HALF-HOURS WITH GREAT SCIENTISTS

HALF-HOURS

GREAT

THE

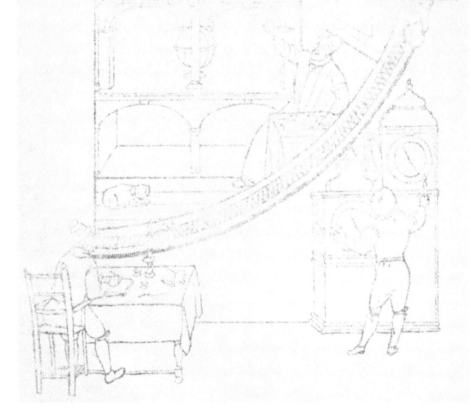

WITH

SCIENTISTS

STORY OF PHYSICS

By
CHARLES G. FRASER
M.A., Ph.D., B.Paed., F.C.I.C.

UNIVERSITY OF TORONTO PRESS

This book is dedicated
to my wonderful wife
to our children
and to all our fellow-citizens
who may derive from it
pleasure or benefit

Preface

THE present age is sometimes called the Scientific Age. This does not imply that every member of the community is an expert scientist—far from it. It does mean, however, that the labours of the scientists have given the age certain features which influence the life of every citizen to some degree. Accordingly it is desirable that as many as possible should have some understanding of the scientists' work, of their aims, their point of view, and their methods.

If we had a wishing-rug or some sort of space-time car that could transport us at will to any place and time, we might visit the scientists of every age, see them at work, listen to their discussions, and even take a hand in the proceedings. The wishing-rug is not available but the literature of science will serve the purpose for anyone who will do the necessary searching, reading, and thinking. Unfortunately, some of that literature is decidedly inaccessible. To meet the difficulty this book has been written in the hope of bringing some of the most important passages of the literature of science within the reach of everyone.

Every part of the vast edifice of science is necessarily the work of some human being, and most of us become more interested in the building, and are able to understand and appreciate it better when we know who were the architects and builders and when, how, and why they did their work. The story of science is a noble epic of the struggle of man from ignorance toward knowledge and wisdom and toward the mastery of nature and of himself.

One purpose of science is to systematize experience, and a knowledge of the story of science has helped many in that process of organization. This book, therefore, offers the reader a cordial invitation to embark on a tour of visits with great scientists to learn from them the parts they played in the advancement of science and of the human race.

As this book has taken shape during an interval of years it has accumulated a heavy debt to many who have assisted the author in various ways. Three librarians, Misses K. L. Ball, M. E. A. Cook, and V. A. Taylor, took a lively interest in tracking down rare books. Some of my revered mentors have given fundamental assistance,

Professors F. B. Allan, Alfred Baker, W. Alexander, B. Bensley, C. Chant, F. B. Kenrick, T. J. Meek; R. B. Thomson and W. Lash Miller of the staff of the University of Toronto, Professor M. A. Chrysler of Rutgers, Principal Chas. G. Fraser, Miss H. Charles, F. C. Colbeck, and Mr. and Mrs. R. Gourlay. I received valuable help and encouragement also from a number of colleagues and class-mates, Professors D. Ainslie, Lachlan Gilchrist, W. H. Martin, and H. Grayson Smith, of Toronto; Professor R. L. Allen of the University of Western Ontario; C. A. Girdler, Miss M. Hawkins, Miss E. Affleck, and E. F. Kingston of the staff of Harbord Collegiate Institute, Toronto; W. F. Bowles; Rev. J. Garden of Mount Royal College, Calgary; J. H. Wilson of Central Scientific Co.; and Rev. J. A. Tuer. Some of my students and friends have been zealous critics and willing test-objects, Professors W. C. G. Fraser of Rensselaer and P. Matenko of Brooklyn College; Miss E. Bott, W. C. G. Fraser, S. Golden, C. Gotlieb, H. R. Hugill; Mr. and Mrs. W. A. Ladd of Columbian Carbon Co., New York; Lt. Com. Leon Leppard, J. Levinson, Mrs. A. McRoberts, A. Rapoport, Mr. and Mrs. H. Rutledge, Mrs. R. Proctor, and F. Molinaro. It is a pleasure to declare my gratitude to these kind and able friends.

To four who helped to make the publication a possibility, I owe a special debt, Professors E. F. Burton and J. Satterly of the University of Toronto, and F. H. Turner and G. G. Hawley of Reinhold Publishing Corporation, New York. I wish also to express my appreciation of the help and counsel of Miss E. T. Harman, Associate Editor and Production Manager of the University of Toronto Press.

Even yet the list is incomplete, as the reading of the text will show.

                                                              C.G.F.

List of Acknowledgements

THE following authors, representatives, and publishers have kindly given permission to reprint copyright material:

Messrs. George Allen & Unwin Ltd. for extracts from *Homeric Hymns*, translated by Andrew Lang.

Messrs. G. Bell & Sons, Ltd., for selections from Aristophanes' *The Clouds*, translated by M. Rogers; and from *The Universe of Light*, *Concerning the Nature of Things*, and *The World of Sound*, by Sir William Bragg.

Sir Lawrence Bragg, for extracts from the above books by Sir William Bragg.

Cambridge University Press, for an extract from the *Collected Works* of Sir James Dewar.

Clarke, Irwin & Company Limited, for selections from *The World of Sound*, by Sir William Bragg.

Messrs. E. P. Dutton & Co., Inc., for selections from *The World of Sound*.

Harper & Brothers, for selections from *Concerning the Nature of Things* by Sir William Bragg.

Harvard University Press, for selections from *Collected Papers on Acoustics*, by Wallace Clement Sabine.

The Loeb Classical Library, for a passage from Plato's *Ion*, translated by Lamb, for one from Ovid's *Metamorphoses*, translated by F. J. Miller, and one from Hesiod's *Cosmogony*, translated by H. Evelyn White.

The Macmillan Company of Canada Limited, and the Macmillan Company, Inc., New York, for extracts from *The Universe of Light*, by Sir William Bragg.

Messrs. Macmillan & Co. Ltd., London, for passages from *Dialogues on Two New Sciences* by De Crew and Salvio; from *The Science of Musical Sounds*, by D. C. Miller; and from *Theory of Sound*, by Lord Rayleigh.

Messrs. Methuen & Co. Ltd., for extracts from *Memories of My Life*, by Francis Galton; and from *Great Physicists*, by Ivor Hart.

Munsell Color Company, Inc., for paragraphs from *A Color Notation*, by A. H. Munsell.

Messrs. Thomas Nelson & Sons, for brief passages from *A Book of Myths*, by Jean Lang.

The *Physical Review*, for an extract from an article by Dr. Charles T. Knipp.

Messrs. G. P. Putnam's Sons, for passages from *A Book of Myths*, by Jean Lang.

W. B. Saunders Company, for a passage from C. N. Camac, *Epoch-Making Contributions to Medicine*, translated by J. Forbes.

The Society for Promoting Christian Knowledge and The Sheldon Press, for paragraphs from *Waves and Ripples in Water, Air and Aether*, by Sir J. A. Fleming.

Messrs. Taylor & Francis, Ltd., for extracts from articles by Lord Rutherford, W. Moseley, J. J. Thomson, and N. Bohr, which appeared in the *Philosophical Magazine*.

The University of Chicago Press, for excerpts from *The Electron*, by R. A. Millikan; and from *Ancient Records*, Vol. I, translated by Luckenbill.

Messrs. Winsor & Newton, for an extract from *Colour Science*, by W. Ostwald, translated by J. Scott Taylor.

Every effort has been made to locate the owners of copyright material. If, however, any acknowledgements have inadvertently been omitted, the author and publishers will be grateful for notification, so that the proper credits may be included in subsequent editions.

Contents

List of Illustrations

THE STORY OF
Mechanics

CHAPTER ONE

Weapons and Implements

IN the long ago, primitive man had to contend with fierce and power-ful enemies such as the sabre-toothed tiger and the mammoth. Against such antagonists, it might seem that the puny human creature could not have even a faint chance of success. But there were some potent factors in his favour. Chief among these was his superiority of brain with all which that implies. There was also another item in his favour—one that is of fundamental importance in the story of mechanics. Related to man's superior ability to think and foresee was his use of weapons, traps, and other mechanical devices. Without these he probably would not have survived and certainly could never have reached his present mastery of the world. His use of weapons and tools was, of course, dependent upon his upright posture and upon his possession of fingers. Man was the "weapon animal."

In the recent world war, the use of weapons and machines was crucial in the survival of our civilization. In the conflict with the Nazis, we encountered the most dangerous kind of foe, one who was thoroughly armed with engines of war and destruction. With more machines and some better ones, with superiority in using them and with some strokes of good fortune, the Allies finally won the struggle.

The science that deals with mechanical devices was christened "mechanics" by Aristotle of Athens about 350 B.C. It is the oldest science except possibly astronomy, which may be called "celestial" mechanics. In any case, mechanics is the fundamental science; and that branch of it which deals with practical applications of mechanical principles in human affairs is called "applied" mechanics.

EARLY RECORDS

The earliest records of the invention and use of simple machines are some of the implements themselves which have survived, or draw-ings made at a time when man had not yet invented writing. The first devices such as clubs, spears, needles, were made of wood or horn or bone. Then in a later age, appliances of stone were made. That age, consequently, is called the Stone Age (fig. 1). During an early period of

the Stone Age, the artifacts were of roughly chipped stone (fig. 5). The period is, accordingly, referred to as the Old Stone Age or palaeolithic era (Gk. *palaios*, old; *lithos*, stone). It was in palaeolithic times that the mammoth of fig. 2 was drawn on a cave-wall in southern France and the bison of fig. 3 painted in four colours on the wall of a Spanish cave. Archaeologists have discovered excellent palaeolithic sketches of reindeer, wild ponies, and mammoths, in the form of carvings on bone or on

FIG. 1. Primitive weapons.

mammoth-tusk ivory. It used to cost a mammoth rather dear to have his picture tattooed on his tusk. The instruments which the artists employed in making such sketches were stone awls (fig. 5).

The artifacts of a later era (probably 20,000 years later or more) were smoother and of greater variety. The period is called the neolithic or New Stone Age. The smoothness of neolithic implements was

FIG. 2. Mammoth in trap.

FIG. 3. Bison on cave wall.

achieved by grinding and polishing. The improved workmanship displayed in these implements shows that men were advancing in their knowledge of the materials with which they were working and in their skill in applying that knowledge. The increased variety of mechanical devices is an indication of progress in invention. Although the palaeolithic and neolithic ages differed in many aspects, yet we have just seen that the division of the Stone Age into periods is based on stages of progress in applied mechanics.   One criterion of the advancement of a tribe or nation is its progress in mechanics and allied sciences.   Excavators who unearth deposits of bones of cave-hyenas and other Stone Age animals often find spear-heads and arrow-heads mingled with the bones.   Not infrequently they find an arrow-head penetrating part way through a bone.   Primitive man evidently killed such animals with weapons for food and clothing or in self-defence.   Arrow-heads would prove the

FIG. 4. Bow and arrow.

existence of the archer's bow even if the wooden bows and shafts and the bow-strings had all disintegrated and vanished.   Fig. 4, however, gives further proof; it is taken from an ancient cave-wall drawing and shows a bow in use.   The bow has been of service to the human race in various ways—in hunting and in fishing, in warfare and athletics, in boring holes, in making fires.   Familiarity with the bow led to the invention of certain musical instruments and the bow itself is used in producing music from stringed instruments.

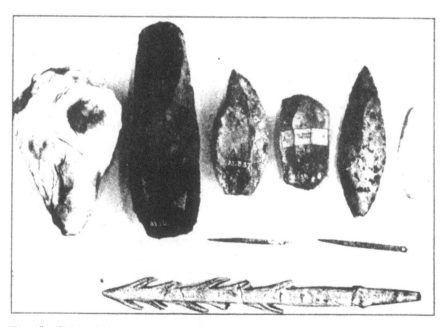

FIG. 5.   Palaeolithic artifacts: fist-hatchet, skinner, arrow-heads, spear-points, awls, and harpoon.

APPLIED MECHANICS IN ANCIENT MESOPOTAMIA

Some relics recently unearthed at Al 'Ubaid near Ur of the Chaldees in Mesopotamia are significant in the story of mechanics. They date about 4000 B.C. Among them are (1) stone hoe-heads, worn smooth at their cutting edges by use; (2) stone nails and pegs; (3) flint saws, resembling modern hack-saw blades; (4) sickles of baked clay, modelled after animal jaw-bones—for example, the kind of jaw that Samson used in arguing with the Philistines; (5) stones, narrowed at the middle and therefore called "waisted," which were used for weighting fishing-nets and loom-warps; (6) smooth one-piece stone bowls.

These articles bespeak such occupations as agriculture, carpentry, building, fishing, weaving, dyeing, pot-manufacture, brewing, and so forth. They tell a great deal about the state of applied mechanics among the people who made and used them in the region of the Tigris and Euphrates rivers some six thousand years ago.

SIMPLE MACHINES

Most of the mechanical devices mentioned thus far are applications of the simple machine called the *inclined plane*. As the name implies, a natural illustration of the inclined plane is a hill-side, of which a gang-plank and a stairway are direct copies. One advantage of the inclined plane as used in the loading device of fig. 6 is that the man can roll the barrel into the van by means of a push that is less than the weight of the barrel. Thus the task becomes easier than by lifting the load directly into the van. The inclined plane increases or multiplies the man's lift or push or *force*. This illustrates a general trend, for mankind continually seeks an easier, quicker way of performing work, provided the new way does the work as well as the old or better. This tendency is referred to as the "principle of ease or economy of effort." Parsimony of effort differs from laziness; it does not shirk work but enables one to do more work in a given time. In other words, it increases our power, for power is rate of doing work. A machine may be defined as an apparatus which is used in applying or modifying a force in doing a certain piece or kind of work.

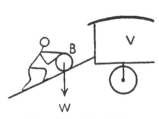

FIG. 6. Inclined plane.

Other applications of the inclined plane of which we make frequent use are the wedge, the winding stairway, the screw, the razor, the shoe-horn, and so on ad infinitum. In the biological world, the inclined

FIG. 7. The shaduf.

plane is amply illustrated by "tooth and claw" for these are wedges or double inclined planes. The wooden clothes-peg is a hollow wedge. Although the inclined plane was the first simple machine to be invented and used, it was by no means the first to be understood theoretically.

The second simple machine to be invented in prehistoric times was the *lever*, a rigid body free to turn about an axis, called its fulcrum. Common examples are the crowbar or "jimmy," the boat-oar, hoe-handle, nut-cracker, and forceps. A pair of scissors consists of a pair of levers whose cutting edges are inclined planes. The legs and arms of vertebrates are natural illustrations of the lever. The shaduf (fig. 7), a lever used in Mesopotamia and Egypt for lifting water in the irrigation of fields, was invented as early as 2000 B.C. and is still in use in those regions. The see-saw or teeter-totter and the balance are also ancient examples of the lever.

A wheel may be thought of as a group of levers fastened together in a circle and free to turn about a common axis, which is usually at the centre. The earliest use of the wheel probably occurred in the Euphrates valley. In 1928 excavations in the royal cemeteries at Ur brought to light a carving that represents a four-wheeled cart (fig. 8). It was a remote ancestor of the automobile. A fragment found by the same archaeological expedition represents a two-wheeled chariot produced by the Sumerian civilization about 2700 B.C. The contribution of the wheel to the advancement of mechanics and civilization has been tremendous.

Pottery discovered at Al 'Ubaid indicates the existence of the potter's wheel as early as 4000 B.C. The design on an old vase (fig. 9) found at Khafaje in Mesopotamia records the use of the wheel about 3000 B.C. The newly domesticated animal, which in this design has for the first time "got into the pictures," was called by the Sumerians "the

FIG. 8. Four-wheeled chariot.

ass of the mountains," but by us, of course, its modern descendants are called horses.

The inclined plane and the lever are classified as simple machines. Two primitive modifications of the inclined plane, developed in antiquity, were the wedge and the screw (foreshadowed by the winding stairway). Two simple derivatives of the lever were the pulley and the wheel-and-axle. These six simple machines were all invented and used in prehistoric times. When Aristotle wrote his book on machines (350 B.C.) and entitled it *Mechanics*, these six were all well known from time immemorial and their origins were already veiled in the mists of tradi-

FIG. 9. Wheel and horse.

tion. The Greeks made no addition to the list. In fact, during the whole historic period up to the present, only one person has ever succeeded in inventing a new type of simple machine: in 1620 Pascal of Paris invented the hydraulic press.

WRITTEN LANGUAGE

The invention of writing was an enormous advance for the human race. It marked the beginning of the historic period. Historians require records or documents of some sort. Documents necessitate writing: but writing is impossible without such instruments as the stylus, brush, pen, or pencil. Since these are all applications of the inclined plane and the lever, it follows that simple machines were necessarily invented in prehistoric times, for the historic period could not begin until after their invention. Then the use of writing was handed down anonymously as a priceless heritage to all succeeding generations. The magnitude of its influence on the advancement of science and of human welfare is beyond all estimation.

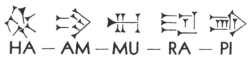

HA — AM — MU — RA — PI

FIG. 10. Cuneiform name of king.

Writing owed its origin in the Euphrates valley partly at least to the need of keeping account of the goods and estates belonging to the temples. Marks were made in moist clay; then the clay was hardened in the sun or in a furnace. In later times, the instrument used for making such marks was a three-cornered stylus of about the size of a crochet-hook. Since its impressions were wedge-shaped, the writing is called cuneiform (Lat. *cuneus*, wedge). Fig. 10 shows a sample of this writing; it spells the name of King Hammurapi, the great law-giver who reigned in Babylon about 1700 B.C.

Although palaeolithic artists had a spoken language, they did not label their drawings; they had no written signs to represent the oral signs called words. The first attempts in that direction were pictographs or ideographs, i.e. pictures to stand for the things or ideas which words represent. A circle, for example, could represent the word "sun" or the word "day"; the drawing of a man, the word "man"; and a man drawn in an appropriate attitude could represent the idea of worship.

The development of machines, in general, was from simple to complex but that of language writing discloses some advancements from complex to simple. Fig. 12 gives histories of the development in the writing of three words.

In Egypt, also, writing originated in the temples. Its invention was ascribed to the god Thoth, the Recorder. In fig. 13, the god Knemu is in the act of fashioning a man out of clay on a

FIG. 11. Pictograph.

potter's wheel. The potter's wheel is operated by a pedal, an example of the lever which unfortunately is eclipsed in the picture. Thoth is counting off the man's allotment of years by notches on a crooked staff from the palm tree, which was often called the "tree of life." This old picture is the first in this chapter that is labelled: the names

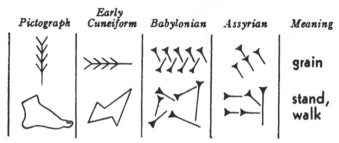

FIG. 12. Development of cuneiform writing.

of the two gods are inscribed in hieroglyphics or priestly writing, and each name is framed by a cartouche. The original cartouche was a circle representing the sun's disc, but later it was modified to the form shown. The character at the right side of these labels indicates the godhead of Knemu and Thoth: it is a picture of Osiris and symbolizes divinity. The illustration, therefore, signalizes the advent of writing.

FIG. 13. Creation of man.

It also indicates that the use of written numbers developed along with that of language symbols. In fact, numbers may have had some priority; they were probably the first of all written language symbols if we exclude pictographs.

APPLIED MECHANICS IN ANCIENT EGYPT

The ancient carvings shown in fig. 14 represent Egyptian hunters or warriors who lived about 3000 B.C. or earlier. Each wore a head-feather, a belt and a loin-cloth. Each was bedecked with a jackal-tail, which in this old picture might easily be mistaken for a wooden war-club. Hunter A brandishes a stone mace, B flourishes a double-headed

Fig. 14. Ancient Egyptian hunters.

stone axe, C wields spear and boomerang, and D shoots a stone-tipped arrow from his bow. The standards borne by A and B are symbols of the god of their tribe.

The branch of applied science or engineering which has to do with the constructing of buildings is certainly of prime importance in human affairs. One of the greatest engineers of all time was the Egyptian Imhotep, who was the first to erect a large building of stone masonry. He was one of the most remarkable of all Egyptians. Fig. 15 gives his name in hieroglyphics. The sign of divinity in that name indicates that after death he was deified, an honour which in the long history of Egypt was accorded to only two commoners. Imhotep was very learned and skilful. About 2800 B.C., he built the step pyramid of Sakkarah as a tomb for Zoser, a Pharaoh of the third dynasty. Imhotep is, therefore, the first contributor to applied mechanics whose name is known; for now the story of mechanics is coming into historical times.

Fig. 15. Imhotep's name in hieroglyphics.

Imhotep's method of constructing a pyramid was an application of the inclined plane on a tremendous scale (fig. 16). The trestle-work of this picture is probably erroneous: the ramps almost certainly were wholly of earth and rubble. The blocks of granite or other rock, some of them weighing fifty tons, were sawn at the quarries by means of bronze saws set with diamonds and other hard jewels. They were transported by barge and on land, over logs or rollers which were

Fɪɢ. 16. Building a pyramid.

primitive applications of the wheel. Hundreds of slaves or conscripts
pulled the papyrus ropes which moved the huge blocks of stone over
the rollers while others carried logs around from behind and placed
them in front.

To dress the blocks the masons had copper or bronze chisels. It is
said that each chisel lost its temper at one blow and had to be removed
for retempering. Yet the blocks were so "truly shaped and fashioned"
that after they have rested in position for some forty-seven centuries
without mortar between, it is commonly impossible to slip a post-card
between them.

> In the elder days of Art,
>   Builders wrought with greatest care
> Each minute and unseen part:
>   For the Gods see everywhere.

When the building was completed, the final operation was to dig
out the pyramid and remove from the landscape the gigantic mound

FIG. 17. Balance.

of rubble which would be, in the case of the Cheops pyramid, about 480 feet high, that is, of the same height as the Canadian Bank of Commerce at Toronto, the tallest building in the British Commonwealth.

The influence of Imhotep affected other departments than mechanics and was indeed so great as to inaugurate in the culture of Egypt a new age of splendour.   In the building of the pyramid we have seen the advent of the copper or bronze age, for although the artisans were working in stone some of their implements were of bronze.

The scene depicted in fig. 17 is from the *Book of the Dead*, which dates about 1300 B.C.   It was compiled from earlier writings called the Pyramid Texts, which were written about 2400 B.C.   The scene portrays an important application of the lever, namely, the *balance*.   The dog-headed god, Anubis, is weighing the heart of a person who has just entered the spirit world.   Thoth is recording the examination results. It may seem that the requirements were not overly drastic for the counterpoise was a feather!   It should be added, however, that the feather was an Egyptian symbol for righteousness, and

FIG. 18. Egyptian cabinet-maker's shop.

lightness of heart was perhaps in this case a desideratum.

Fig. 18 represents a scene which was painted on the wall of a tomb in the Nile valley about 2500 B.C.   It shows the interior of a cabinet-maker's shop.   Among the tools in use are shown a mallet, a chisel of stone or bronze, a copper saw, planes, and a bow-drill employed in boring a hole.   The mallet is a descendant of the ancient war-club.

All the rest are applications of the inclined plane. The bow-drill is also
an example of the wheel used as a *pulley*. Implements from different
ages are seen here together; the mallet, from the wooden age, the chisel
from the stone age, and
the saw from the cop-
per or bronze age.

FIG. 19. Plough and clod-breaker.

Another scene from
the same period (fig.19)
shows a plough and a
clod-breaker. The lat-
ter was a large form of hoe from which the plough was developed. Both
hoe and plough are applications of the inclined plane and the lever.

At the left side of this painting stands a scribe to keep tally of the
work done. On his report the taxes will be based. His standing aside
may suggest that his popularity is about the same as that of a modern
income-tax officer. The inscription gives the badge of his profession,
consisting of a palette, an ink-pot, and a reed-pen (fig. 20). His writing
materials devised at the dawn of the historical period—papyrus (Egyptian "paper"), ink, and pen—are still more commonly used throughout
the world than any others, for nothing better has ever been invented.
Even the printing-press and the typewriter are essentially elaborations
of this same old Egyptian writing equipment. Many palettes were
ornamented, some with maxims, prayers, charms, or incantations, and
some with elaborate carvings. The Egyptian hunters of fig. 14 are
taken from a palette carving that portrays
a lion-hunt.

In the palette were two sockets or wells,
one for black ink and the other for red. If
you were to enter a modern office and there
ask for any sign of Egyptian influence about
the place, you would likely receive an astonished negative. But look at the desk set
—two ink-wells, one black and one red.
Look at the typewriter—just paper, ink,

FIG. 20. (*a*) Symbol of
scribe's profession. (*b*) A
scribe's equipment — pal-
ette, ink-pot, reed-pen.

and a complex pen. Or the ribbon—upper side black, lower side red.

MATHEMATICS—"THE LANGUAGE OF SIZES"

At first only a few priests could write. Later this skill was communicated to the privileged order of the scribes. To the uninitiated,
writing was incomprehensible, and their fear of the unknown made
them shun it. To them it seemed sinister, savouring of magic and
black art. Through sixty centuries, however, literacy has grown until

now, in advanced nations, about ninety per cent or more of the people enjoy the conveniences and pleasures of written language. No longer does writing seem to the majority something difficult or to be shunned, though many evince a lethargy toward letter-writing.

A similar development has been taking place, though more slowly, in the language called mathematics, which is particularly facile in the discussion of quantities. Thoth, or his priests, in listing the temple belongings expressed such quantities as

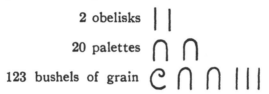

and so forth. Ordinary writing served to tell the kind of article or the unit of measurement, but special symbols were invented for telling the number of each kind, i.e. the measure of the quantity. The symbols, 1, 2, 3 or I, II, III, etc., are called digits (Lat. *digitus*, finger) because counting was often done, and is still sometimes done, on the fingers. Mathematics has also been shunned by the uninitiated, but steadily there has been an increase in the proportion of the people who enjoy the advantages of this language of sizes or magnitudes.[1]

Among those who have profited by the use of mathematics are the students of mechanics. In fact, the two sciences have continually rendered mutual assistance in their progress. So much so, indeed, that the physicist is glad to acknowledge that what was a luxury has become a necessity and that he could not carry on his work efficiently without mathematics.

Another kind of writing that has been of inestimable assistance in the progress of mechanics and geometry is that of diagrams. Such writing is related, of course, to the picture-writing which was the earliest form of language symbols. Diagrams can be exceedingly expressive. One Chinese writer has gone so far as to say, "One diagram is better than a thousand words." The first diagrams were maps and building-plans. Accordingly, fig. 21, the oldest known map, is an important source in mechanics. It was drawn on a clay tablet about 2300 B.C.[2] The cuneiform inscription at the top (fig. 22) means "east," that at the bottom "west," and that on the left, "north." The central circle is an estate and the other three are cities. A mountain range lies

[1]L. Hogben, *Mathematics for the Million* (Norton, 1937) gives an excellent discussion (p. 15).

[2]Discovered at Nuzi in 1931 by Professor T. J. Meek of the University of Toronto.

along the east border and another on the west. Through the valley-plain between them runs an irrigation canal which is labelled "The Fructifier." Its waters flow southward and are drawn through three

FIG. 21. The oldest known map.

FIG. 22. Key for map.

canals from a body of water at the north whose waves are represented by dashes. This is probably the first mention of the important term, *wave*, in the literature of the world.

Some clay tablet inscriptions are building-plans, primitive, of course, but sufficiently like modern architects' blue-prints to be immediately recognizable.

FINDING THE AREA OF A CIRCLE

A knowledge of the properties of the circle is important in mechanics and in other branches of physics, for instance, in the study and use of pulleys and in connection with engines that have cylinders and pistons. In particular it is frequently necessary to know the area of a circle. This area is more difficult to measure than that of a square or rectangle, for when one attempts to mark it off in square inches or square centimetres, difficulty is encountered at its curved boundary or circumference which cuts through a number of squares and includes various fractions of squares which are not easy to estimate. Archimedes of Syracuse discovered about 240 B.C. that the area can be obtained for any circle with an error of not more than 1 in 3000 by multiplying the area of the square on the radius by $3\frac{1}{7}$ or 3.141. . . (fig. 23, p. 18).

Egyptian priests, surveyors, and architects, however, had another method of making this calculation as early as 1800 B.C., sixteen centuries before Archimedes, and possibly even ten centuries earlier, in the days of Imhotep. This is attested by the following quotations from an old papyrus scroll now in the British Museum. It is called the Rhind Papyrus[3] after a Scottish pioneer archaeologist, Alexander

[3]Washington Congressional library has an edition of this papyrus. by Professor E. Peet of Oxford.

Rhind, who purchased it and later presented it to the Museum. It bears the intriguing title, *How to Know All Dark Things, Every Mystery and Every Secret.* It was written at Heliopolis about 1800 B.C.

Behold this roll was written in the likeness of a writing, handed down from antiquity [!]. It was the scribe Ahmose [or Ahmes] who made this copy.

The date makes Ahmose a possible contemporary of Joseph, the Hebrew Chancellor of the Exchequer who cornered the wheat market for Pharaoh about 1800 B.C.

Ahmose gives the following rule for finding the area of a circle.

Method of reckoning a circular plot of ground of diameter 9 khet [unit of length]. What is its land area? You are to subtract one-ninth of it, namely, 1; remainder, 8. You are to multiply 8 times. It becomes 64. This is the land area.

The doing as it occurs [solution]:

Let 1 represent	9 [diameter]	
1/9 of it	1	
Subtract from it;		
Remainder	8	
Let 1 represent	8	
Then 2 represents	16	
" 4 "	32	
and 8 "	64.	

This old Egyptian rule can be expressed by an equation:

Eq. 1. $A = \left(\dfrac{8}{9} d\right)^2 = \left(\dfrac{8}{9}\right)^2 (2r)^2 = \left(\dfrac{16}{9}\right)^2 \cdot r^2 = \left(\dfrac{256}{81}\right) r^2 = 3.16\, r^2.$

If we measure the diameter *d* of a circular block, and its circumference *c*, and similarly for a number of blocks, it is found that the *ratio* between *c* and *d*, i.e. the fraction *c/d*, has the same value for all circles. It is called a *constant*. Some quantities remain constant for a limited time but this one never changes. It is constant forever. Its value is more than 3 and less than 4. Some men have spent most of their lives in computing the value of this important constant to hundreds of decimal places and have used their results as their epitaphs. Here it is to 27 places: 3.141592653589793238462643383 ... Some who wish to assist their memories use the following rather languid mnemonic to eleven places. The numbers of letters in the words give the required digits,

> See, I have a rhyme assisting
> My feeble brain its tasks resisting.

Since this result never terminates or repeats, no number in our system can represent the constant exactly. Consequently, Leonard Euler, the blind mathematician of St. Petersburg, Russia, proposed in A.D. 1750

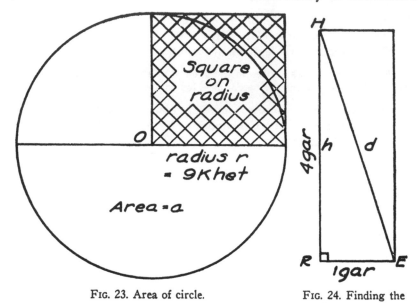

FIG. 23. Area of circle.   FIG. 24. Finding the
                                            hypotenuse.

that it be represented by the Greek letter π (pronounced *pie*). His suggestion has long since been universally adopted. Then

Eq. 2. $\dfrac{c}{d} = \pi$, or $c = \pi d = 2\pi r$

where r is the radius. The same constant, π, occurs in Archimedes' rule for finding the area of a circle from its radius. It is written thus:

Eq. 3. $A = \pi r^2 = (3.141\ldots)r^2$ (fig. 23).

By comparison with *Eq.* 1 (p. 17), it is seen that this value of π, namely, 3.141 . . . differs from the Egyptian value 3.16, by less than 1 per cent—actually by 2 in 300 or about 0.6 per cent! Using this rule we obtain the area of Ahmes' circle:

$A = \pi r^2 = 3.141\left(\tfrac{9}{2}\right)^2 = 63.6$ units (square khet).

In comparison with this result the Egyptian answer, 64, as we have just seen, was about 1 per cent high. This achievement of the Egyptian priests who wrote the Pyramid Texts (2400 B.C.), is nothing short of astonishing. According to the Hebrew Scriptures, the Phoenician brass-moulders in King Solomon's day (1000 B.C.) were less accurate on this point. They used for π the value 3.0, which is 4 per cent below 3.14 and has, therefore, an error four times greater than the Egyptian.

In sharp contrast to such advanced knowledge, is the fact that Ahmes apparently could not multiply 8 by 8 directly, but was obliged to perform the computation by three major onslaughts with the 2-times table. One dark mystery that remains is how the ancient Egyptians discovered their rule. Shall we say, "by an infinite amount of trial and error"? Ahmes shows how to apply his rule but gives no light on how it was derived. That was simply another dark secret.

These excerpts from the Rhind Papyrus are the first example in this chapter of a source in mechanics which is in the form of a quotation from a treatise. They indicate that mechanics had reached the documentary or historical stage in Egypt by at least 1800 B.C. and possibly several centuries earlier. The map from Nuzi (fig. 21, p. 16) marks a similar transition in the Euphrates valley about 2300 B.C.

Since astronomy and mechanics are the oldest sciences, it is not surprising to find that they have long since developed to a quantitative stage and are therefore much in need of help from algebra and geometry. Accordingly, it is plain why a history of physics must include some items from the story of mathematics.

SIDES OF A RIGHT-ANGLED TRIANGLE

Another mathematical principle of great utility in mechanics is the famous theorem ascribed to the Greek philosopher Pythagoras, who lived at Crotona in southern Italy in the sixth century B.C. The Pythagorean theorem calculates the length of the hypotenuse or longest side of a right-angled triangle from the other two sides by adding their squares and extracting the square root of the sum.

In the museum at Berlin, Germany, there are (or were) some cuneiform fragments which show that several centuries before Pythagoras Babylonian mathematicians could make this calculation for at least some types of the problem. They used a system of numbers called sexagesimal because it was based on the number 60. In other words, the radix of their numbers was 60, whereas the radix of our number system is 10. Hence our numbers are decimal (Lat. *decem*, ten) and we have noted the same in Egyptian numbers. The influence of the sexagesimal system is still seen in our division of the hour into 60 minutes and the minute into 60 seconds, and also in the division of the circle into 360 degrees, each degree into 60 minutes, and each minute into 60 seconds of arc. In the following fairly free translation of one of the Berlin inscriptions, sexagesimal numbers have been replaced by decimals for the reader's convenience. The unit of length employed was the *gar*. The time and place was Babylon, 1800 B.C., about a century before Hammurapi.

To find the diagonal of a door [fig. 24] whose height $h = 4$ gar and width $w = 1$ gar. The width, 1 gar, is to be squared: result, 1. This result is to be divided by the height, 4 gar. Result, $\frac{1}{4}$. Half of this result, $\frac{1}{8}$, is to be added to the height. This gives $4\frac{1}{8}$ or 4.125 as the length of the diagonal.

The Pythagorean solution is as follows: Let d be the diagonal; then

Eq. 4. $d = \sqrt{(1)^2 + (4)^2} = \sqrt{17} = 4.123$ gar.

Hence the error in the Babylonian value was actually only 2 in 4000, i.e. 1/20 per cent or 0.05 per cent. To solve this problem with such accuracy and without the use of square root[4] was an astounding achievement.

The Babylonian method is briefly expressed in algebra thus:

Eq. 5. $d = h + \dfrac{w^2}{2h}$

where d is the diagonal of the door and h and w its height and width. Since the right side of this equation is the sum of two terms, let us express the right side of *Eq. 4* as a series by means of the binomial theorem. We obtain,

$$d = \sqrt{h^2 + w^2} = \left\{ h^2 \left(1 + \frac{w^2}{h^2} \right) \right\}^{\frac{1}{2}} = h \left(1 + \frac{w^2}{h^2} \right)^{\frac{1}{2}}$$

$$= h \left\{ 1 + \frac{1}{2} \cdot \frac{w^2}{h^2} + \frac{\frac{1}{2}\left(-\frac{1}{2}\right)}{\underline{|2}} \left(\frac{w^2}{h^2} \right)^2 + \ldots \right\}$$

$$= h + \frac{w^2}{2h} - \frac{1}{8} \cdot \frac{w^4}{h^3} + \ldots$$

If we discard the third term of this series which equals

$$-\frac{1}{8 \times 64} = -\frac{1}{512} = -0.002 \text{ (approximately)}$$

we obtain $d = h + \dfrac{w^2}{2h}$

which is the Babylonian result. The third term reduces the Babylonian answer by 0.002, the very excess we found in it over the Pythagorean (4.125 − 4.123 = 0.002).

Again the mystery confronts us, "How did these mathematicians derive or discover their rule?"—for, like Ahmose, they give no hint of explanation. That they should have discovered by trial and error even the first two terms of this binomial expansion or series is a great credit to their astuteness and perseverance. If they had shown a knowledge of the third term, we should have been forced to suspect that they knew something about the binomial theorem.

[4]Discussion of Babylonian knowledge of square roots and logarithms is given in *Mathematical Cuneiform Texts* by Otto Neugebauer and A. Sachs (American Oriental Society and American School of Oriental Research, New Haven, 1945), p. 33.

Consider a door that is 3 gar by 3 gar. By the Pythagorean theorem,

$$d = \sqrt{(3)^2+(3)^2} = \sqrt{9 \times 2} = 3\sqrt{2} = 1.414 \times 3 = 4.242.$$

By the Babylonian method the result is $d = 3 + \dfrac{3^2}{2 \times 3} = 4.50$ gar which is 6 per cent too high. So the acute old mathematician did not choose that example, or at least did not publish his results as far as we know.

From the pyramids it is seen that the Egyptians could draw a right angle with remarkable accuracy. This practical problem in surveying was handled by a special cult of priests who were called "rope-stretchers" or, in Greek, *harpedonaptai*. They used a rope loop divided by knots into three parts whose lengths were in the proportion 3:4:5. They stretched it by three stakes at the knots to obtain on the ground a triangle. Since $3^2 + 4^2 = 5^2$, it is seen that the rope-stretchers knew in a practical way at least one example of the Pythagorean theorem, and this probably as early as Imhotep and the pyramid of Sakkarah (2800 B.C.). The rope-stretcher method of drawing a right angle is still used sometimes by carpenters in making small buildings.

THE MECHANICS OF FLUIDS

Most of the early mechanical appliances that have been mentioned thus far were in the solid state of matter. But there were also in early times mechanical devices which made use of the properties of fluids,

FIG. 25. Babylonian reed-raft.

i.e. of liquids and gases. Typical of these were the boat and the pump. The raft and the "dug-out" were certainly of prehistoric origin. Even the Babylonian reed-raft of fig. 25, and the Assyrian "boat of skins" (fig. 26), though primitive, were by no means the original raft and boat. The date of the Egyptian vase with boat design represented by fig. 28 is probably about 4000 B.C. The standard on the boat is like those of fig. 14, p. 11. There is a well-known picture of an Egyptian boat transporting a 350-ton obelisk down the Nile from the quarry at Aswan during the reign of Queen Hatshepsut (1450 B.C.). Her diary

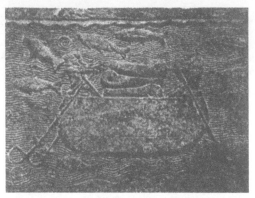

FIG. 26. Assyrian boat of skins.

FIG. 27. Drinking "through straws."

says that the quarrymen completed the obelisk only seven months after she placed the order.

The convivial and romantic scene of fig. 27 shows an old Mesopo-

FIG. 28. Ancient boat design.

tamian custom of using a straw or reed for sucking up a beverage from a vessel. It also shows that the economical plan of two straws to a glass, one for each participant, has very ancient precedent and did not originate north of the Clyde. The behaviour of the liquid in this circumstance depends on the action of atmospheric pressure and foreshadows the invention of the pump, which occurred long before Aristotle wrote his book on mechanics in 350 B.C., and long before mankind had any idea of atmospheric pressure. In this instance practice preceded theory by centuries.

HOW ASSYRIA APPLIED MECHANICS

The following excerpt is from cuneiform inscriptions on clay tablets deposited in the Temple of Adad at Ashur by the Assyrian king, Tiglath-Pileser, about 1100 B.C. It refers to a number of mechanical devices and especially to that branch of applied mechanics which has to do with the constructing of buildings.

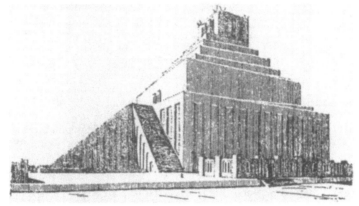

FIG. 29. Reconstruction of temple tower.

Tiglath-pileser, the mighty king, king of the universe, who is without rival, king of the four quarters of the world, lord of lords, king of kings whose splendour overwhelms the quarters of the world.

I took my chariots and my warriors and over the steep mountains and through their wearisome paths I hewed a way with pickaxes of bronze. I crossed the Euphrates in vessels of skins. At the bidding of [the god] Ninurta who loves me . . . with my mighty bow and with my iron spear and with my sharp darts, in the desert, I killed four aurochs which were mighty and of monstrous size [11 feet tall at the shoulders]. I delighted the hearts of the gods.

I put the ploughs to work throughout all Assyria and heaped up grain in greater quantities than any of my fathers.

The temple of Ishtar . . . I rebuilt completely. I made bricks. I dug down and laid the foundations upon the firm bed of the mighty mountain. The whole of that chasm I filled with brickwork like unto an oven. I built two mighty tower temples. The interior thereof I made to shine like the heart of heaven: I made beautiful its walls like the brightness of the rising stars. The towers thereof I raised to heaven. The holy temple which shone as a star in the heavens, being richly adorned by the skill of my craftsmen, I planned with care.[5]

Another edifice that was "raised to heaven" was the Tower of Babel in the "land of Shinar" (Sumer), as may be judged by the size of the human figure in the foreground of fig. 29. The exteriors of some Assyrian temple-towers displayed remarkable winding stairways in which inclined planes alternated with horizontal promenades.

[5]Luckenbill, *Ancient Records* (Univ. of Chicago Press, 1927), I, 86.

Sometimes ancient machines were misused by unscrupulous tyrants to cause dreadful suffering. This will be seen in the cruel words of Ashur-nazir-pal, King of Assyria in 870 B.C.:

In the valour of my heart and with the fury of my weapons, I, Ashur-nazir-pal, stormed the city of Shuru. I built a hollow brick pillar over against its city gate. I flayed alive all the chief men and I covered the pillar with their skins. Some I walled up alive within the pillar; some I impaled on stakes. I cut off the limbs of the officers. From many, I cut off their hands. Of many, I put out the eyes. In the city of Shuru ... I increased the tribute and taxes imposed upon them. I fashioned a heroic image of my Royal Self; my power and my glory, I inscribed thereon. In the midst of their palace, I set it up. I fashioned memorial monuments and inscribed thereon my glory and my power. Fifty cities, I captured. I slew their inhabitants. The cities, I destroyed, I devastated, I burned with fire. The terrifying splendour of my dominion, I poured over them.[6]

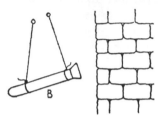

FIG. 30. Battering-ram.

Ashur-nazir-pal's ability to overpower any city in spite of its walls was largely due to a mechanical invention. In a previous reign, an engineer had devised the battering-ram. It was essentially a huge hammer (fig. 30) and a forerunner of the pile-driver. It was also related to the *pendulum*, for the metal-tipped beam *B* swung at the ends of two ropes. The ram was moved into position by means of an armoured car or "tank," whose wheels are shown in fig. 31. It is smashing a breach in a city wall and the bricks are flying in all directions. The towers and human figures are drawn to different scales, a very convenient style of art—for the artist.

Of the rich heritage bequeathed to mechanics by the ancient civilization of Mesopotamia only a meagre sketch could be given in these

FIG. 31. Assyrian tank with battering-ram.

[6]Ibid., p. 145.

few pages. The tireless archaeologists, moreover, are perpetually adding items to the list. Among other appliances devised by early inventors in the Euphrates valley were the sling (fig. 32), the fish-hook (fig. 33), the gate-hinge, a water-clock, the sun-dial, a seeder or drill for planting grain, standard weights (fig. 34), the tripod, and many others. Our clocks and protractors declare their Sumerian origin by their sexagesimal division of time and angles. Our calendar also bears witness to the labours of the Chaldean astronomers.

Fig. 32. Slinger.

Another important contribution of Mesopotamia to mechanics and to the world was that of libraries; for modern libraries are the lineal descendants of those at Nineveh and Babylon. It would be difficult to overstate the importance of libraries in the progress of mechanics. In libraries we have a second criterion of the progress and advancement of a nation, the criterion of mechanics having been mentioned previously.[7]

A third criterion of any civilization or culture—and probably the most searching of all—is the kind of people it produces. Throughout

Fig. 33. Mesopotamian
fishhook.

Fig. 34. Armenian royal
standard weight.

this book we shall be observing a certain group of people, produced by succeeding cultures—the scientists.

MECHANICS AMONG THE HEBREWS

The Hebrew Scriptures, though concerned primarily with ethics and religion, contain, nevertheless, many passages that refer to mechanical appliances. They sometimes reflect the influence of Egypt, Babylon, or Phoenicia. A few of these will be quoted here and refer-

[7]The library of the British Museum today is the largest which has ever existed. In it there are over five million books and fifty-five miles of shelves. The beautiful Library of Congress at Washington, with three million books, is the largest in America and third largest in the world. In some stages of a scientific research, the library is a better milieu than the laboratory.

ences given to some others which the reader can readily consult for himself. The first is from the prophecy of Isaiah, who lived in Jerusalem about 760 B.C. It mentions four applications of the inclined plane:

And they shall beat their swords into ploughshares and their spears into pruning-hooks; nation shall not lift up sword against nation, neither shall they learn war any more.—Isaiah ii.2.

Reference is made to the axe, the hammer, and a winding stairway in 1 Kings vi. 7 (1000 B.C.); indirectly to spades and pickaxes in Deuteronomy viii.6 (1450 B.C.); and to the adze, saw, plane, and trowel in 2 Samuel v.11. The balance is discussed in Leviticus xix.36. The bow is part of Jacob's equipment in Genesis xlviii.22 (1700 B.C.), and also of Jonathan's accoutrement in 1 Samuel xvii.58 (1050 B.C.).

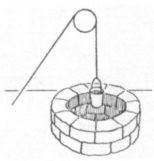

FIG. 35. Simple pulley.

David's use of a wheeled ox-cart is recorded in 2 Samuel vi.2, and chariot wheels come prominently into the picture in Exodus xiv.21. The only scriptural reference to a pulley occurs in Ecclesiastes xii.6, in the expression "or the wheel broken at the cistern" (fig. 35).

In 1 Kings vii.13, the sacred writer used the Phoenician value of 3.0 for π, the ratio between the circumference and the diameter of a circle:

And King Solomon sent and fetched Hiram from Tyre. . . . And Hiram made a molten sea, ten cubits from brim to brim; it was round and a line of thirty cubits did compass it about.

Instead of the measurement of 30 cubits in this passage, a more accurate value would have been 10 × 3.14 or 31.4 cubits. A cubit, the length of the forearm, was two spans or about 18 inches. Hence in these measurements of Hiram's "molten sea," there was an absolute error of about 1.4 cubits or nearly two feet, and a per cent error of about five.

A deadly expert use of the sling is recorded in the following well-known passage:

And there went out a champion out of the camp of the Philistines, named Goliath of Gath, whose height was six cubits and a span. And his spear's head weighed six hundred shekels of iron. And the Philistine said, I defy the armies of Israel this day. And David put his hand in his pouch, took from it a stone, slang it and smote the Philistine on the forehead . . . . and he fell to the earth upon his face.—1 Samuel xvii.4.

In this encounter David's weapon was symbolic of the Stone Age while Goliath's was of the more advanced Iron Age. Yet the champion of the Stone Age won. It has just been noted that the span or spread of the hand was about nine inches and the cubit, or two spans, about a foot and a half. So Goliath stood about three yards tall in his stocking-feet. As the shekel was about half an ounce, Goliath's spear weighed some eighteen pounds!—a nasty weapon to throw at an opponent.

The sling is a simple form of the appliance called a *centrifuge*. When a chestnut is swung around at the end of a string or when any body moves along a curved path, it exerts an outward force away from the centre of motion. This was named centrifugal (Lat. *fugio*, flee) force by Christian Huygens in A.D. 1673. The force exerted by the string on the chestnut toward the centre of motion is called centripetal force (Lat. *peto*, seek). Any contrivance which makes use of centrifugal or centripetal force may be called an application of the centrifuge. More particularly the name is used in reference to such devices as the cream-separator, shirt-dryer, or Watt's steam-engine governor. Other centrifugal applications are the top and gyroscope and the banked curve of a highway or race-course. A celestial example is the moon revolving around the earth—much like the stone in David's sling before he let fly and gave Goliath the impression that the moon had hit him. In the sequel we shall find the centrifuge of great importance in the theory of mechanics.

A tiny one-passenger boat is described in Exodus ii.2 with reference to a historic embarkation that occurred about 1570 B.C. Genesis gives the following specifications for a large one-door boat with one window:

> And God said unto Noah . . . Make thee an ark of gopher wood . . . Thou shalt pitch it inside and outside with pitch. . . The length shall be three hundred cubits [450 feet!], the breadth of it, fifty cubits and the height of it, thirty cubits. A window shalt thou make in the ark . . . and the door shalt thou set in the side.—Genesis vi.13.

CHAPTER TWO

The Classical Period

THE advent of the Greeks in science was heralded by the work of the philosopher Thales. He was of Phoenician extraction and lived at Miletus at the mouth of the Meander in Asia Minor about 600 B.C. His name stood first among the Seven Wise Men of Greece. His greatest renown arose from his knowledge of astronomy, which he derived partly from Mesopotamian sources.

Though a physicist, he took an active part in the practical affairs of his community. By cornering the olive market in a good season, he had become wealthy and was, therefore, able to devote much leisure time to study. Since the Greek word for leisure is *schole*, he was called a scholar or man of leisure. A scholar, then, is essentially one who enjoys leisure and a school is, etymologically at least, a place of leisure. When a student modestly refers to himself as "a gentleman and a scholar," this derivation helps us to appreciate the remark.

After Amasis, a Pharaoh of the Twenty-sixth Dynasty, had opened the gates of Egypt to tourists, Thales took a business and pleasure trip to that country. He was entertained by the priests, who were the teachers in that land. By keeping their knowledge to themselves, they secured for their cult great power and prestige. Even Pharaoh was pretty well under their thumb. Thales found that the annual shifting of landmarks by the floodwaters of the Nile had led the priests to the practice of land-surveying, and from them he acquired a knowledge of that subject. His attitude, however, was somewhat different from theirs. They dealt with such practical quantities as area and volume. Thales, on the other hand, found interest in the subject beyond its immediate practical value; he studied such ideas as arise in considering lines and points. With the imagination and power of abstraction characteristic of the best of the Greeks, he laid the foundations of the science of *geometry*. This name he derived from two Greek words, *ge*, land, and *metron*, measure. It records, therefore, the practical origin of the science in Egypt. Here are five of Thales' theorems which are useful in mechanics:

1. Where two straight lines cut each other, vertically opposite angles are equal [fig. 36].

2. Angles at the base of an isosceles triangle are equal [fig. 37].

3. The sum of the angles of any triangle is two right angles [fig. 38].

4. The angle in a semicircle is a right angle [fig. 39].

5. In similar triangles, corresponding linear dimensions are proportional [fig. 40].

Proofs of these theorems are given in elementary geometry texts. The enunciations are cited here for use in our discussions. They signify progress in geometry—a science which is so useful in mechanics

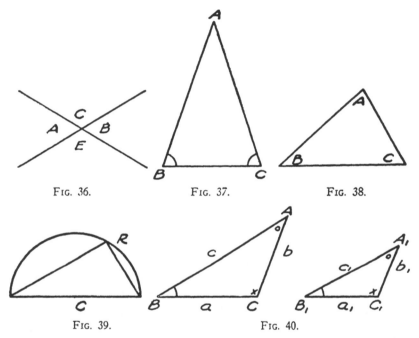

FIG. 36.   FIG. 37.   FIG. 38.

FIG. 39.   FIG. 40.

that any advance in it is potentially an advance in mechanics. In theorem 3 above, and throughout his work, Thales used a characteristic sentence in the language of mathematics, namely, the *equation*. It should be recalled, however, that Ahmes also used equations in his treatise on *All Dark Things*, twelve centuries before Thales. In theorem 5 Thales used a *proportion*, a statement of equality between two fractions or *ratios*. In fig. 40, $\triangle$ ABC is similar to $\triangle$ $A_1B_1C_1$ ($\triangle$ ABC $|||$ $\triangle$ $A_1B_1C_1$).

Thales was a great teacher and a pioneer in the Ionian school of culture, if not its founder. Among his pupils was his companion Anaximander (611-547 B.C.), who was the first Greek to propose the idea that the earth is poised in space. He considered its shape cylindrical.

As a summation of his observations of Nature, Thales proclaimed that she is not capricious nor subject to the whims of gods or goddesses, but governed by fixed and immutable *natural law*, sometimes referred to as the law of cause and effect. This fundamental tenet in science, he expressed in the famous saying: *Necessity governs all.*

THE PYTHAGOREAN CULT

A few decades after Thales, the Greek philosopher Pythagoras founded a famous school or college at Crotona in southern Italy. The college was primarily a secret religious society which aimed first at improvement of the individual and second at world reformation. Many of its votaries were ascetics who bestowed their worldly goods on the college and devoted themselves to the "rapturous pursuit of knowledge and wisdom." In time the society became a wealthy and powerful institution.

Pythagoras was a noble citizen with a magnificent mind and a very friendly personality. He was the most learned Greek of his day and travelled extensively. During his stay in Egypt, he improved his geometry. His belief in reincarnation and transmigration of souls may well have been a reflection of his sojourn in India. He studied to the age of fifty before he felt himself properly prepared to teach. He then taught to the age of ninety-five, thirty years beyond our superannuation age!

His marriage with his beautiful pupil Theano was ideally happy though of the type styled "May and December." Theano wrote a biography of her great husband but unfortunately, it, like many other priceless books, was destroyed in one of the world's worst disasters— the destruction of the Library of Alexandria (A.D. 640).

There are no writings of Pythagoras extant, for, like Thales, Socrates, Christ, and some other great thinkers, he left no writings. To keep the secrets of the society inviolate, he allowed no text-books in his college. All instruction was oral and no student was allowed to take notes at a lecture on pain of death. Might there be a modern professor who would welcome a revival of that rule?

In the Pythagorean curriculum geometry was an important subject, and one kind of writing that was permitted was the making of diagrams. What those diagrams were like is concealed in mystery, with one remarkable exception, and that exception is actually the secret symbol or crest of the society, shown in fig. 41. It is a five-point star, drawn with a continuous line, i.e. with no breaks or lifting of the stylus. Its vertices were labelled with the letters of a Greek word for good health, *hygeia*, from which comes our word hygiene. One may well wonder how this diagram came to be made public, for it would seem the least likely to be divulged. Indeed it was forbidden to disclose the symbol to the uninitiated on pain of death except in a mortal emergency. The story goes that a certain member of the cult was overtaken by a fatal sickness while travelling far from home. He summoned his innkeeper and explained that he had brought along enough money and goods for living expenses but not enough for dying. He gave his host a card bearing the sacred star and assured him that if he would let the funeral costs be "on the house," and would then expose the strange design at the portico, he would in due time be amply repaid. Presently a brother Pythagorean happened along who, on seeing the secret symbol boldly exposed to view, nearly jumped out of his mantle. He hastily entered and inquired, then paid the landlord handsomely, urging him to go fetch the crest and tarry not on the manner of his going.

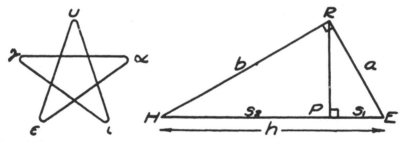

FIG. 41. Pythagorean emblem.   FIG. 42. Pythagorean theorem.

To us the design is interesting chiefly because it is the first known diagram which had the corners labelled with letters. This method of labelling has now long since become standard in geometry and physics.

Although the proof of the Pythagorean theorem given in geometry texts has admirable elegance and rigour, nevertheless, it seems rather cumbersome in comparison with the following algebraic proof which Pythagoras probably used:

In figure 42, HE is the hypotenuse h of a $\triangle$, right-angled at R. The perpendicular RP divides HE into the segments s_1 and s_2. By Thales' theorem of similar triangles,

$$\frac{s_1}{a} = \frac{a}{h} \text{ and } \frac{s_2}{b} = \frac{b}{h}$$

$$\therefore \ hs_1 = a^2 \text{ and } hs_2 = b^2$$

Eq. 1. $\therefore h(s_1 + s_2) = a^2 + b^2.$ $\therefore h^2 = a^2 + b^2.$

Tradition says that when Pythagoras first hung out his shingle as a teacher he could find only one person who would consent to listen to his lectures, and that one agreed only on the stipulation that he would receive at the end of each lecture a small coin as payment for listening.　After a number of sessions, the philosopher observing how interested his "audience" had become, refused to continue.　Then the pupil, importuning him, offered to pay a small coin at the beginning of each lecture.　Pythagoras thereupon resumed his lectures, thus affording a classical example of the modern terms "strike" and "chiselling"; also a horrible precedent for indecently low pay to teachers. The story, it must be confessed, does not march too well with the fact that Pythagoras was independently wealthy by inheritance.

Pythagoras is said to have coined the word "mathematics," which means literally the science of learning, and also the word "philosopher." A student once asked him, "Master, art thou a sophist?" (literally, a wise man).　Pythagoras replied, "I pray thee, call me not a sophist but rather a *philo*sophist ($\phi\iota\lambda\acute{o}\sigma\sigma\phi\sigma$)," i.e. one who loves wisdom. Pythagoras held with Anaximander that the earth is poised in space, but he considered it spherical; many Pythagoreans held also that it rotates on its axis.

From his study of numbers and of Nature, Pythagoras came to the general conclusion that the universe is constructed on a mathematical basis or, in other words, the Deity is a Supreme Mathematician. This idea he couched in the saying: *All things are numbers.*

ZENO'S PARADOXES

About a century after Pythagoras, there lived at Elea in southern Italy a philosopher named Zeno who is famous for his paradoxes. Some of these are especially interesting in mechanics because they discuss motion and because they show that some Greeks at that time were beginning to wrestle with the problem of infinitely small and infinitely large quantities, a topic which was peculiarly foreign to the Greek attitude of mind.　Such considerations lead to the department of algebra called the infinitesimal calculus, or briefly, the calculus. Here are five of Zeno's paradoxes:

1. If a bushel of wheat turned out upon the floor makes a noise, each grain must likewise make a noise; but in fact it is not so.
2. So long as any body is at one and the same place, it is at rest. Hence an arrow is at rest at every instant of its flight and therefore, also during the whole flight: but in fact this is not true.
3. Two bodies moving with equal speeds traverse equal spaces in equal times. But when they move with the same speed in opposite sense, each passes the other in half the time required if the other were at rest. [Thus the time required to pass each other at a "given speed" has two unequal values.]
4. Before a body can reach a distant point, it must first traverse the half distance. Before it can go the half distance, it must go the quarter distance; and

so on ad infinitum. Hence a body to pass from one point to another at a finite distance must traverse an infinite number of distances. But an infinite distance cannot be traversed in a finite time. Consequently, the goal can never be reached; but in fact this is not so.

5. If the tortoise has the start of Achilles, he can never overtake the tortoise: for while Achilles goes from his starting-point to that of the tortoise, the latter advances a certain distance: and while Achilles traverses this distance, the tortoise makes a further advance and so on ad infinitum. Consequently, though Achilles run ad infinitum, he cannot catch up with the tortoise: but as a matter of fact this is not true.

In 4 and 5 can be seen the germ of the important idea of a *variable* quantity *approaching a fixed limit*. The theory of limits is fundamental in the infinitesimal calculus. Zeno seemed to miss the fact that the sum of an infinite number of infinitely small quantities may be finite. Divide an hour into minutes, these into seconds, and these into an infinite number of infinitely small parts. Their sum is a finite quantity— one hour.

A CORPUSCULAR HYPOTHESIS

A prominent philosopher of the Ionian school, Anaxagoras (500- 425 B.C.), on whom the mantle of Thales had fallen, brought the spirit of scientific enquiry westward to Athens—to be later repaid for his labours by exile. Among his friends and disciples were Pericles, Euripides, and Socrates. One important idea that he advanced was that bodies are composed of infinitely small *particles*. The same opinion was held by Leucippus of Miletus and his disciple, Democritus of Abdera (460-370 B.C.), who was sometimes called "the laughing philosopher" because of his jovial banter. To Democritus goes the chief credit for developing and promulgating the particle view of matter in ancient times.

Abdera was a commercial town and its inhabitants in general considered trade and commerce the only pursuits worth while. So when Democritus spent his whole patrimony on travel and study, the townsmen thought he must be demented. A group of Abderites, although it was really none of their business, sent for the greatest of Greek physicians, Hippocrates himself, to diagnose the case, to find out what cogs were loose, and to prescribe. But the doctor who came to treat a madman remained to converse with a savant and to become his disciple. He reported to the money-grubbing Abderites that apparently there was considerable insanity in Abdera but it was not in Democritus.

Among Greek physicists Democritus ranks second only to Archimedes, though decidedly a distant second. He supposed that the ultimate particles of matter are indivisible. Hence he called them *atoms* (Gk. *a*, not, and *temno*, cut or divide) and his assumption is referred to as an atomic or corpuscular hypothesis (Lat. *corpus*, body, and *-ulus*,

little). Incidentally, the Latin analogue for the Greek *atomos* (ἄτομος)
is our word "individual" (Lat. *individuus*, not divisible), which we
apply to a person, meaning that individuals are the ultimate units of
which a community is composed.

The writings of Democritus covered a wide field but only his treatise on ethics is extant. Of his discussion of the atomic hypothesis,
we have only a number of quotations found in the works of other
authors. Of these the following are typical:

From Aristotle's *Metaphysics:*

Leucippus and his companion say that the Full and the Empty are the
primordial elements. They affirm that . . . the Full and the Solid constitute
Being, and the Void constitutes Not-Being.

From Aristotle's *De generatione et corruptione:*

They say that bodies are composed of indivisible particles . . . which are
infinite in number.

From Diogenes Laërtius:

Nature is composed wholly of atoms and the void. . . . The atoms are infinitely small and infinitely numerous: they move about everywhere in the universe,
executing whirling movements. It is of them that the components, fire, water,
air, and earth are formed.

From Simplicius' *De caelo:*

Leucippus and Democritus say . . . that the fundamental particles which they
call atoms or indivisibles, are . . . indestructible, because they are solid and without pores . . . and separated from each other in the infinite void. As they encounter each other abruptly they come into collision. Democritus considers
them so small that they escape our senses.

To our senses, even with the aid of the microscope, a piece of "solid
silver" (or iron) seems to consist wholly of silver (or iron) at every
point within its boundaries. It seems uninterrupted by pores or spaces,
and as such would be called a *continuum*. The atomic hypothesis of
Democritus, however, assumed in spite of appearances that the silver,
like all matter, is discontinuous or has an interrupted structure—
that it is made up of lumps and holes somewhat like lace or raisin
cake, or like a cloud of gnats. The solid parts Democritus called
atoms and the spaces he assumed to be vacuum. In a later age, we
shall see further development of this hypothesis when much pertinent experimental evidence will have become available.

If, on encountering a new phenomenon, a person's curiosity is
aroused, he seeks an explanation. He ranges through his experience
until he recalls another event which seems analogous and which he
thinks he understands. He then says that he has explained the new
event in terms of the old. Since mechanics is the fundamental science

and the one in which most people have the greatest knowledge and experience, he is apt to seek the required analogy in that science. Thus Democritus, to explain the fact that wet clothes hung up in a breeze become dry, assumed that the water particles or atoms flew into the air. He found his explanation by imagining a system of small bodies obeying the laws of mechanics. In other words, his explanation was by means of an imagined mechanical picture or model devised to suit the case.

Most advancements in science are either the discovery of new facts or the invention of new instruments, if we include in the latter new theories or ways of interpreting facts. Some theories are powerful instruments and the most successful have great scope and applicability. As we proceed on our journey we shall see that the most prominent landmarks are of this type—hypotheses and theories.

ARISTOTLE'S *MECHANICS*

The atomic hypothesis was looked upon with disfavour by the philosopher Aristotle (384-322 B.C.), who lived at Athens about a century after the days of Democritus. Aristotle was Dean of the Lyceum, a college situated in a park about two bow-shots outside the eastern city-gate and surrounded by a garden. He was the greatest of Plato's disciples in the Academy, where his nickname was "The Brain." He was one of the greatest thinkers of all times. It is to be doubted if any other man has ever recorded so many facts and principles in so many subjects as Aristotle. His writings were the encyclopaedia of the ancient world. In the Middle Ages, unfortunately, they were viewed by the church as infallible and final. Such an attitude was a serious hindrance to progress and, indeed, was quite foreign to that of Aristotle himself.

Because Aristotle and his students habitually paced to and fro in the gardens of the Lyceum as they discussed their subjects, they were dubbed "Peripatetics." To this day, certain lecturers and teachers exhibit vestiges of the same peripatetic habit when speaking to their audiences.

Aristotle's father was the court physician of King Philip of Macedon and Aristotle's chief scientific excellence was in biology. He was a tolerable mathematician but as a physicist, unfortunately, he was a much better biologist. In examining his work in physics, we shall see him at his worst, but even so, he displays brilliant flashes of genius. The historian is obliged to paint the picture, as Cromwell said, "warts and all."

To his treatise on machines, Aristotle gave the title *Mechanics* (ἡ Μηχανική), and thereby christened the fundamental science with the name which it has retained ever since. In his *Mechanics*, Aristotle stated the law of the lever quite clearly, but the discussion by which he attempted to derive it is utterly unsatisfactory. In a practical way, Imhotep undoubtedly knew this law as well as did Aristotle or even better. Here is Aristotle's so-called argument. It refers to fig. 43, which is Aristotle's own diagram, labelled, as you see, in Pythagorean style:

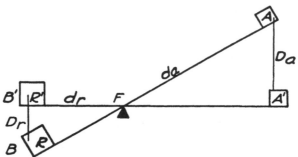

FIG. 43. Aristotle's diagram of the lever.

Let AB be a lever; R, the weight to be lifted, A, the acting weight, and F, the fulcrum. Now since [the end of] a longer radius moves faster than that of a shorter one, and since a lever has three elements, namely, a fulcrum F, and two weights, the mover A, and the moved R, therefore [?] as the resisting weight R is to the acting weight A, so inversely are their respective arms FA and FB to one another.

The farther one is from the fulcrum F, the more easily will one raise the weight R, the reason being . . . that the point which is farther from the centre of motion describes a greater circle. The original cause of all such phenomena is the circle, the perfect curve. There is nothing strange in the circle being the origin of any and every marvel.

The last paragraph sounds like an annotation by a mystic commentator of the Middle Ages. The law can be expressed in algebra thus:

Eq. 2. $$\frac{R}{A} = \frac{d_a}{d_r} \quad \text{or}$$

Eq. 3. $$R \cdot d_r = A \cdot d_a$$

where A and R are the acting and resisting forces and d_a and d_r the corresponding arms of the lever. The·product $R \cdot d_r$ is called the *moment* of R about F and $A \cdot d_a$ is the moment of A about F. *Eq. 3* is a statement of the *law of moments* for the simple case of two forces: when two forces act on a lever to produce equilibrium or balance, their moments are equal and opposite and their algebraic sum is zero.

If there is one word in physics more important than all others, it is probably the term *energy*, which means ability to do work. Aristotle did not employ the concept of energy, although he most certainly knew the meaning of work. Yet in his own diagram, the idea of energy was staring him in the eye, almost seductively ogling him. Instead of thinking always and only of the arms of the lever, of the products $R.d_r$ and $A.d_a$, and the equation $R.d_r = A.d_a$, if he could only have "come around the corner" to thinking of the distance D_r through which R is lifted, and D_a through which A descends, he could soon have arrived at the product $R.D_r$,—the work obtained

FIG. 44. Aristotle.

from the machine—and $A.D_a$, the work put into the machine. The same triangles would have shown that $R.D_r = A.D_a$. This is a simple illustration of the *law of conservation of energy*, which may be stated thus: Whatever energy one body gains some other body or system loses. "System," by the way, is a term applied to a body when we think of it as being made up of parts: thus a clock is a system of wheels, springs, and so on. Since it took the physicists, even including Archimedes and Newton, twenty-one centuries to round that corner of the diagram, we can hardly blame Aristotle for not succeeding. The schools, in making Aristotle the final authority, forced all students to think about this subject in the orthodox way, i.e. in terms of moments. Such compulsion is inimical to progress.

At this juncture we might broaden the definition of the term *machine*. Any body or system that is used to transport energy from one place to another or to transform energy from one kind to another may be called a machine. We may now have carried the broadening too far for convenience and some narrowing may be necessary later. Thus our definitions develop with experience.

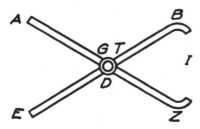

Fig. 45. Greek dental forceps.

Aristotle's *Mechanics* contains the following description of a double lever, which incidentally throws some light on the state of dentistry in those times:

Why is it that dentists pull teeth more easily by applying the additional force of the tooth-puller than with the bare hand only? The truth is that the forceps consist of two levers, opposed to one another, with the same fulcrum at the point where the pincers join. Hence they use the instrument for the extraction in order to loosen the teeth more easily.

Let *A* be one end of the forceps, *B* the other end which draws the tooth, *ADZ* one lever, *BGE* the other, and *GTD* the fulcrum. Let the tooth whose resistance is to be overpowered be at the point *I* where the two levers meet. Holding the tooth with both *B* and *Z* at the same time, he pulls and when he has loosened it, he can take it out more conveniently with his fingers than with the instrument.

Fig. 45 is Aristotle's own diagram. His article may have been written from bitter experience. Ancient Egypt, too, had her dentists, whose skill included the technique of putting gold fillings and even jewels in tooth cavities. The callousness of a geometrical account of the pulling of a tooth may recall the instructions for ridding a farm of "potato-bugs" without the use of poison—simply by placing the insect on a block *A*, superposing block *B* and pressing *B* firmly with the foot. To give the fulcrum three letters as Aristotle does in his diagram seems rather heavy-handed; perhaps he did it for emphasis. The same treatise discusses over a dozen applications of simple machines—the rudder, capstan, windlass, double-pulley, yoke, steelyard, and nut-cracker. The number of mechanical inventions was growing. Aristotle could have mentioned the lift-pump, for it was in use in Greece in his day.

ARISTOTLE ON INERTIA

From riding a bicycle or from any study of moving bodies, a person soon reaches the conclusion that in order to set a body in motion or to bring it to rest, force is required. Prehistoric man was undoubtedly aware of this idea in a practical way. A spear when hurled continues moving until stopped by some target. The latter applies force to the spear and reduces its velocity. What is more important practically is that the spear exerts force on the target. We all make application of this principle every day, for example in shovelling coal into a furnace, in stamping snow from our shoes, in throwing or catching a ball, and so forth. The property which every body has of opposing any change in its velocity is called *inertia* (Kepler, A.D. 1608). That Aristotle recognized inertia as a property common to all bodies is shown by the following sentence from his treatise called *Physics:*

Every body has the tendency to resist being forced.

i.e. to resist having its velocity changed. In our visits with Galileo and Newton, we shall see that this quotation foreshadows the First Law of Motion, sometimes called the inertia law.

The following famous passage is also from Aristotle's *Physics*. It deals with free fall, i.e. the motion of bodies that fall to earth without support (other than that of the air).

Bodies that have greater weight fall in less time than lighter ones through equal distances and in the ratio of their magnitudes [weights].

This passage has generally been interpreted as claiming that a 2-lb. block falls twice as fast as a 1-lb. block. The serious error in this statement laid Aristotle open to severe attack; and although he had to wait centuries for that attack, it certainly came at last. Apparently Aristotle did not stoop to test his conclusion experimentally, for simply to drop a stone and a pebble to the ground is enough to disprove it. One of Galileo's most blistering and satirical attacks on the peripatetics was his experimental and jubilant manhandling of this passage. It is of considerable historical importance, for by disproving it Galileo struck off some of the mediaeval shackles and gained for us greater freedom of thought. Nevertheless, in this year of grace, there are still people who instinctively subscribe to the old Aristotelian error.

Continuing his discussion, Aristotle drew another inference which has not marched well with subsequent research. His argument hinged partly on the opinion that the motion of any body B is caused by the medium in contact with it, e.g. the air. This was because it was commonly denied that any body could exert force at a point where it was not.

Therefore, bodies would also move through the void [vacuum] with this ratio of speeds. But this is impossible, for why should one move faster when surrounded by nothing? Therefore, all will possess equal velocity. But this is impossible. It is plain, therefore, . . . that Vacuum does not exist either by itself or as a component of bodies.

In the hands of Aristotelians, this passage and others like it became the much-used formula, "Nature abhors a vacuum." The idea was often referred to as the "horror of the void," or *horror vacui*.

In opposition to Democritus, Aristotle considered that a material body is a continuum. There could be a similar difference of opinion about a pencil-mark which has no gaps. One person might think of it as being infinitely divisible. No matter how often its parts were subdivided, every part could still be divided (with a sufficiently sharp knife) into two shorter lines. A line which is thus infinitely divisible is called a continuum; every part of it shades into the next part so that there is no telling where any part begins or ends. If, however, we say the parts are geometrical points, i.e. with zero length, we run into a Zenoid paradox, for the sum of all their lengths would be zero; "but as a matter of fact it is not so."

One might also think of the line as a series of small dots or points somewhat like a chain of minute beads or particles. Such a line would be made of indivisible *discrete* parts and could not be divided more finely than into its component dots; it would not be a continuum. With a sufficiently good microscope, it would be possible to distinguish where any dot began and ended.

In such a controversy, that opinion is preferable which agrees better with the experimental facts. The topic was not profitable for debate between Democritus and Aristotle or their disciples, for they lacked important facts which have since been discovered. The only logical verdict in their day was one of suspended judgment until enough pertinent facts were available.

Aristotle considered a solid bar of pure silver, for instance, as composed wholly of silver throughout. The space enclosed by its surfaces he held to be completely filled with silver only. He denied that solid silver contained even infinitesimal pores or interstices. Hence he called it a *plenum* (Lat. *plenus*, full). Since this was his conception of all bodies, he claimed that Nature also is a plenum: we have just read his denial of the existence of vacuum. Similarly, water in a bowl, he considered, could be divided into two bodies of water, each of these into two and so on forever if a sharp enough knife could be obtained.

Here is a passage from Aristotle's treatise *Concerning the Sky*, which discusses the topic. Again there is a note of mysticism. This

time it concerns the magic of the "perfect number 3" which was held
in veneration by the Pythagoreans:

Physics studies bodies, . . . their properties and the laws that they obey.
Now a continuum is infinitely divisible; hence a body is a continuum. A magnitude, if divisible in one direction, is a line; in two dimensions, a surface; and in
three, a body [volume]. No magnitude can transcend these three because there
are no more than three dimensions. For three is the perfect number [!]. As the
Pythagoreans say, the universe is based on the number three which has a beginning, a middle and an end. . . . Thus body [volume] is the only magnitude which
is perfect for it alone has three dimensions.

THE SECOND LAW OF MOTION

Aristotle's most brilliant feat in physics was his partial anticipation
of Newton's second law of motion, which was first clearly stated by
Galileo nineteen centuries after Aristotle. Here is the passage:

If the acting force F has moved a body B through a distance S in a time T,
then in the same time the same force will move $\frac{1}{2}B$ through $2S$, twice the
distance.[1]

The two cases can be synopsized thus:

$$(a)\ F,\ B,\ S,\ T$$
$$\text{and } (b)\ F,\ \tfrac{1}{2}B,\ 2S,\ T.$$

Thus if F and T are constant, $S \propto \dfrac{1}{B}$. Aristotle also stated that if B
and T are constant, $S \propto F$.

$$\therefore \text{ if } T \text{ is constant, } S \propto \frac{F}{B} \text{ or } F \propto BS.$$

Aristotle did not use Galileo's term *acceleration* (symbol a) which
means change of velocity per sec. When two bodies begin from rest
and move with uniform accelerations a_1 and a_2, the distances they
traverse in equal times are to one another as $a_1 : a_2$. Or briefly, $S \propto a$.
If we substitute weight w for body B and acceleration a for distance S,
then Aristotle's statement, $F \propto BS$, becomes

$$F \propto wa \quad \text{or} \quad F = kwa.$$

This was Galileo's form of the second law of motion. In this matter,
therefore, and to the degree indicated, Aristotle was far ahead of his
time. Then he got to thinking about some of the implications of his
statement and ran into what looked like a Zenoid paradox. One consequence, for example, is that any force, however small, should be able
to move any body, however large. For instance, a boy should be able
to move a ship. This deduction shook Aristotle's faith in his theorem.
He should have stood his ground, for he was right, but who could blame
him for wavering? In the case of the ship, a new force, liquid friction

[1]Aristotle *Physics*.

or viscosity, comes into play and neutralizes the boy's push as soon as
he begins to move the ship. Remove or counteract this force and he
can move the ship, though with relatively small acceleration and
velocity.

ARISTOTLE ON VECTORS

There are certain quantities which are not completely described
until their directions (and senses) are specified, e.g. displacement,
velocity, acceleration, force, etc. Such quantities are called directed
quantities, or, briefly, *vectors*. It is frequently necessary in mechanics
to calculate the combined effect of two forces which act on a body in
different directions. In other words, we wish to find the sum of two
vectors. That sum is called the *resultant* (**R**) of the two vectors and
each is called a *component* of **R**. If they are at right angles to each
other, they are called *rectangular components*. Aristotle was the
first to show how to calculate the resultant of two simultaneous displacements if they concern the same body and take place at right
angles to each other (fig. 46).

Now if the two displacements of a body A are in any constant ratio, the
resultant displacement must be along a straight line, namely, the diagonal of
the rectangle formed by the lines drawn in that ratio. [2]

Conventionally, a vector is indicated by writing a bar above or
below the symbol, or, in printing, by bold face type. Thus in fig. 46,

Eq. 4. $c_1 + c_2 = R$ (a vector equation)

If we divide the displacements by the times, which in this case are
equal, we obtain a corresponding theorem for velocities. If we substitute vector for displacement in Aristotle's statement, we obtain the
well-known theorem called the *triangle of vectors* for the special case
of rectangular components. If we consider only the magnitudes of these
vectors as represented by the lengths of the three lines, e.g. in the case
of velocities, disregarding direction and considering only the speeds,
then

Eq. 5. $R = \sqrt{c_1^2 + c_2^2}$ (*Pythagorean theorem*)

The process of combining two vectors or replacing them by a single
equivalent vector (their resultant), is referred to as composition or the
compounding of vectors. The opposite process is called resolution or
the resolving of a vector, i.e. replacing a vector **R** by two or more components c_1 and c_2, whose vector sum is **R**. Using fig. 46, Aristotle
showed how to resolve a displacement along a straight line into two
rectangular components, thus:

[2]Aristotle *Mechanics*.

The converse is also true. If a point is moved along the diagonal by two displacements, it is moved necessarily according to the proportion of the sides of the rectangle.

Then Aristotle made an excellent preliminary step by these words:

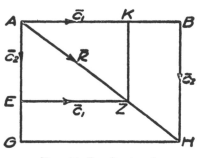

FIG. 46. Resultant and components.

If the two displacements do not maintain a fixed ratio during any interval, a curve is produced.

This passage contains the germ of a method for dealing with bodies that move along circular or other curved paths, for example, projectiles moving along parabolic trajectories or the moon on its orbit around the earth.

As we say au revoir to this wonderful old Greek we hear a student ask him the question, "Master, where dwell the Muses?" and the master's reply, "The Muses dwell in the souls of those who love *work*."

THE FIRST PROFESSOR OF GEOMETRY

One of Aristotle's old pupils, Alexander, the Macedonian conqueror (or marauder), founded the city of Alexandria at the mouth of the Nile in 332 B.C. and modestly gave the city his own name—a compliment which he also vouchsafed to some sixteen other cities that he overpowered. His ablest general later became Pharaoh of Egypt as

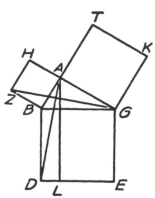

FIG. 47. Euclid, Book I, Proposition 47.

Ptolemy I, and used his great power to foster the advancement of science. Inspired by the great library that he had seen in Babylon, he established the Library of Alexandria, which grew to be the largest of the ancient world containing some 700,000 scrolls, many of them single copies. Its nucleus was the private library of Aristotle. Ptolemy also founded at Alexandria the first university, as we understand the term, and it became the greatest of the ancient world, with an enrolment of about 14,000 souls. It was dedicated to the Muses and called, therefore, the Museum. Its first professor of geometry was the celebrated Euclid (315-250 B.C.) whose famous text, *Elements of Geometry*, has been the supreme marathoner of all text-books, for it has remained in use for over twenty centuries. The meaning of the word Museum has changed somewhat. In the modern sense it suggests a collection of fossils, but in the Alexandrine sense it suggested a group of professors.

During the centuries when Euclid's text was used, the science of geometry was commonly called "Euclid"; this man is the only one after whom a science has been named. His contribution to the human

Fig. 48. Euclid.

race has been great. His book has given training in exact reasoning to the youth of sixty generations throughout the civilized world. His influence is seen in the work of others, for instance, in the writings of Archimedes, Apollonius, Galileo, Newton, and Huygens. Like Bach in music, he has been a master of masters. Fig. 47 is the well-known diagram for Euclid's classic demonstration of the Pythagorean theorem. We shall use several of Euclid's propositions in our arguments.

Like many great teachers, Euclid is said to have had at times a rather caustic tongue. At the end of a demonstration, a pupil once asked him, "What money shall I gain by learning this?" Euclid summoned his slave, that is, in modern terms, his demonstrator, and said, "Give this youth threepence, since he must make money by what he learns."

Tradition says that on a certain occasion King Ptolemy, who found by experience that the study of geometry entails considerable mental exertion, asked Euclid for an easier method of mastering the subject. Euclid's reply is famous: "There is no royal road to Geometry."

THE FATHER OF EXPERIMENTAL SCIENCE

In the realm of physics and mathematics, Aristotle and Democritus were far surpassed by Archimedes of Syracuse (287-212 B.C.), the greatest physicist of the ancient world and one of the greatest mathematicians of all time. He was of noble birth, a relative of King Hieron of Syracuse. His father, Pheidias, who was an astronomer and geometrician, sent him to the University of Alexandria, where he studied physics and geometry. Ultimately he became his alma mater's most distinguished alumnus. He may have had the privilege of taking lectures from the great Euclid himself during the last year of the latter's professorship, or if not, at least his instructors were men who had studied under Euclid. In any case, there is no doubt that Euclid had a strong influence on Archimedes.

After a brilliant course at Alexandria, Archimedes returned home to Syracuse, where for the rest of his life he carried on research in geometry. His kinsman, King Hieron, realized that Archimedes was an unusual genius, but bemoaned the fact that a man of such superlative intellect should potter about night and day with cones, spheres, cylinders, and "sich like theoretical twaddle," when the world was fairly bristling with important practical problems crying for solution. It happened that Hieron had sent his jeweller a weighed quantity of gold to be made into a new crown. When the crown was delivered and duly tested on the family scales, it was found to have the correct weight. Yet Uncle Hieron was haunted by a lurking suspicion that the rascally goldsmith had substituted silver for gold in the interior of the crown. So he offered his nephew the nice task of solving the mystery. Archimedes was not to mar the precious bauble by so much as a scratch, but was just to tell his king the crown's true inwardness without looking inside.

Ordinarily Archimedes considered it beneath the dignity of a pure geometrician to have any truck with so sordid a subject as physics, but the difficulty of this problem intrigued him. So he undertook the investigation. It is said that one day as he was walking to the public baths, influenced probably by the plunge he was about to take, he kept thinking about a fact that had been known to every swimmer since prehistoric times, namely, that a stone or a body immersed in a river is easier to lift than on land. What was not known before Archimedes' day was *how much* lighter the stone is in water. The attendant had filled the bath-tub brimful of water and as Archimedes hopped in and submerged, water flowed over the brim. While the attendant was soaping, rubbing, and perfuming the philosopher, the latter was absorbed in his problem. Suddenly the idea struck him that the weight he lost on submerging was equal to that of the water which brimmed over.

In his joy at making this discovery, tradition says that he jumped out of the tub and, absentmindedly forgetting his clothes, raced down the path, wrapt in frenzy and exclaiming "Eureka, eureka"—a rather risky thing to do even in the balmy Syracusan climate.

Arriving home without arrest, he performed a series of experiments to test his surmise. At last he had captured the idea which solved the puzzle, and which has ever since been named in his honour "Archimedes' Principle." He found that when a body is immersed in a fluid, the apparent loss in its weight equals the weight of the fluid displaced, or in other words, the buoyancy of a fluid on any body immersed in it equals the weight of the fluid displaced. Archimedes now wrote his famous treatise, *Concerning Floating Bodies* (περὶ ὀχουμένων), which is a marvellous piece of work judged even by modern standards. Thus Archimedes brought the science of hydrostatics from a qualitative to a quantitative status. He has been called the "Father of Experimental Science," although Pythagoras and Euclid were both experimenters.

Here is a literal translation of Archimedes' statement of his hydrostatic principle as it appears in his treatise. The influence of his tub experiment can be plainly seen in the wording.

Theorem vii. Any body which is denser than a fluid, if let down into the fluid, will sink to the bottom and will be lighter in the fluid by the weight of the fluid which has the same volume as the immersed body.

This differs somewhat from the statements commonly given in modern texts.

Postulate I of the same treatise will assume special interest when we come to discuss the work of Pascal, for it comes close to what is called Pascal's Principle.

Postulate I. Let it be granted that a fluid is of such a nature that its parts lying evenly and being continuous, any part of it which is at less pressure is driven along by one that is at greater pressure, and that each of its parts is thrust by the fluid vertically above it, if the fluid is in a container and subjected to pressure.

It is seen that Archimedes agreed with Aristotle in considering a body of water a continuum. That he should be of the Aristotelian school is not surprising if we recall the chain of influence suggested by the names, Aristotle, Alexander, Alexandria, Euclid, and Archimedes. Thus science too has her chain of apostolic succession.

Theorem ii of the same book points out the essential characteristic of a liquid and incidentally shows that Archimedes, like Pythagoras, considered the earth a sphere.

Theorem ii. The surface of any liquid at rest is the surface of a sphere which has the same centre as the earth.

Let us take a glance at the calculations by which Archimedes analysed the crown. But first, did it have any silver in it? Optimist! Suppose the crown weighed 684 grams in air and 644 gm. in water. The density of gold is 19.3 gm. per cc. and of silver 10.5.

Loss of weight $= 684 - 644 = 40$ gm.

$\therefore$ weight of water displaced $= 40$ gm. (Archimedes' Principle)

$\therefore$ volume of crown $\qquad = 40$ cc. (density of water 1 gm. per cc.)

Crown's weight if pure gold $= 40 \times 19.3 = 772$ gm.

$\therefore$ shortage in weight $\qquad = 772 - 684 = 88$ gm.

Substituting 1 cc. silver for gold reduces the weight by
$$19.3 - 10.5 = 8.8 \text{ gm.}$$

$\therefore$ number of cc. silver in crown $= \dfrac{88}{8.8} = 10$ cc.

Similarly, no. of cc. gold $\qquad\qquad = 30$ cc.

$\therefore$ by volume, the crown was ¾ gold and ¼ silver, or 75% and 25%.

Archimedes' treatise *On the Lever* has not come down to us, for it perished in the destruction of the Library of Alexandria by the Mohammedans in A.D. 640. While Archimedes was writing the treatise, Uncle Hieron managed to get himself into another of his practical dilemmas. He had built an unusually large trireme as a present for Pharaoh, so large, in fact, that his workmen could not drag it into the water. As usual, he appealed to his theorizing nephew to extricate him from his trouble. Archimedes' reply is his most famous short saying:

Give me where to stand and I will move the earth.

Hieron doubtless thought his nephew's head was quite swollen but he swallowed his irritation since he yearned greatly to have his ship launched. Archimedes meant merely that if a suitable fulcrum is furnished and a long enough and strong enough lever, any weight what-

ever can be lifted. We have seen that the pulley is a development of
the lever. With a combination of pulleys which we call a pulley-block
or a block-and-tackle, and which Archimedes devised for the occasion,
he moved the ship with ease. He also let Hieron pull the rope, to see
how easily a practical man can solve his problems when he has a
master of theory to help him.

Archimedes' discussion of the lever referred repeatedly to the equal-
arm balance and particularly to the condition of rest or "balance"
which it reaches when the two weights are equal. The word that suggests the equality of the two forces and the resulting condition of
balance is the word *equilibrium* (Lat. *libra*, scales). This idea is of
tremendous importance in science. The greatest legacy bequeathed to
mechanics by the ancient world was Archimedes' treatise *On Equilibrium*. Because of it, he is called the "Founder of Statics," the department of mechanics which deals with systems at equilibrium.

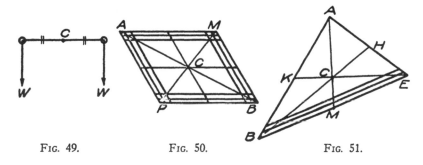

Fig. 49.   Fig. 50.   Fig. 51.

In that treatise Archimedes introduced the term *centre of gravity*
(or centre of mass). That centre is a unique point in, or related to,
any body or system, through which the resultant of the weights of all
its parts always passes no matter how the system is turned. Here are
a few of Archimedes' fundamental propositions on *centre of gravity*
(κέντρον βάρεος).

Postulate I. Equal weights at equal distances are in equilibrium [fig. 49].
Proposition iv. If two equal weights have not the same centre of gravity,
then the centre of gravity of the system is at the middle point of the straight
line joining their centres of gravity [fig. 49].
Proposition x. The centre of gravity of any parallelogram is the point of
intersection of its diagonals [fig. 50].
Proposition xiv. The centre of gravity of any triangle is the intersection
of its medians [fig. 51].

This point in a triangle is called the *centroid* and frequently the term
centroid is used as a brief synonym for the longer expression, centre
of gravity, especially among engineers.

Archimedes' treatise *The Sand-Reckoner* was addressed to Prince

Gelon, the son of King Hieron. It deals with relatively huge numbers and contains the germ of the ideas of indices and logarithms. Its title intimates that it proposes to express the number of sand-grains in a sphere as large as the solar system.

In *The Sand-Reckoner* there is an interesting reference to another treatise which unfortunately is no longer extant—another lamented casualty in the destruction of the Library of Alexandria. This treatise was written by one of Archimedes' contemporaries, the astronomer Aristarchus of Samos, who proposed a *heliocentric* theory which was an anticipation of the Copernican theory of the solar system. This is the reference.

> Aristarchus of Samos published some writings advancing certain hypotheses. . . . He assumed that the fixed stars and the sun remain at rest and that the earth revolves around the sun on the circumference of a circle with the sun at the centre of its orbit.

It is worth noting that Samos was also the home of Pythagoras who was exiled because of his fearless questioning of some accepted ideas. The same insensate hatred which has been displayed so often by narrow ignorant office-holders against their intellectual and personal superiors cast Aristarchus into prison on a false charge of impiety and well-nigh cost him his life.[3]

THE VALUE OF π

Archimedes once said to some friends that if ever his grave had a tombstone, he wished that the sculptor would ornament it with a design of sphere and cylinder, because he thought these would symbolize his best work. In this opinion history has shown that he was right, and Hieron, who contemned such theoretical studies, was wrong.

When the fall of Syracuse was imminent, the Roman general, Marcellus, gave orders that the life and property of Archimedes were to be spared, for he had come to hold unbounded admiration for the sage whose inventions had kept a Roman army at bay for three years. During the sack of the city, however, an ignorant soldier slew Archimedes. Marcellus deplored this tragic blunder and throughout his triumph at Rome lamented the death and absence of Archimedes.[4] He erected a splendid monument over Archimedes' grave, sculptured according to the sage's wish. In 75 B.C., Cicero found the grave in disrepair and had it rehabilitated, as befitted the memory of so great a man.

To obtain the value of π, Archimedes inscribed and escribed regular 96-gons in and about a circle and calculated their perimeters.

[3] Oswald Spengler, *Decline of the West* (Knopf, 1926), I, 9 and 68.
[4] Plutarch *Marcellus*.

In making these calculations he used the Pythagorean theorem and the third proposition of Euclid's Book VI. Taking the circumference as intermediate between the two perimeters and dividing by the diameter, he obtained the value π between limits. He published his result in the treatise *Measurement of the Circle:*

> Thus the ratio of the circumference to the diameter is greater than 3 10/71 but less than 3 10/70.

That is to say, $3.1428\ldots > \pi > 3.1408\ldots$ The result can be written $\pi = 3.1418 \pm 0.001$. (Archimedes' remarkable calculations in this research are given in greater detail in Appendix A.)

Such then was the first determination of the value of π with an accuracy of about 1 in 3000. We saw that Ahmes' value of π was about 1 per cent too great. An outstanding difference between the two treatments was that the Greek scholar showed how he obtained his rule, whereas the Egyptian kept his method a dark secret and quite possibly did not know how it had been derived by some earlier mathematician.

To give some idea of the difficulties under which the Greeks laboured in making calculations, the Greek numerals are tabulated here:

1	2	3	4	5	6	7	8	9
A	B	Γ	Δ	E	F	Z	H	Θ
10	20	30	40	50	60	70	80	90
I	K	Λ	M	N	Ξ	O	Π	Q
100	200	300	400	500	600	700	800	900
P	Σ	T	Υ	Φ	X	Ψ	Ω	&

There was no zero symbol.

Let us try a subtraction problem and then a piece of division.

$$
\begin{array}{cccc}
32 & \Lambda B & 17)\overline{323}(19 & IZ)\overline{TK\Gamma}(I\,\Theta \\
17 & IZ & 17 & PO \\
\hline
15 & IE & 153 & \overline{PN\Gamma} \\
& & 153 & PN\Gamma
\end{array}
$$

Note the inconvenience of having unrelated symbols for such a sequence as 4, 40, 400. As a special flutter we might try to extract the square root of 3 using Greek or Roman numerals. As we struggle with the intricacies of this computation, our admiration of the mental power of Archimedes, "The Reckoner," increases by leaps and bounds; especially when we learn that he likely carried out the process without writing, or, as we say in school, by "mental" arithmetic, or "in his head." He expressed his result thus:

$$\frac{1351}{780} > \sqrt{3} > \frac{265}{153}$$

or in decimals, $1.7320512 > \sqrt{3} > 1.7320261$. The average of these limits, $1.7320386 \ldots$ would be low by about 2 in 170,000 for $\sqrt{3} = 1.7320508. \ldots$ Just how Archimedes managed to make this calculation is not definitely known. He often checked his results by physical experiments, for example, the area of a circle by cutting one from metal foil and weighing it and a square of the same material. But such a method would be difficult to apply in this case.

In finding the area of a circle, Archimedes employed the famous method of Eudoxus, called *exhaustion*. To prove, for instance, that $a = b$, Eudoxus would show that a is not greater than b $(a \ngtr b)$ and

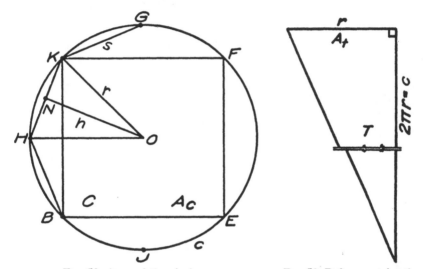

FIG. 52. Area of the circle.   FIG. 53. Reference triangle.

that a is not less than b $(a \nless b)$. Then a must equal b. The procedure is similar to Sherlock Holmes's method of deduction by elimination.

In fig. 52, C is a circle with centre O, circumference c, radius r, and area A_c. T is a $\triangle$ with sides about its right-angle respectively equal to r and $c = 2\pi r$. $\therefore$ its area $A_t = \frac{1}{2}cr = \frac{1}{2} \cdot 2\pi r \cdot r = \pi r^2$ (fig. 53).

It is claimed that $A_c = A_t = \pi r^2$. If not, then (i) $A_c > A_t$, or (ii) $A_c < A_t$. Suppose $A_c > A_t$ and that $(A_c - A_t) = e$, a constant. Inscribe in C a square $KBEF$. Its area $A_q < A_c$ by the four segments that it subtends. Bisect the arcs KF, KB, etc., at G, H, J, etc., and join these points to the vertices to form a polygon whose area $A_p > A_q$ but $< A_c$ by the segments that it subtends. As this procedure is carried on, A_p approaches A_c, or, in symbols, $A_p \rightarrow A_c$. Continue the process until $A_c - A_p < e$, i.e. until A_p is nearer A_c than is A_t. Then $A_c > A_p > A_t$. The perimeter of the polygon $P < c$.

FIG. 54. Archimedes.

Let the polygon have n sides each of length s. Then $P = sn$. Let ON, the perpendicular on any side from $O = h$ ($<r$).

$$\therefore A_p = n \cdot \tfrac{1}{2}hs = \tfrac{1}{2}(sn)h = \tfrac{1}{2}Ph.$$
$$\because A_t = \tfrac{1}{2}cr \text{ and } \because P<c \text{ and } h<r$$
$$\therefore A_p < A_t.$$

This is absurd since $A_p > A_t$ by hypothesis,
$$\therefore A_c \not> A_t.$$

Similarly by means of an escribed polygon, it is found that $A_c \not< A_t$

$$\therefore \text{ by the method of exhaustion, } A_c = \pi r^2.$$

By a more complicated use of exhaustion, Archimedes found that the surface of any sphere is four times its greatest circle. If an orange is cut in two by a knife blade that passes through the centre, the area

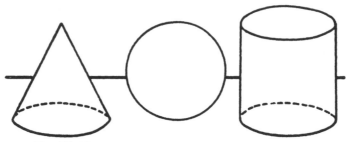

Fig. 55. Volume of right cone, sphere, and cylinder.

of the dome of each hemisphere has twice the area of the circle exposed (πr^2). Then, if A_s is the area of the surface of a sphere,

Eq. 6. $$A_s = 4\pi r^2.$$

Archimedes also showed that if a right cone, a sphere, and a cylinder have equal diameters and heights (fig. 55), their volumes are in the proportion 1:2:3, for their values are respectively $(\tfrac{1}{3})2r \cdot \pi r^2$, $(\tfrac{4}{3})\pi r^3$, and $2r \cdot \pi r^2$, which are in the proportion $(\tfrac{2}{3})\pi r^3 : (\tfrac{4}{3})\pi r^3 : (\tfrac{6}{3})\pi r^3 = 1:2:3$. These are all volumes and accordingly r is in the third power in each.

We have seen that in some of his work Archimedes used the principles of limits, which are fundamental in infinitesimal calculus. In some calculations he really used the calculus—nineteen centuries before Newton and Leibnitz. Great then must be our veneration for this lone pioneer who stands in splendid isolation, far above all his contemporaries.

Archimedes considered it beneath a pure geometrician to write about such practical things as machines. Accordingly, we find in his treatises hardly any references to his famous mechanical inventions. These included grappling irons which capsized a Roman trireme, a catapult that sank another, a combination of pulleys that moved a large ship with ease and the Archimedean spiral pump which seems to cause water to run uphill, and which is still used in irrigation.

The famous painting of Archimedes (fig. 54) gives a powerful impression of his tremendous concentration on a problem. We see him about to die as he had lived, engrossed in mathematical contemplation. In our subsequent discussions, we shall often refer to him, for his work has endured through the centuries. Though dead, he yet liveth.

MECHANICS UNDER ROME

Among the Romans, science made less progress than with the Greeks. Romans in general were not equal to the Greeks in imagination and abstract thinking; their bent was more practical and commercial. They were, however, excellent warriors and builders, and the constructing of buildings and of engines of warfare belongs to the province of applied mechanics and engineering. The magnificence and grandeur of Roman architecture expressed the power and dominance of the Roman empire.

About the time of Julius Caesar, there lived at Rome a remarkable man who was both scientist and poet—Lucretius Carus. As a scion of one of Rome's wealthiest patrician families, he received the best education obtainable at Rome and in all probability continued his studies abroad in Greece. He was deeply impressed by the teachings of Epicurus and of Democritus, the chief proponents of the atomic hypothesis. The influence of that theory is quite patent throughout the famous poem, "Concerning the Nature of Things" (*De rerum natura*), which Lucretius published in 55 B.C., the last year of his life. In the following three passages from that poem, Lucretius addresses to his friend Memmius an invitation to join him in the pursuit of science; but the scope of the invitation is universal:

Now apply your mind, I pray, to a true reasoning; for a truth wondrously new is struggling to reach your ears, and a new view of creation to reveal itself. . .

Nor am I unaware of the difficulty of expounding in Latin verses the abstruse discoveries of the Greeks, especially when many of the notions necessitate the coining of new words because of the poverty of our language and the novelty of the ideas. . . .

But your merit and the anticipated pleasure of your sweet friendship urge me to undertake any toil and induce me to keep vigil through the silent nights, searching by what words, yea and in what measures, to spread before your mind a bright light by which you may examine deeply into hidden truths. . . .

Many scientists have accepted this friendly invitation, among them Sir William Bragg and the German philosopher Immanuel Kant. They have found Lucretius a sincere and stimulating friend and have declared with enthusiasm their indebtedness and gratitude. The second quotation emphasizes the fact that the study and use of language is an important factor in the progress of science. Words are instruments just as truly as calipers, balances, and clocks, and they should be used with the same meticulous care and precision. A considerable part of science is word-work.

The third excerpt gives a true picture of the attitude of a great teacher. Perhaps you have noticed that among the great scientists whom we have met thus far a high proportion are teachers, for example, Thales, Pythagoras, Anaxagoras, Aristotle, and Euclid—and so it will be to the end of the story.

Fig. 56. Lucretius.

In the following passage about empty space, Lucretius opposes the Aristotelian principle of the horror of the void, and the idea that the universe is a plenum:

And yet not all bodies are held close-hemmed on every side by matter: for there is empty space in bodies. . . . There is accordingly Void, mere space, intangible and vacant. For if there were not, by no means could any body move; for the property of matter of resisting and hindering [other matter from occupying the same space] would perpetually confront all bodies. Nothing, therefore, could advance, since no body could be first in yielding place.

The argument here assumes that matter is impenetrable, that is to say, if a certain body, e.g. a chair, occupies a certain position, another body, e.g. another chair, cannot occupy the same position at the same time. The first body fills the position to the exclusion of other bodies. Most people instinctively assume the impenetrability of matter, having abstracted the idea unconsciously from dozens of experiences that come to all. For example, in a dark room, we learn that if a doorpost occupies a certain place, a nose or toe cannot occupy any of the same space even though brought forcibly into close contact with the post. Any apparent success we may enjoy in this project merely results in the removal of part of the post.

ATOMIC THEORY

Lucretius' enthusiastic adoption of the atomic hypothesis is plainly seen in the following typical excerpts:

> There exist then certain minute bodies which consist of no parts. They are the smallest particles in nature. . . . All nature is composed of two things; for there are atoms and vacuum in which the atoms are situated and in which they move hither and thither.

Since Lucretius spoke of coining Latin words to translate Greek terms, one might expect him to introduce the important term *atomus* from the Greek *atomos*, meaning indivisible. In this however, one is disappointed. In fact, the first Roman to use the term *atomus* was Cicero, a contemporary of Lucretius.

Lucretius both supports and attacks Aristotle's error about free fall. He claims that *in vacuo* all heavy bodies will descend with the same acceleration.

> For whatever bodies fall down through air, . . . the air, though of tenuous consistency, cannot retard all bodies equally, but recedes more rapidly when displaced by a heavier body. But on the other hand, empty vacuum can never resist any body in any direction. Wherefore, *all bodies must needs fall through inert vacuum with equal acceleration even though they are of unequal weights.*

Later we shall see Aristotle's error attacked by Galileo experimentally, and therefore more forcibly. But Lucretius was no experimenter. He observed Nature and drew his inferences. Among those who have followed this method, it is to be doubted whether there has been in the whole story of science a keener or more astute investigator. Sometimes Lucretius' prophetic insight was downright uncanny, for instance, when he surmised the existence of disease germs and thus foreshadowed the wonderful work of Louis Pasteur, Robert Koch, and modern bacteriologists.

VARIABLES AND CONSTANTS

Any quantity that changes value during a certain time is called a varying quantity, or, briefly, a *variable*, during that interval. The temperature of the air, for example, frequently varies: the lengths of day and night continually change from season to season, and the price of butter usually rises. The poet has said, "Change and decay in all around I see."

We saw that any quantity which retains the same value for a certain interval of time is called a constant quantity during that time, or, briefly, a *constant* (p. 17). Thus the velocity of an automobile may remain unchanged for several minutes. The height of the house in which you live has probably remained fairly constant for several years. We have found that the ratio between the circumference of a circle and its diameter, $\pi = 3.14159.\,.\,.$, is a constant forever. Constants are much rarer than variables. One chief occupation of the scientists seems to be a perpetual hunt for a constant in a world full of variables. A facetious philosopher once said, "The only constant in the world is change." More than one scientist has gained fame by the discovery of a new constant or by a more accurate determination of the value of an old one. Thus Archimedes' determination of π was an important advance on the work of Ahmes.

Lucretius declared the significance of constants and pointed out two important examples in the following passages:

For it must be that something remains changeless, or else the whole creation would be utterly annihilated. . . .

Wherefore, the atoms are now in motion with the same motion as they had in ages past and to all eternity will continue to move in the same way. . . .

The constancy referred to in these words foreshadows the discovery of the principle called the *conservation of momentum*, which we shall discuss when dealing with the work of Galileo and Newton.

Toward the end of the eighteenth century, the great French chemist Lavoisier won fame by his experimental derivation of the law of *conservation of mass*, which may be stated thus: Regardless of what processes occur in any isolated system, its mass or quantity of matter remains the same. If the whole experiment is carried out at the same place, as it usually is, the term weight can be substituted for mass in the statement, for in that case the mass of every body is proportional to its weight. In fact, mass is commonly measured by weight, i.e. by means of a balance; and the units of mass and of weight, though quite different, are spelled the same, thus: pound (mass) and pound (weight) or gram (mass) and gram (weight). (It has been suggested that the abbreviation for pound [force] be *pd.*, and for pound [mass] *lb*. These abbreviations will be used henceforth in this book.) The weight of a body is the force of attraction between it and the earth. The mass of a body or its quantity of matter is not a force nor does it depend on the earth. When coal burns in air, the total mass (or weight) of the chimney gases and the ashes produced is the same as the initial mass

of the coal and oxygen used up in the process. If this is true throughout the universe, the principle can be stated thus: The mass of the
universe is constant. The following passage shows that eighteen centuries before Lavoisier, Lucretius, by his remarkable power of induction,
his poetic flight, and his use of the atomic hypothesis, had reached the
same idea without experiments:

> Things cannot be produced from nothing, nor, likewise, when created, can
> they be annihilated. . . . No agent can destroy the atoms, the ultimate consti-
> tuents of bodies, for by their solidity they persist forever. . . . Not a single
> body, therefore, vanishes into nothing. . . . The universe is seen to remain con-
> stant, since all atoms that depart from any body diminish that from which
> they depart and produce an equal increase in those to which they are added. . . .

· FRICTION

Wherever two solid surfaces rub against each other, the force of
friction is manifested and particles are separated from the surfaces.
This wearing away of a surface where there is friction is called *attrition*.
In sharpening a knife, we make use of attrition. But when we are
obliged to buy new brake-bands or new shoes, we are apt to wish
vainly that there could be friction between solids without attrition.
The purpose of lubricants is to reduce friction and attrition. Lucretius
used the facts about attrition to show that atoms must be exceedingly
minute.

> Nay more, as the sun's year rolls around again and again, the ring on one's
> finger becomes thin by wearing. Drops of falling water can wear holes in stone. The
> curved iron ploughshare, cutting hiddenly through the earth, becomes worn
> away. We see the pavement stones of the streets worn away by the feet of the
> multitude. Again, by the city gates, the brazen statues reveal that their right
> hands are wearing thin through the hand-clasps of the visitors who greet them
> ever and again as they pass on their way. All things, then, we see grow less as
> they are rubbed away. Yet what particles leave them at each moment, the vexa-
> tious limitations of our sight prevent us from seeing.

Lucretius has a friendly word of encouragement to any who feel
that they have not sufficient mental power to study science: "Even a
shallow pool of water on the road can reflect the vast heights of
Heaven."

ALEXANDRINE SCHOOL

A century after the days of Lucretius, an Alexandrine physicist
named Hero (or Heron) published some books on mechanics. In these
he described a number of appliances invented by himself and by a
predecessor named Ctesibius. Among the latter's inventions were a
water-clock or *clepsydra*, a form of catapult, a hydraulic pipe-organ,
the force-pump, and a fire-engine composed of a pair of force-pumps
(fig. 57). One of Hero's devices is called Hero's Fountain (fig. 58). The
following description of it is taken from his book *Pneumatica*. Hero

had a special fondness for any contraption that was mystifying to the layman.

Upon a base $\alpha\beta$ a small satyr is fixed with a water-skin on his shoulder and a small basin is placed near by. The base is sealed water-tight. It is divided by the partition $\gamma\delta$ into two compartments. Through the partition runs a tube $\epsilon\xi$ which reaches nearly to the roof. Through the top of the box is passed a tube $\eta\theta$ which is soldered to the roof of the reservoir as well as to the partition. Similarly another tube $\kappa\lambda\mu$ passes through the roof and down nearly to the partition $\gamma\delta$.

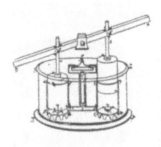

FIG. 57. Ctesibius' fire-engine, a double force-pump.

FIG. 58. Hero's Fountain.

FIG. 59. Penny-in-the-slot machine.

FIG. 60. Tantalus cup or intermittent siphon.

This tube conducts a stream of water to the basin. The chamber $\alpha\delta$ is filled with liquid through any opening ν and ν is then closed. If now liquid is poured into the basin, it flows through the tube $\eta\theta$ into the chamber $\beta\gamma$ from which air escapes through the tube $\epsilon\xi$ into the compartment $\alpha\delta$ and forces the liquid in it through the tube $\kappa\lambda\mu$ into the basin. This water then runs again into the chamber $\beta\gamma$ and in the same manner, drives air from it which in turn forces water from $\alpha\delta$ into the basin. This process continues until all the water is driven from $\alpha\delta$. The tube $\mu\lambda\kappa$ should be of narrow bore in order to make the show last longer.

The first penny-in-the-slot machine was described and probably invented by Hero. His diagram (fig. 59) shows how the device operated.

It was used in the temples to furnish each pilgrim with a spoonful of holy water (for a consideration). The same book discusses also the *siphon* and the Tantalus cup or intermittent siphon (fig. 60).

The most eminent of all Greek astronomers was Hipparchus of Rhodes, who flourished a century after Archimedes (fl. 146-126 B.C.). He obtained some knowledge of astronomy from Eudoxus of Cnidus (408-355 B.C.) and some from Chaldean sources, for instance through a school founded in Aristotle's day by Cidenas of Sippra on the Euphrates (343 B.C.). All such knowledge Hipparchus verified experimentally. By his life's work he established astronomy on a sound geometrical foundation.

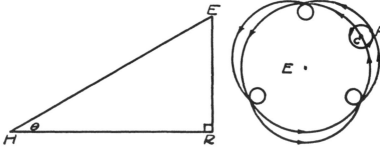

FIG. 61. Right-angled triangle.   FIG. 62. Epicycle.

Among his contributions to science was his use of longitude and latitude in locating points on the earth. From the idea that the world was longer from east to west than from north to south, distance east and west was called longitude (Lat. *longus*) and north and south distance, latitude (Lat. *latus*, broad).

Another great contribution of Hipparchus was his founding of the science of trigonometry which deals with measurement of triangles, as its name implies. (Greek *tri*, three; *gonu*, knee). As a gesture to the work of Hipparchus and for future use in our arguments, let us express the Pythagorean theorem in trigonometrical notation, though not in that used by Hipparchus. In the right-angled triangle HER (fig. 61) the ratio ER/HE is called the sine of the angle θ and is written sin θ Cosine θ is defined as the ratio HR/HE and is written cos θ. Thus in the right triangle of the rope-stretchers sin $\theta = 3/5$ and cos $\theta = 4/5$. Let EHR be any right triangle. Then by the Pythagorean theorem, $ER^2 + HR^2 = HE^2$.

$$\therefore \quad (ER/HE)^2 + (HR/HE)^2 = 1$$

Eq. 7. $\therefore \quad \sin^2\theta + \cos^2\theta = 1.$

Although the writings of Hipparchus (except one book) were unfortunately lost at the burning of the Library of Alexandria, yet we know something of their contents through the celebrated physicist Claudius Ptolemy (A.D. 100-160), who studied them at that library. Later he published a famous book entitled *Syntaxis* (σύνταξις τῆς ἀστρονομίας), to which the Arabians, about six centuries afterward, gave the name *Almagest* meaning "the greatest book"; now it is often called redundantly "The Almagest." It accounted for the motions of the planets by the assumption of *epicycles*,[5] an idea which Hipparchus had obtained partly from the Alexandrine geometer Apollonius of Perga. The name planet means literally "wanderer" and was given to describe the peculiar paths of such bright stars as Venus, Mars, and Jupiter through the fixed pattern of the other stars. Hipparchus and Ptolemy

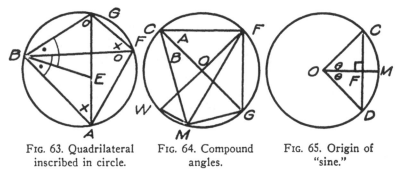

FIG. 63. Quadrilateral   FIG. 64. Compound   FIG. 65. Origin of
  inscribed in circle.   angles.   "sine."

assumed that the orbit of a planet is an epicycle, i.e. a circle whose centre *C* (fig. 62) revolves on a larger circle around the earth. Beneath all this was the fixed idea that the motion must be circular since the circle was perfect. Thus their theory of the solar system, which became known as the Ptolemaic system, was geocentric. It was in general acceptance for about fourteen centuries until discarded in favour of the less complicated heliocentric theory of Aristarchus as revived and developed by Copernicus (A.D. 1543). Here is a brief excerpt from *Almagest:*

It must be admitted that the sky is of spherical form and rotates as a sphere: also that the earth has the shape of a sphere: that it is poised in the heavens as a centre of revolution and that by reason of its size and distance from the fixed stars, it is merely a motionless point.

Ptolemy divided the hour into 60 minutes and each minute into 60 seconds and by translation into Latin his names for these units of time were rendered as *partes minutae primae* and *partes minutae secundae.* This was the source of our words "minute" and "second." Thus the Babylonian sexagesimal system took firm root in Greek astronomy and mathematics.

[5]F. Cajori, *A History of Physics* (Macmillan, 1933), p. 31.

Where we use the idea sin $\theta = CF/OC$ (fig. 65), Ptolemy would double the angle (COD) and consider the arc CMD and the corresponding chord CFD, i.e. the chord of the double arc. Thus in his table of sines he gives for 20°30' the length of CD as 21,12,12 in sexagesimals. This equals $\dfrac{21}{60} + \dfrac{12}{(60)^2} + \dfrac{12}{(60)^3} = 0.35588$ in decimals. Hence his value for the chord CF is 0.17794 and if the radius OC (hypotenuse) is unity, his value for sin $\theta = 0.17794$.[6] Our tables give sin 10°15' = 0.17794. We shall see later how these ideas, employed by Hindu and Arabian mathematicians, led ultimately to the term "sine" in the twelfth century, whereas the term "cosine" was first introduced in England and not until 1620.

Another important contribution of Ptolemy was his discovery of a geometrical theorem that is named after him: If a quadrilateral is inscribed in a circle (fig. 63), the product of its diagonals equals the sum of the rectangles formed by its opposite sides. His elegant argument based on the principle of similar triangles is given in *Almagest*. It shows that $AG.BF = AF.BG + AB.FG$.

This theorem enables trigonometricians to handle *compound angles* (fig. 64).

Let $\angle GCF = A$ and $\angle GCM = B$.

Bisect CG at O. With centre O and radius OC, describe the circle CGF.

Produce FO to cut the circle at W. Join FG, GM, and MW.

By Ptolemy's theorem,

$$MF.CG = FG.MC + CF.MG$$
$$\therefore \quad \frac{MF}{CG} = \frac{FG}{CG}\cdot\frac{MC}{CG} + \frac{CF}{CG}\cdot\frac{MG}{CG}$$
$$\therefore \quad \sin(A+B) = \sin A \cdot \cos B + \cos A \cdot \sin B.$$

These passages from *Almagest* are the last in our story of mechanics which were originally written in Greek. They therefore mark the decline of the Greek schools and forebode the close of the classical cycle of culture and civilization.

[6]F. Cajori, *History of Elementary Mathematics* (Macmillan, 1896), p. 88.

CHAPTER THREE

The Mediaeval
and Early Modern Period

SINCE mechanics needs calculations in its work, the introduction of a better method of reckoning is an important event in the story of mechanics. The invention of the system of numbers that we commonly use is ascribed to a school of mathematicians which flourished at Ajmir in India about A.D. 600. One of the most noted members of this school was Brahmagupta. Hindu numbers may, however, have come originally from Tibet. Their cradle history may be epitomized as follows:

Brahmagupta represented each integer up to nine by as many vertical or horizontal strokes as it contains units, somewhat as in the match-patterns we make in kindergarten (fig. 66a).

FIG. 66a. Early numbers.

Since the symbols for seven, eight, and nine, were confusingly alike, he substituted for seven the symbol 7 taken from an ancient inscription, and by adding two strokes, he obtained for nine the symbol 9. For speed and convenience, then, these characters became modified to their modern forms (fig. 66b).

123456789

FIG. 66b. Hindu or arabic numerals.

NUMERALS

Later, these symbols, called digits, were used by the calculators of the Arabian Empire in the Middle Ages. Hence they came to be called Arabic numerals. The fact that they are called digits reminds us of the primitive method of counting on the fingers and even on toes, vestiges of which procedure can still be seen now and then, for example when some speakers are announcing firstly, secondly, thirdly, and so forth.

The advantage of Hindu numbers over Greek numerals has already been noted during our visit with Archimedes. Their advantage over Roman numerals is just as great, as those who attempt to extract the square root of three by Roman numerals will be willing to admit a long while before they finish the calculation.

A good deal of argument about the origin of the Hindu numbers is on record. There is a question whether the credit for inventing them should go to Brahmagupta or to his school, and further whether the school produced Brahmagupta or vice versa. Let us adopt an attitude of compromise. Indeed, throughout the whole history of science, while we give full credit to the leaders, we acknowledge their debt to their predecessors and to their contemporaries, whose names are often unknown to us.

That most numeral systems, Egyptian, Hebrew, Greek, Roman, and Hindu, are on a basis of ten, or, as the mathematician says, have the *radix* 10, obviously resulted from our pentadactylity (Gk. *penta*, five, and *dactyl*, finger). Counting was often done by using pebbles as counters. This is recorded in the word "calculate," which comes from the Latin *calx*, meaning pebble. To this day, chips are used as counters in certain games. Another method of calculation was by means of an *abacus* (fig. 67), a counting-board with beads, of which we see examples to-day in the kindergarten abacus and the swanpan used in Chinese laundries. The hand was the original abacus. The readings of an abacus form a discontinuous or incremental series: they jump from 1 to 2 and from 2 to 3 with no intermediate values.

Some Roman numerals are derived from pictographs of counting on the fingers, thus, /, //, ///, ////, ∨///. The symbol V for 5 is from a picture of the open hand. It is not from the Latin word for 5 (*quinque*), nor is it related to the letter V which stands for the first sound in the word Victory. Similarly, the symbol X for 10 is the picture of two open hands: it is two V's. It is not derived from the Latin word for 10 (*decem*), nor does it represent the final sound in radix. Roman numerals were peculiar in employing the idea of subtraction as in IV and XC. This feature made calculations more difficult.

The word "score," meaning 20, suggests that counting was sometimes done not only on the fingers but also on toes. The number 20 also occurs in several tables of measures, e.g. 1 ream = 20 quires.

Twelve is a better radix than 10, because of its greater factorability. Whereas 10 is prime to all smaller numbers except 2 and 5, the number 12 has factors in common with 2, 3, 4, 6, 8, 9, and 10. The advantage of 12 as a radix is attested by the fact that much counting is done in dozens and gross; also by the frequent recurrence of 12 in tables of units, e.g. 1 foot = 12 inches; 1 penny-weight = 24 grains; 1 day = 24 hours; 1 shilling = 12 pence; 1 quire = 24 sheets. *Dozens* of other examples could be cited. When our nation becomes more scientific, it will adopt the radix 12 for its numbers, but that event seems to be well in the future. If the human race had only been

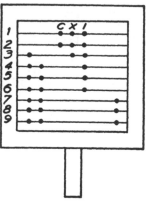

Fig. 67. Abacus or swanpan.

six-fingered, we should have had a better number system without the need of a world-shaking reform.

SPECIFIC GRAVITY

After the fall of the Roman empire during the fifth century, the Mohammedan culture of the Arabians became the torch-bearer of learning for several centuries. In 1025 one of their physicists who lived at Khiva in Persia and who rejoiced in the name Abu R Raihan Muhammad ibn Ahmad al-Biruni,[1] invented the "overflow-can" method of finding the *density* of a body, i.e. its weight or mass per unit volume. Archimedes, of course, was quite familiar with the notion of density.

This excerpt from Biruni's report has a laboratory flavour (fig. 68).

I persevered in setting up one apparatus after another, correcting in each the defects of its predecessors until finally I contrived a conical vessel, broad at the base *a*, and narrow above. From there upward extended a cylindrical neck. At the middle of the narrow part, . . . I made a small hole and soldered to it a tube *d*, of the same bore and curved downward. . . . At the time of an observation, its end *e* stood over the pan of the balance *f*. I broke each sample of metal into larger and smaller pieces. The larger ones were not larger than the width of the neck of the vessel and the smaller ones had the size of a millet seed. With these pieces, I proceeded as follows. First, I dropped the large ones into the mouth *c*, since this disturbed the water considerably and raised it higher than was necessary. There was no error in this procedure if I let the tube drain. Then, as I

[1]G. Sarton, *Introduction to the History of Science* (Williams, 1927), I, 707.

dropped in the small pieces with tweezers, the technique was less violent. . . .
Obviously the water rises in accordance with the volume of the bodies submerged
and hence there flows from the spout a volume which is equal to that of the
bodies introduced into the vessel.[2]

FIG. 68. Al Biruni's
overflow-can.

Having weighed a piece of gold and determined its volume by his overflow-can method,
Biruni found the density as follows: If w is
the weight and v, the volume, then by definition, the density

Eq. 1. $d = \dfrac{w}{v}$.

The result would evidently depend on the
units of weight and volume employed. Thus
the density of water is 62.4 lb. per cu. ft. or
10 lb. per gal. or 1 gram per cc. Biruni, of
course, used Persian units which he knew were unfamiliar to other
scientists of the empire.

In order to express his results independently of the units used,
Biruni introduced the idea of *specific gravity*, the ratio between the
weight of a body and that of an equal volume of water. An algebraic
expression of the definition is

$$s = \frac{w_b}{w_w} \qquad \frac{(weight\ of\ body)}{(weight\ of\ equal\ volume\ of\ water)}.$$

Regardless of what unit is used for weight and what unit for volume
in this determination, the ratio would evidently be the same for any
given substance.

If we divide each term of the fraction by v, the volume of the body
or of the water ($\because v_b = v_w$), we obtain,

$$s = \frac{w_b/v_b}{w_w/v_w} = \frac{d_b}{d_w}.$$

This equation says that the specific gravity of a substance is the
number of times its density is greater than that of water. Again it is
seen that this ratio is independent of the units used in the determination. No choice of units could influence the number of times a substance
is denser than water. Since 1 cc. water weighs 1 gm., the denominator
of the last fraction, d_w, equals 1, in the metric system. Consequently,
in that system, the specific gravity of a substance and the numerical
coefficient of its density are the same. This is one of the many advan-

[2]E. Weidemann, *Verhandlungen der deutschen physikalischen Gesellschaft*, 1908,
x, 340.

tages of the metric system of units, which was devised with a view to gaining maximum convenience.

Al Biruni's value for the specific gravity of gold was 19.06; since the value now accepted as standard is 19.3, his error was about 3 in 190 or about 1.5 per cent. His values for copper (8.7) and for a score of other metals and gems were all within 3 per cent of modern values. With the apparatus at his disposal his performance was, therefore, very creditable. Indeed, the overflow-can method is still used in elementary classes in physics. The influence of Archimedes' tub experiment is quite apparent in the work of this Persian investigator; in fact, the overflow-can is sometimes called a "eureka-can."[3]

When we were visiting Brahmagupta's school in India we might have noted that some Hindu mathematicians, especially astronomers, in discussing Ptolemy's diagrams (figs. 64 and 65), applied to the chord *CD* the name *jiva*, which was their word for the cord of a hunter's bow —obviously a very apt term. In Biruni's day Arab mathematicians in dealing with such diagrams placed more emphasis on the arc *CMD* and applied the term *dschaib* meaning "breast." In the twelfth century Plato of Tivoli, translating into Latin a book on astronomy by the Arabian Al Battânî, rendered the term *dschaib* by a Latin word for a curved surface—*sinus*. Thus we have traced to its source the term sine.

FRIAR BACON

Our itinerary leads again westward, carries us through two centuries in time, and brings us to Paris in the thirteenth century. Here we consider the work of an English friar, Roger Bacon (1214-1294), who was both prophet and martyr in the cause of science. The brilliance of his mind is attested by his sobriquet, "Doctor Mirabilis." He was much influenced by the teachings of four Arabian philosophers, Al Biruni, Alhazen, Averroës, and Avicenna. He realized that the peripatetic schoolmen of his day were making little progress, and he saw the cause of the stagnation. Their studies had lost touch with the practical affairs of the world and remained centred in unprofitable discussions which belonged to the past. The chief value to be derived from tracing the history of physics is the understanding and help we gain in dealing with the problems of to-day and to-morrow. The schoolmen's highly developed powers of logic, for example, were being frittered away in disputing such unprofitable topics as "How many angels could stand on the point of a needle?" Bacon realized that if progress was to be made, new experiments should be performed and discussed and the findings applied. There is no doubt that Aristotle himself, the object

[3]J. A. Cochrane, *Readable School Physics* (Bell, 1922), p. 14.

of so much monkish veneration, would have been the first to agree
with Bacon's contention.

Bacon had the courage of his convictions. He attacked ignorance
and smugness entrenched in high places of power; and dearly did he
pay for his bravery. He was persecuted, exiled from England, and cast
into prison in Paris, which by that time had become the hub of the
world of schooling. At the request of Pope Clement, however, he published, in 1266, his famous book, *Opus Majus*, which proclaimed the
fundamental importance of experimentation in science. Like most
treatises of the Middle Ages, it was written in Latin; but it is unusual
in being a scientific book written with a religious purpose. Here is a
translation of one of its passages.

Now there are four chief obstacles that hinder every man, however learned,
from grasping truth . . . namely, submission to ignorant authority, influence of
customs, popular prejudice, and concealment of our own ignorance.

There are four great sciences. . . . Any person who is familiar with these can
make glorious progress in pursuit of learning. Of these sciences, the gate and
key is mathematics, which the saints discovered[!] at the beginning of the world.

I now wish to unfold the principles of experimental science, since without
experience nothing can be definitely known. . . . For if a man is without experience that a magnet attracts iron, and has not heard from others that it attracts,
he will never discover this fact without an experiment. Since this Experimental
Science is wholly unknown to the rank and file of students, I am, therefore, unable to convince people of its utility without first describing its characteristics
as well as its excellences. This science has three chief features. The first is that
it investigates by experiment the important conclusions of all other sciences. . . .
This queen of the speculative sciences alone is able to give us important truths
within the confines of the other sciences which those sciences can learn in no
other way. . . . But there is a third dignity of this science. It arises from those
capacities in reference to which it has no connection with other sciences but by
its own power investigates the secrets of nature.

Thus Bacon's prophetic insight was keen enough to give him some
fairly clear glimpses of the future of experimental science three centuries or more ahead of his times.

Since some have credited Bacon with the invention of gunpowder,
his reference to fire-crackers is of special interest. He does not claim
the invention as his own.

We have an example of this in that children's toy which is made in many
parts of the world, namely an instrument as large as the human thumb. From the
force of the salt called saltpetre, so horrible a detonation is produced at the
bursting of so small a thing, namely a piece of parchment, that we perceive that
it exceeds the roar of a thunder-clap [!] and the flash exceeds the greatest brilliance of lightning.

Evidently this writer suffers no deficiency of imagination. One might
wish to read his description of the explosion of an atomic bomb!

We could hardly expect Bacon to foretell the use of gunpowder in firearms at the battle of Crécy (1344), or to predict the great influence of the "children's toy" on civilization. Firearms outmoded mediaeval castles and armour, and contributed to the ending of the feudalism of Bacon's day.

In the optics section, Bacon reports his use of a reading-glass and his prediction of the telescope and the microscope. In estimating his work we must consider not only his inventions but also the advances made because of his influence on others: the more important part of his contribution was indirect.

It was one of the tragedies of science that despotic authority was able to fetter and cripple the genius of this man. In bitter dejection, he closed his life with these words: "It was on account of the ignorance of those with whom I had to deal that I have not been able to accomplish more."

THE MOST VERSATILE GENIUS

The long jump of two centuries which we took from Al Biruni to Roger Bacon is indicative of the stagnation and aridity against which Bacon agitated. It indicates that scientists and contributions to science were few and far between. Because of the lack of freedom under autocratic authority during the Middle Ages, and the resulting failure of science to advance, that period has been called by some the Dark Ages, although the term would not apply to the Arabian empire of that epoch, which, in fact, kept alight the torch of knowledge.

Now another gap of two centuries brings us to the work of one of Italy's greatest sons, the celebrated painter Leonardo da Vinci. He was in all probability the most versatile genius known to history. If Roger Bacon typifies the passing of the Middle Ages and the early dawn of the modern era, then da Vinci is the radiant rising sun. His insatiable pursuit of knowledge, both practical and theoretical,

Fig. 69. Leonardo da Vinci. (By gracious permission of H. M. The King.)

led him into every department that was even remotely related to painting. He seemed to have a golden touch which at every turn yielded valuable results: great paintings, music, sculptures, canals, buildings, the parachute, the camera, a speedometer, the lamp-chimney, a hygrometer, the anemometer, a flying-machine, a diving-suit, the law of moments, interpretation of fossils, the phenomenon of capillarity, Newton's third law of motion, and so on. What a man!

Da Vinci's first patron was Lorenzo "The Magnificent," a member of an ancient and wealthy family, the Medici of Florence. The evil deeds of some Medici have lived after them, and their good deeds have so often been interred with their bones, that a word about their patronage of art and applied science may be appropriate here. Of importance in our story was the assistance and encouragement which Lorenzo gave da Vinci at the beginning of his career.[4] Two other Medici who on various occasions gave da Vinci the employment he needed were Giuliano de' Medici and his brother Giulio, who became Pope Leo X. Da Vinci later had other patrons, but in the closing decade of his life he returned to the protection and assistance of his old friends the Medici.

Leonardo's enthusiasm for mechanics he declared in these words:

Mechanics is the paradise of mathematical science, for here we reach the fruit of mathematics.[5]

Or, in the original Italian:

La meccanica e il paradiso delle scienze matematiche perche co quella si viene al frutto matematico.

This quotation marks the beginning of the end for the mediaeval custom of using Latin in scientific writings. Since, however, da Vinci left his writings in manuscript, it is not certain that he intended to publish a treatise.

While the alchemists of the Middle Ages were searching for the "philosopher's stone" that would transmute lead into gold, other mystery seekers were striving to invent a "perpetual motion" machine that would yield its owner a perpetual supply of work for nothing and thus bring him to the blissful state of not having to work for a living. Da Vinci's sagacity is seen in his opinion that such a pursuit is futile. This is how he expressed it:

O ye speculators on perpetual motion; how many will-o'-the wisps ye have fashioned! Go and rub noses with those who dream of creating gold.[6]

This denial contains the germ of the law of conservation of energy, according to which no energy can be created in any of our experiments or machines. Even to this day, however, one of the signs of spring

[4] Lorna Lewis, *Leonardo the Inventor* (Nelson, 1937), p. 40.
[5] Kensington manuscripts.
[6] Ibid.

seems to be an announcement in a newspaper that someone else thinks he has at last discovered the secret of a perpetual motion machine that will give out more work than is put into it.

Napoleon brought some of da Vinci's manuscripts to France from Italy as spoils of war. They were placed in the library of L'Institut de France at Paris, which is the second largest library in the world. Indeed, for rarity of contents, some consider it to be the greatest in the world. In 1797, the physicist Venturi published a booklet on the Paris manuscripts of da Vinci. Yale University has a copy of this brochure. In it one finds da Vinci's statement of the law of the lever in terms of moments. (Da Vinci realized that the human skeleton is a system of third-class levers, and accordingly he must needs make a study of levers.) Da Vinci states that the moment or turning-effect of a force

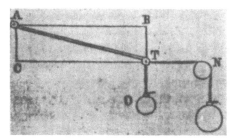

Fig. 70. The law of the lever in terms of moments.

about a fulcrum or axis is the product of the force and the perpendicular distance from the fulcrum to the force's line of action. This distance da Vinci called the "potential lever":

Let AT be a lever . . . its fulcrum at A, the weight O suspended at T, and the force N, which must produce equilibrium with the weight O. Draw AB normal to OB and AC normal to CN. I call AT the real lever but AB and AC the potential levers. We have the proportion,

$$N : O = AB : AC.$$

This relation can be expressed conveniently in products thus

$$N \times AC = O \times AB.$$

$N \times AC$ is the moment of N about A and $O \times AB$ is the moment of O about the same axis. Thus if two forces keep a lever in equilibrium, their moments or turning effects about the fulcrum or about any axis are equal and opposite. Their algebraic sum is zero. The law holds for any number of forces. In this form, the statement is an improvement on Aristotle's expression of the law. Da Vinci makes no attempt to explain why the law holds true.

Da Vinci's manuscripts are difficult to read. They are scribbled in old Italian, often irregularly, are full of abbreviations and emendations, and, to cap it all, are written from right to left. Some think that this was because da Vinci was left-handed. From the lines of shading in some of his drawings, it can be inferred that he favoured the left hand. The general opinion, however, is that he wrote in this fashion as a code to keep the words secret. One need only recall the maltreatment of Roger Bacon to realize the danger in those days of expressing one's thoughts aloud, to say nothing of leaving them around in writing. Even in our day, the cynic says, "Do right and fear no man: don't write and fear no woman." Venturi says that da Vinci's writing is best read with a mirror; but Richter's opinion is that it is more convenient in the long run to learn to read it directly, somewhat as a type-setter learns to read reversed type.

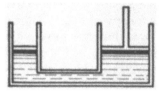

FIG. 71. Liquid seeks its own level.

At one time, da Vinci served his patron as an engineer in charge of the construction of a canal. In characteristic style he made a study of hydrostatics and wrote a treatise on that subject. In it occurs a sentence which gives at least one source of the common saying that "a liquid seeks its own level." Here is the passage:

The levels of all [uniform] liquid bodies, at rest and joined by an intercommunicating tube, are always at the same height [fig. 71].[7]

Da Vinci also discovered an important exception to this rule, namely, the phenomenon of *capillarity*. A glass tube whose bore is about as fine as a hair is called a capillary tube (Latin *capilla*, a hair). If a clean capillary tube with a bore of diameter 1 mm., open at both ends, is dipped in pure water, the liquid is drawn up in the tube to a height of about 3 cm. or $1\frac{1}{4}$ in. This explains how blotting paper soaks up ink and a towel dries a wet surface. Two common applications of capillarity are a candle-wick and a split pen-nib.

From his study of the flight of birds, da Vinci reached the invention of the parachute, as may be seen in the following excerpt from the manuscript at Milan, called the Codex Atlanticus. The same passage shows a recognition of Newton's third law of motion, the reaction law.

An object offers as much resistance to the air as the air does to the object. You may see that the beating of an eagle's wings against the air supports its heavy body in the highest and rarest atmosphere close to the sphere of elemental fire.

If a man have a tent-roof of caulked linen, 12 bracchia [yards] broad and

[7]Quoted in H. Grote, *Leonardo da Vinci als Ingenieur*.

twelve bracchia high, he will be able to let himself fall from any height without danger to himself.

The first parachute descent was made in 1783 by Lenormand.

In the story of optics, we shall consider da Vinci's invention of the pin-hole camera. For the present, the interview closes with a saying of his which intimates the high level at which he lived: "As a day well spent gives joyful sleep, so does a life well spent give joyful death."

THE HELIOCENTRIC THEORY

In 1543, the Polish churchman and astronomer, Dr. Nicolaus Kupernik, or Copernicus, of Frauenburg, published an important Latin treatise entitled *Concerning the Revolutions of the Celestial Orbs.* He was an able mathematician, a thorough scholar, and altogether an admirable man. His busy life was filled with philanthropic work among the poor, especially in giving free medical assistance to those who could not afford it. His manuscript was virtually completed in 1530, but knowing the dissension and persecution which it would likely occasion, he hesitated to publish it. At the insistence of his friend Rhaeticus, however, he finally consented. The treatise was published in the last year of his life. In fact, he was on his death-bed when the first copy was hurried to him and placed in his hands. By death he escaped persecution, but not so some others who championed his theory, for example Bruno, who was burned at the stake in 1609 by the Inquisition, and his friend Galileo, who was imprisoned by the same tyranny for saying that the earth rotates on its axis and revolves around the sun.

In the following quotation Copernicus proposed the heliocentric theory of Aristarchus. He developed that hypothesis and showed its superiority over the more cumbersome Ptolemaic system. Now it is universally accepted by astronomers.

Every observed change of position is due either to the motion of the observed body or of the observer or to the motions of both. Since the planets appear now nearer, now farther from the earth, this shows necessarily that the centre of the earth is not the centre of their circular orbits.

He still holds with Aristotle that the orbit must be a circle, "the perfect curve":

If one admits the immobility of the sun and transfers the annual revolution from the sun to the earth, there would result, in the same manner as actually observed, the rising and setting of the constellations and the fixed stars, by reason of which they become morning and evening stars. Finally, one will be convinced that the sun itself occupies the centre of the universe, that is, if one only, so to speak, will look at the matter with both eyes!

In advising Copernicus to publish his book, Rhaeticus (Georg

Joachim) was not preaching what he would not practise, for he was an independent thinker. He was first to break away from the ancient custom of considering trigonometric functions with reference to the arc of a circle. Since Hipparchus, the sine had always been defined with reference to a right triangle placed in a circle with the angle in question at the centre. Rhaeticus discarded the circle and dealt solely with the triangle. What had been a radius was now a hypotenuse. This lead him to the ratio called secant and written

$$\sec \theta = \frac{hypotenuse}{base}.$$

He prepared a table of secants.

When Rhaeticus discarded the classical circle and arc for defining trigonometrical ratios, he probably contributed to the subsequent introduction of the ratio called cosine (abbrev. cos.) This step was taken eight decades later (1620) by the Englishman Edmund Gunter, who was a contemporary of Napier. In the right triangle HER (fig. 61), $\angle EHR = 30°$, and $\angle HER$ ($= 60°$) is complementary. Gunter defined cosine as the ratio, base/hypotenuse, i.e. $\cos 60° = \frac{1}{2}$ and $\cos 30° = \frac{1}{2}\sqrt{3}$. But $\sin 30° = \frac{1}{2}$ and $\sin 60° = \frac{1}{2}\sqrt{3}$, or, in general, $\cos \theta = \sin (90 - \theta)$. The name cosine suggests that the sine of an angle is the cosine of its complement. There is the same kind of relation between secant and cosecant and between tangent and cotangent. Thus

$$\operatorname{cosec} \theta = \sec (90 - \theta)$$
and
$$\cot \theta = \tan (90 - \theta)$$

STEVINUS AND THE INCLINED PLANE

Although the inclined plane was the first simple machine employed in prehistoric times, the law of its action was not discovered until eighteen centuries after Archimedes wrote his treatise on the lever. The delay suggests that the problem was difficult. Victory over this difficulty was achieved by the Flemish scientist Simon Stevin of Bruges (1548-1620), who is often referred to by his Latin name Stevinus, in accordance with mediaeval custom. Stevinus began his career as a clerk in a dry-goods store in Amsterdam. By dint of industry and intelligence, he rose to be the King's Master of Ordinance and Overseer of Canals and Fortifications in Holland. He certainly contributed effectively to the advance of science and his name deserves to be better known.

In 1586, Stevinus published a book on statics and hydrostatics which introduced the principle of triangle of vectors and thereby gave a new impetus to the science of statics.[8] He showed that Aristotle's rectangle of displacements and velocities is a particular case of a more general law, the parallelogram or triangle of vectors. Whereas Aristotle showed how to combine rectangular components and to resolve a vector into rectangular components, the work of Stevinus led to the

FIG. 72. Stevinus.

knowledge of performing these operations for any practical value of the angle between the components. The argument is as follows:

[8]A French translation of Stevinus' book was published at Leyden in 1634 by the physicist E. Girard. The University of Michigan possesses a copy of this rare volume. Grateful acknowledgment must be made here to Dr. Warner Rice, the chief librarian, for his generous co-operation in lending such a book on the Interlibrary Exchange Plan. By that plan, an investigator, at the nominal cost of postage and insurance one way, has at his disposal what is potentially almost the Library of North America. The librarians who have conceived and established this excellent service rightly deserve gratitude and praise.

In fig. 73, $\mathbf{R} = \mathbf{c}_1 + \mathbf{c}_2$ (addition and resolution of vectors).

By the Pythagorean theorem,

$$R^2 = OH^2 + HE^2 = (c_1 + c_2 . \cos \phi)^2 + c_2^2 . \sin^2\phi$$
$$= c_1^2 + c_2^2 . \sin^2\phi + c_2^2 . \cos^2\phi + 2 c_1c_2 . \cos \phi$$
$$= c_1^2 + c_2^2 (\sin^2\phi + \cos^2\phi) + 2 c_1c_2 \cos \phi$$
$$= c_1^2 + c_2^2 + 2 c_1c_2 \cos \phi. \quad \text{(Hipparchus, } Eq. \text{ 7, p. 60)}$$

$\therefore$ the length of $R = \sqrt{c_1^2 + c_2^2 + 2 c_1c_2 \cos \phi}$.

If $\phi = 90°$, then $\cos \phi = 0$

$\therefore R = \sqrt{c_1^2 + c_2^2}$. This is Aristotle's theorem ($Eq.$ 5, p. 42).

If the vector $\mathbf{OB} = \mathbf{c}_2$ is moved parallel to itself to the position $\mathbf{AE}$, then the $\triangle OEA$ is a triangle of vectors. Its sides express the magnitude,

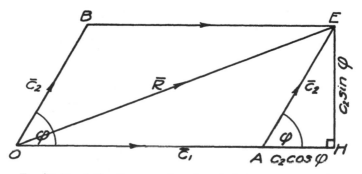

FIG. 73. Resolution for any value of angle between components.

direction, and sense of the three vectors, c_1 and c_2 two components, and $\mathbf{R}$ their resultant.

In the same book Stevinus gave his discovery of the hydrostatic paradox which we shall discuss when considering the work of Pascal.

A rather amusing invention made by Stevinus was the first horseless carriage. It was propelled by sails and carried twenty-six passengers. On more than one occasion Stevinus took his patron the Prince of Orange and a party for a ride in that carriage along the shore of the Zuyder Zee, faster than a horse could run.

In 1605, Stevinus published the book of which the title-page is shown in fig. 74. It suggests that Stevinus considered his work with the inclined plane his best. Three years later Willebrord Snell, professor of physics at Leyden, published a Latin translation with the title *Hypomnemata Mathematica* ("mathematical studies").[9] In the following derivation of the law of the inclined plane (fig. 75), taken from

[9]A copy is in the library of the University of Chicago.

these books, the author makes a clever use of the denial of perpetual motion:

> To show that as $AB : BC$ (2:1), so is the action of the weight E to that of D. To the perimeter of the triangle ABC, apply a cycle of 14 spheres, equal in weight and size and equidistant from each other, such as $D, E, F, \ldots Q, R$, strung upon a line passing through their centres so that there can be 2 spheres on the side BC and 4 on BA, and so that the system can revolve without friction along the sides AB and BC.
>
> If the pull of the weights D, R, Q, P, be not equal to the force of the two spheres E and F, one will pull harder. First, let the four, D, R, Q, P, pull harder. The four, O, N, M, L, are equal to the four, G, H, I, K. Hence the side of 8

FIG. 74. Title-page of Stevinus' book,
published 1605.

> spheres, $DRQ \ldots ML$ will move downward. Presently D will arrive where O now is and likewise the others. Consequently, the cycle of spheres will have the same configuration as at first. Therefore, the motion would have no end; but this is absurd. The demonstration for the other side will be the same.
>
> The part of the cycle, $DR \ldots L$, will, therefore, be in equilibrium with the part $E \ldots K$. Remove $ONML$ and $GHIK$. The remaining $DRQP$ will be in equilibrium with the part EF. Hence E will exert a pull which is twice that of D.
>
> Therefore as $BA : BC$ so is the action of E to that of D.[10]

The *mechanical advantage* of a machine is defined as the ratio of the resisting force R to the acting force A when the system is *at equilibrium*, or algebraically, $M = R/A$. From Stevinus' theorem $M = R/A = HE/EG = \operatorname{cosec} \theta$. This is a statement of the law of the in-

[10]E. Mach, *Science of Mechanics* (Open Court, 1942) deals interestingly with this topic (p. 24).

clined plane; its mechanical advantage is the cosec of the angle of slope or inclination (*GHE*). See fig. 76.

Stevinus was first to distinguish among three kinds (or degrees) of equilibrium, namely, stable, neutral, and unstable. These can be illustrated by the three possible positions of a cone. Because of his work in this department of mechanics, he was called the "Father of Modern Statics," whereas Archimedes is called the "Founder of Statics."

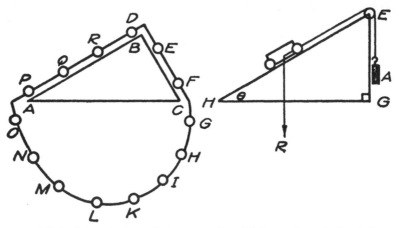

FIG. 75. Deriving the law of the inclined plane. FIG. 76. Law of the inclined plane.

As early as the twelfth century, decimal fractions were used by some mathematicians for the extraction of square roots.[11] Indeed, they were known to some Chaldeans. It was Stevinus who introduced them into common use, a contribution about which some boys and girls seem none too enthusiastic, wishing that Stevinus had found other employment for his great talents. He did not use the decimal point. For example, the number 42.354 he wrote as $42_0 \, 3_1 \, 5_2 \, 4_3$, in which the subscripts are evidently the indices of the denominators,

$$\text{for } 42.354 = \frac{42}{10^0} + \frac{3}{10^1} + \frac{5}{10^2} + \frac{4}{10^3} \cdot$$

To bring the advantage of decimals to the people, Stevinus published in 1585 an explanatory pamphlet on the topic in Flemish, the language which he thought would some day become the universal language of the world. His pride in his native language resembled that of certain folk who believe that Gaelic is the language of Heaven. Later, in order to reach a larger number of readers without waiting for the spread of

[11]Cajori, *A History of Elementary Mathematics.*

Flemish, Stevinus translated his pamphlet into French. In this case, therefore, he was also a pioneer in breaking away from the mediaeval use of Latin in scientific treatises.

In the history of decimals we have an example of how slowly progress may occur, for after Brahmagupta had shown the possibilities of a system of numbers based on 10 and its powers with positive indices, one might expect that an extension to negative indices would soon follow. But as we have just seen, it required at least ten centuries to take that step.

Stevinus pointed out the advantages of using the decimal idea in tables of units and predicted decimal coinage and the metric system of units.

INVENTION OF LOGARITHMS

For the priceless convenience of logarithms and the slide-rule, we are deeply indebted to John Napier, Laird of Murchiston, Scotland. Most of Napier's life was spent, not inappropriately for a scientist and a Scotsman, in contriving economies of time and labour in calculations. In astronomical computations logarithms reduce the time of operations at least fifty per cent, or as Laplace once expressed it, "they have doubled the life of the astronomer."

In Archimedes' *Sand Reckoner* it was seen that a convenient way to deal with large numbers is to handle them as powers. Thus 10^6 is a more convenient symbol than 1,000,000, and, in the ancient problem from the chessboard, the notation 2×10^{64} is decidedly more convenient than a 2 followed by sixty-four ciphers. Since logarithms are really indices or exponents they replace multiplication by addition, and division by subtraction. Thus $16 \times 32 = 2^4 \times 2^5 = 2^{4+5} = 2^9 = 512$. The only operation in this case was the addition of 4 and 5; after that, by consulting a suitable table opposite 9, we obtain the value 512. The crux of the whole scheme is the preparation of the table. In the relation $2^4 = 16$, the index 4 is called the *logarithm* of 16 to the base 2. Napier coined the term for the purpose. Similarly the logarithm of $1,000 (= 10^3)$ to the base 10 is 3. In general if a number $n = b^L$, then L is the logarithm of n to the base b, or in symbols, $L = \log_b n$. Furthermore, n is called the *antilogarithm* of L.

If the problem is to divide 12 by 4 the table will need to have fractional numbers. Thus $\log_{10} 12 = 1.0792$ and $\log_{10} 4 = 0.6021$. And $\therefore 12 \div 4 = 10^{1.0792} \div 10^{0.6021} = 10^{0.4771} = 3.0$. Some amazing cuneiform inscriptions have been discovered recently which date about 1800 B.C. They give tables of numbers expressed actually as powers with fractional indices. These show that thirty-four centuries before Napier Babylonian mathematicians were at least close to the idea of logarithms. "There is no new thing under the sun," said Solomon in his pessimism.

By years of computations that were ingenious, intricate, laborious, and tedious, Napier succeeded in preparing a table of numbers which met especially the needs of astronomers. They are more like our logarithmic sines than any other of our modern tabled numbers. Following Napier's lead, mathematicians have prepared other logarithmic tables of greater general convenience. Common or Briggsian logarithms are to the base 10, the radix of our number system. Mathematicians generally use what they call natural logarithms. They were first introduced by John Speidell of London in 1619. These are to the base $\epsilon = 2.718281828459. . .$, which at first glance may appear to be a very unnatural choice; but on closer examination, the epithet proves to be quite appropriate. Sometimes natural logarithms are erroneously called Napierian. It is true, however, that they can be calculated from Napier's table.

In 1614, Napier published his famous book, *Description of the Wonderful Rule of Logarithms*. Here is a translation of one of its paragraphs:

From these preliminaries, the erudite will realize what advantages logarithms bring them; since by adding them, multiplication is avoided, by subtracting them, division is avoided, by dividing by 2, extraction of square root, by dividing by 3, cube root is avoided and by other easy transformations, all the more laborious operations of computations.[12]

The part of a slide-rule that is used for multiplication and division is essentially a convenient working-model of the logarithm table. The same is true of the scales that are used for finding squares and cubes and for extracting square and cube roots. The slide-rule saves time and labour by doing away with the need of writing during the calculations. It was invented about 1625 by Wm. Oughtred (of England).

Of all books that have been written by British scientists, Napier's is considered second only to Newton's *Principia*. It resembles Archimedes' writings on the circle and sphere in that it anticipates Newton's use of the theory of limits and adumbrates his development of the calculus. Coming events cast their shadows before them. Napier's treatise makes rather difficult reading and affords additional illustration of the general truth that a first solution of a problem is apt to be relatively cumbersome. Afterwards simplifications are made which allow solutions that are briefer, more elegant, and easier, in accordance with the principle of economy or parsimony of effort and thought.

Napier derived his logarithms from ideas belonging to mechanics

[12]*Mirifici logarithmorum canonis descriptio.* A copy is in the library of the United States Military Academy, West Point.

and especially to the study of motion. A particle or point M (fig. 77) moves from O toward F, a fixed point, which is at a distance c ($=10^7$ cm.) from O. M's velocity v is proportional to the distance $MF = y$. Since $v \propto y$, $\therefore v = ky$. Let $k = 1$. Then

Eq. 2. $$v = y.$$

Thus at the initial instant $v_0 = MF = 10^7$ cm. per second $= c$ cm. per second. As M proceeds, its speed decreases but the distance OM grows larger approaching the limit OF. $OM \rightarrow OF$ and $MF \rightarrow 0$.

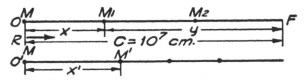

FIG. 77. Napier's diagram.

Another particle M' begins at the same instant at O' and moves with uniform velocity, c cm. p. s. $\therefore$ the distances M' traverses in 1, 2, 3, 4, 5, ... sec., or in intervals proportional to these, are in Arithmetical Progression (A.P.)

Let M move from O to M_1 during a brief interval dt ($=1/c$ sec.). with speed c. The interval dt is small enough that the speed may be treated as uniform. The distance M traverses $x_1 =$ speed $\times$ time

$$= c \cdot dt = c\left(\frac{1}{c}\right) = 1 \text{ cm.} \quad \therefore y_1 = M_1F = c - x_1 = c - 1 = c\left(1 - \frac{1}{c}\right) \text{ cm.}$$

$$\therefore \text{ speed at } M_1 = c\left(1 - \frac{1}{c}\right). \quad (Eq.\ 2.)$$

During a second interval dt, M goes from M_1 to M_2, a distance of

$$x_2 = \text{speed} \times \text{time} = c\left(1 - \frac{1}{c}\right)\left(\frac{1}{c}\right) = \left(1 - \frac{1}{c}\right) \text{ cm.}$$

$$\therefore y_2 = M_2F = M_1F - x_2 = c\left(1 - \frac{1}{c}\right) - \left(1 - \frac{1}{c}\right) = \left(1 - \frac{1}{c}\right)(c - 1)$$

$$= c\left(1 - \frac{1}{c}\right)^2 \text{ cm.}$$

Similarly $M_3F = c\left(1 - \frac{1}{c}\right)^3$, etc.

Hence it is seen that the distances of M from F in dt, $2dt$, $3dt$ ... sec., are in Geometrical Progression (G.P.) with first term $c\left(1 - \frac{1}{c}\right)$ and common ratio $(1 - 1/c)$. Similarly the numbers 1, 10, 100, 1000 ... or

10^2, 10^3 . . . are in G.P. whereas their logarithms to the base 10, namely, 0, 1, 2, 3, 4 . . . are in A.P.

Napier defined OM' $(= x')$ as the logarithm of MF $(= y)$ and thus elaborated his table of logarithms from the relation $x' = \log y$ (Napierian). As a number decreases its Napierian logarithm increases.

The approach to logarithms by means of indices is probably the easiest and most commonly used in elementary work. But it was not Napier's approach, for the simple reason that he actually had comparatively little knowledge of indices. Stevinus had attempted with little success to introduce exponential notation. Indeed, one of the most remarkable features of Napier's monumental work is that he achieved his goal without indices.

In his preface to the *Descriptio*, Napier, with the modesty which is the besetting weakness of his countrymen, begs the reader to pardon his errors. He says, "Since the calculation of this table, which ought to have been accomplished by the labour and assistance of many computors, has been completed by the strength and industry of one alone, it will not be surprising if many errors have crept into it." On the contrary, it was remarkably accurate and had far fewer errors than one would expect. What he did not say was that, feeling the hand of death upon him, he hastened while there was yet opportunity to place his offering, though imperfect, on the altar of science.

Napier was a precocious youth—he was only sixteen years younger than his father. He was sent to Paris to receive advanced education. Yet it is not altogether easy to harmonize his contribution with his environment, which was not remarkable for either its mathematics or its science. Some of Napier's neighbours, not understanding his abstruse studies, reached the profound conclusion that he was practising Black Art and was in league with the Devil, "Auld Nickie-Ben." His familiar, it was rumoured, was an old black rooster. Once when some of Napier's silver spoons had disappeared, the Laird powdered the rooster with lampblack and placed him in a dark room. Each servant was obliged to enter the room in turn and stroke the rooster which, it was given out, would crow when the culprit approached. Throughout the ordeal, the rooster did not crow. Roosters do not crow in dark rooms. But Napier noted that one servant emerged from the room with no lampblack on his hands. Thus the thief was detected and not without some Black Art.

Two years after Napier's death, his son Robert, who had studied with him, published the posthumous book, *Construction of Logarithms*.[13] Napier's canny plan had been to send Part I, the *Descriptio*, into the

---

[13]*Mirifici logarithmorum canonis constructio.* Also in the library at West Point.

intellectual market as a sample or feeler. Then, if it were well received, he could return on a better footing with Part II, the *Constructio*. As a matter of fact, both books were received enthusiastically. This is a sentence from the second book:

If two numbers with known Logarithms be multiplied together forming a third, the sum of their Logarithms will be the Logarithm of the third. Also, if one number be divided by another, producing a third, the Logarithm of the second subtracted from that of the first leaves the Logarithm of the third.

Another device which Napier invented was a set of small rods or strips of bone inscribed with numbers so as to make a simple multiplication machine. The instrument became well known throughout Europe and was dubbed "Napier's bones." Indeed, it gained him more renown in non-scientific circles than did logarithms. To tell the truth, the "bones" were of more bother than worth to anyone who knew the multiplication table up to 9 times 9. From this, we can gain an inkling of the arithmetical status of the rank and file in those days, or in other words, how rank their arithmetic was.

That Napier was a pioneer in using a point to mark the beginning of a decimal fraction, we are reminded by the following sentence from his second book:

In numbers thus divided by a period, whatever is written after the period is a fraction whose denominator is unity with as many ciphers after it as there are figures after the period, thus 25.803 means the same as 25 $\frac{803}{1000}$.

Napier also used the decimal point in his discussion of decimals in his *Rabdologia*, published in 1617. This improvement upon Stevinus' more complicated notation affords a further example of the development from cumbersome to elegant, from complex to simpler, which we have met several times. Strange to say, after Napier's time, other notations were adopted which were less simple. Finally the pendulum swung back so that now the decimal point is, of course, in universal use. In this case, therefore, the path of progress was oscillatory, from complex to simpler then back to clumsy, and finally back to optimum simplicity, ease, and elegance, with maximum economy of work and time. We saw this, too, when Aristarchus' heliocentric theory was superseded by Ptolemy's complicated system of epicycles, and this, in turn, by the simpler heliocentric theory of Copernicus.

A closer look at natural logarithms would bring us the profitable experience of meeting the peculiar number which is their base. It is represented by the Greek letter epsilon, ϵ, which, like π, is an eternal constant. It is possibly even more remarkable than π. An old mathematician once remarked that he could easily imagine the destruction of the universe with the exception that π and ϵ would survive all cataclysms and still remain imperishable. It is clear that he took but little stock in the physicist's principle of conservation of matter.

To obtain the value of ϵ approximately in decimals, let us expand the binomial $\left(1 + \dfrac{1}{n}\right)^n$,

$$= 1 + n\left(\frac{1}{n}\right) + \frac{n(n-1)}{\underline{|2}} \cdot \frac{1}{n^2} + \frac{n(n-1)(n-2)}{\underline{|3}} \cdot \frac{1}{n^3} + \ldots$$

$$= 1 + \frac{1}{\underline{|1}} + \frac{\left(1 - \frac{1}{n}\right)}{\underline{|2}} + \frac{\left(1 - \frac{1}{n}\right)\left(1 - \frac{2}{n}\right)}{\underline{|3}} + \ldots$$

ϵ is the fixed limit approached by this sum as n, the number of terms, grows indefinitely large ($n \to \infty$).

$$\therefore \; \epsilon = \operatorname*{Lim}_{n \to \infty} \left(1 + \frac{1}{n}\right)^n$$

$$= 1 + \frac{1}{\underline{|1}} + \frac{1}{\underline{|2}} + \frac{1}{\underline{|3}} + \frac{1}{\underline{|4}} + \ldots$$

$$= 1 + 1 + 0.5 + 0.1666\ldots + 0.0416\ldots + 0.0083\ldots + \ldots$$

$$\therefore \; \epsilon = 2.72 \text{ approximately, or about } 2\tfrac{3}{4}.$$

KEPLER'S LAWS

If we imagine a visit to the ancient city of Prague in the year 1609, the most interesting man there, to the student of mechanics, is certainly Johann Kepler (1571-1630), Astronomer Royal to Rudolph, King of Bohemia and Emperor of Germany. The title sounds impressive, but we find poor Kepler retained mainly as an astrologer, his chief occupation the casting of horoscopes which he despises, and the intervals of his time filled with an embarrassing struggle to collect his meagre wages. He does not believe in astrology, but is forced to keep on with the horoscopes as his sole means of postponing death from starvation. One encouragement that helps to keep his spirits up is his correspondence with a brilliant Italian physicist named Galileo, whose keen wit and biting satire provide Kepler with many a laugh. He replies in kind. The two physicists "shake their sides with uncontrollable laughter" at Galileo's description of an old fossil who refused to look through the telescope for fear he might see among the heavenly bodies something that would disturb his religious faith.

The circumstances attending Kepler's birth and early years were far from propitious and those of later years were even worse. At four, smallpox left his hands and eyes permanently impaired, and all his life he was beset by misfortune. Yet there must have been at his nativity at least one good fairy, for in spite of all his troubles we find him a kindly, sincere, and generous man, often cheery and jolly and sometimes even hilarious. He takes it on the chin and laughs. In spite of his handicaps he gets much fun out of life for he is a tremendous worker and finds diversion and relief by immersion in the study of science and mathematics.

FIG. 78. Kepler.

In 1609, he published a Latin treatise, *The Motions of Mars*, from which the following excerpt is taken. It declares in vigorous terms a modern scientist's attitude towards truth and dogma. It indicates a growing spirit of independent thought abroad in the world, quite different from mediaeval slavish submission to dogma. The translation used here was made in 1661 by Salusbury.

It must be confessed that there are many who are devoted to Holiness, that dissent from the Judgement of Copernicus, fearing to give the Lye to the Holy Ghost in the Scriptures if they should say that the Earth moveth and the Sun stands still. But he who is so stupid as not to comprehend the Science of

Astronomy or so weak and scrupulous as to think it an offence of Piety to adhere to Copernicus, him I advise, that leaving the study of Astronomy and censuring the opinions of Philosophers, he betake himself to his own concerns and that, desisting from further pursuit of these intricate studies, he keep at home and manure his own ground. In Theology, the weight of Authority but in Philosophy, the weight of reason is to be considered. But to me more sacred than all these is Truth.

In this connection perhaps Pontius Pilate's difficult question comes to mind, "What is truth?"

Aristotle said, "Every body has the tendency not to have its velocity changed," or , in other words, the property of resisting change of velocity. It was Kepler who gave a name to this important property of matter. In the following passage from his *Motion of Mundane Bodies*, published in 1617, he calls it *inertia:*

Now the heavenly bodies are neither heavy nor light but being divinely adapted for circular motion, offer no resistance to motive impulse. Thus on the one hand, any celestial globe is not heavy as a stone on the earth is said to be heavy, nor light as in the case of flame: but on the other hand, the stone by reason of its material, has a natural reluctance about moving from place to place. It has a natural inertia or sluggishness by which it tends to remain at rest in any place where it is situated undisturbed.

Here you see that even as independent and fearless a thinker as Kepler cannot always free himself from the trammels of tradition. For instance, he holds the mediaeval opinion that the circle is the perfect curve and therefore (!) the heavenly bodies being perfect must move on circular orbits: also that the heavenly bodies do not obey the same laws as those on earth. Soon Kepler was compelled by his own observations and calculations to state that these orbits are ellipses.

When Newton, half a century later, was wrestling with the same problem, he took a different attitude. He said, "Could not the moon and the stone be essentially alike? Could there not be the same laws for both?" As Dr. Robert Hooke said, "More laws are vain where less will serve." May it not be the same force which attracts both stone and moon?

Kepler's greatest contribution to science was his discovery of the laws of planetary motion. These three laws are justly named in his honour and constitute his monument. He enunciated them in his *Harmony of the Universe* (1619) from which the following famous passages are taken. The first one helped mightily in toppling Aristotle's writings on physics from the throne of authority which they had held for centuries.

Law I. I have shown that the orbit of a planet is elliptical . . . and the Sun is at one focus of the ellipse.

(To keep our perspective true, however, it should be added that if the earth's orbit, or the moon's, is drawn to scale, it looks like a circle and measurement is necessary to make certain that it is not a circle.)

Law II. I have shown in my Commentaries on Mars from the exceedingly accurate observations of Brahé, that equal diurnal arcs on the same orbit are not traversed with equal velocities but that ... the times being supposed equal as for instance, one natural day in each case, the true diurnal arcs corresponding to them are inversely proportional, to the two distances from the Sun.

This law is often expressed by saying that the areas swept out in equal times by a straight line joining the planet and the sun, are equal.

Law III. After I had by unceasing toil discovered the true distances of the orbits, at last, the true relation ... by a new onset, overcame by storm, the shadows of my mind. The principle is unquestionably true that the periodic times of any two planets are

FIG. 79. Tycho Brahé.

to each other as the cubes of the square roots of their median distances.

Algebraically, the third law may be expressed simply as follows:

Eq. 3. $p^2 \propto r^3$ or $p^2 = k \cdot r^3$ where k is a constant.

(Kepler's initial is chosen for the constant.)

Many of the observations on which Kepler's calculations were based were obtained by his Danish friend and predecessor, Tycho Brahé, who was a rather peculiar character. A physiognomic idiosyncrasy of his was a gold nose. The nose which had belonged to him by nature was cut off in a duel, conducted in the dark with knives. Those were the quiet old days when people had little excitement—beyond that of a duel in pitch dark where at any moment a knife might suddenly slice off one's nose or ear or what not. The story has been told that on cold days Tycho found his gold nose uncomfortable and liked to take it off and put it in his pocket, but his friends beseeched him to wear it, no matter what the weather.

Brahé lacked Kepler's gifts as a mathematician, but he had a great fondness for making accurate observations of the positions of the stars and recording them and their times. He used a huge protractor twice his own height and sighted the stars through two small apertures about two metres or yards apart (fig. 80). He had no telescope, of course.

In 1620, Kepler dedicated one of his books to Napier, since he had used logarithms in it. Also, not knowing of Napier's death three years earlier, Kepler wrote him a beautiful letter full of praise and gratitude for the great help Napier had given him by his toilsome invention of logarithms.

Kepler's joy at having his arduous labours crowned with success he expressed in the following paean of triumph. Well might Kepler exult, for few indeed are the scientists who, labouring under such adverse circumstances, have accomplished so much of lasting value.

That for which I joined Tycho Brahé and to which I have devoted the best part of my life, at length I have brought to light. Nothing holds me. I will indulge my sacred fury! If you forgive me, I rejoice; if you are angry, I can take it. The die is cast. The book is written, to be read either now or by posterity. I care not which. It may well wait a century for a reader, now that God has waited six thousand years for an observer.

FIG. 80. Tycho Brahé's Protractor.

BACON AND THE INDUCTIVE METHOD

An important English contemporary of Kepler was Francis Bacon, Lord Verulam, Viscount St. Albans, and Lord Chancellor of England (1561-1626). He made no direct contribution to mechanics and was not a mathematician; he was not even well posted in mechanics; but as a philosopher he exerted a strong influence on many scientists, for example, Descartes and Boyle, whose work we shall consider later. By his able pen, Bacon did much to usher into the world the present scientific age. He advocated the founding of the Royal Society and the establishment of National Research universities.[14] In 1620, he published a famous Latin treatise entitled *Novum Organum*, literally, "The New

[14]*The New Atlantis.*

Instrument." The implication of the title was that the book was an advance on a group of Aristotle's writings on logic and method in science, which were referred to as the "Organon." Bacon's book exposed the faults of the Aristotelian schools of the day, and, by emphasizing the importance of experimentation (as did his namesake Roger Bacon) and of the *inductive* method of reasoning, pointed the way of escape from the fogs of the Middle Ages. Aristotle, of course, needs no defender; yet it may be said that in spite of Bacon's attacks, he still remains a master of logic and easily the peer of Bacon. It is true that Aristotle made conspicuous use of deductive reasoning; nevertheless, he was completely familiar with the inductive method.

Here is a quotation from Bacon's *Novum Organum* that discusses the printing press and two other inventions.

Now the empire of man over things is founded on the arts and sciences only. For nature is not governed except by obedience. It is worth while noticing . . . the consequences of discoveries. These appear more manifestly in none more than in those three, unknown to the ancients, yet whose origin, although recent, is obscure and inglorious: Printing, Gunpowder and the Mariner's Needle. For these three have changed the face and state of things in all the world; the first, in letters, the second, in war, and the third, in navigation . . . so that no empire, no sect, no star seems to have exercised a greater command and influence over human affairs than have these mechanical discoveries.

Bacon's advocacy of the inductive method or induction is illustrated by a passage from *Novum Organum:*

The Logic which is in vogue is rather potent for confirming and fixing errors . . . than for the investigation of Truth. . . . Our only hope is a true Induction. . . . Those who have handled the Sciences have been either Empirics or Dogmatists. The Empirics, like the Ant, only amass and use: the latter, like Spiders, spin webs out of themselves: but the course of the Bee lies midway. She gathers materials from the flowers of the garden and field; and then by her own power, turns and digests them. From the closer conjunction of the experimental and the rational, good hopes are to be entertained. Our hints for the Interpretation of Nature embrace two parts differing in kind; the first is on the inducing of Axioms from experience; the second on deducing new experiments from Axioms. The derivation of principles or axioms by means of true Induction is certainly the proper remedy for driving out false ideas. First, a Natural and Experimental History must be prepared, for we must not fancy but must discover what Nature does. Tables of Instances are to be formed with such arrangement that the Intellect may be able to act upon them. In this manner, legitimate and true Induction, the very key of Interpretation, is to be applied.

THE FOUNDING OF ANALYTICAL GEOMETRY

The science of analytical geometry, which is of inestimable utility in mechanics, was founded about 1637 by the celebrated French savant, René Descartes (or Des Cartes), whose Latin name is Cartesius. His mother died shortly after his birth in 1596, and his physique was never robust, but by careful attention to the laws of hygiene his health improved and he lived to the age of fifty-four. In sharp contrast to poor Kepler, he enjoyed an independent income of two thousand dollars a year. Three of his friends were Blaise Pascal, Père Marin Mersenne of Paris, and Christian Huygens of The Hague, the inventor of the pendulum clock.

Descartes was a very independent thinker and a philosopher of great influence. He openly acknowledged his indebtedness to Bacon, but to few others. Like Bacon, he became dissatisfied with the Aristotelian schools of the day. He wiped the slate of authority with the sponge of doubt and tried to think out the problem of the universe for himself. His explanations of physical phenomena were all based on mechanics. His starting-point was his most famous short sentence, *Cogito ergo sum* ("I think, therefore I exist"). He gave much thought to education and the best methods of training young scientists. He gave this advice, "The aim of all studies should be to guide the mind to form true and sound judgments on all problems presented to it."

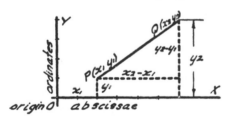

FIG. 81. Cartesian co-ordinates.

His *Discourse on Method*[15] gives a brief sketch of his invention of analytical geometry:

Perceiving further that in order to understand these geometrical relations, I should sometimes have to consider them separately and sometimes to grasp them in the aggregate. ... I thought that in order to deal with them better individually, I should consider them located with reference to two straight lines OX and OY, [fig. 81]. But to keep them in mind or comprehend a group at once, I must designate them by the briefest symbols possible. Thus I could borrow all that was best from geometrical analysis and algebra and correct the defects of each by the help of the other.

The lines OX and OY are called Cartesian axes. Their intersection O, Descartes called the *origin* of the diagram. In Descartes' treatise *La Géometrie*, the X-axis OX is vertical and the Y-axis OY horizontal. The location of the point P is indicated as P (2, 1), meaning that if a point moves from O along OX 2 units and then upwards 1 unit, it arrives at the point P. Similarly Q is the point (6, 4). Distances along OX are called abscissae and along OY, ordinates. The symbol $P(x_1, y_1)$ indicates that a point P has the abscissa x_1, and the ordinate y_1, and $Q(x_2, y_2)$ says that the point Q has the co-ordinates x_2, y_2. The abscissa

[15]A translation is given in the Harvard Classics, Vol. XXXIV.

and ordinate of a point are referred to collectively as the Cartesian co-ordinates of the point. This method is really used every day. If we ask the druggist to direct us to the nearest pawnshop, he may say, "Go two blocks east, then one block south, and there you are." His store is the origin, and he gives us the Cartesian co-ordinates of our destination. The same plan is used in maps where the equator is the X-axis, or axis of longitudes, and the Greenwich meridian is the Y-axis (axis of latitudes). The origin in this case is the Bay of Congo. Hence Descartes' plan harks back as far as Hipparchus. The ideas of Descartes' method were known to earlier mathematicians, but he employed them so effectively that he ranks as one of the first outstanding modern mathematicians. Since analysis is an old name for algebra, the name analytical geometry suggests a combination of algebra and geometry. Twenty centuries before Descartes, Pythagoras made a fruitful combination of geometry and arithmetic.

Descartes' treatise *Geometry* is not very easy reading for most people and deliberately so, as the author warns the reader in the preface, for he was throwing down the gauntlet to conservative mathematicians who opposed the innovations he was advocating. He was saying to them, "Put this

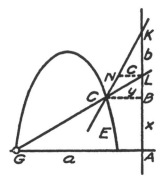

Fig. 82. The hyperbola.

in your pipe and smoke it. May it do you good." He begins with deriving the equation of the hyperbola, a piece of work that occurs much later in most texts on conic sections. Fig. 82 is a copy of his diagram.

The graph EC . . . I consider as described by the intersection of the ruler GL and the rectilinear plane figure $CNKL$ whose side KN is produced indefinitely toward C and which, being moved so that its side KL always coincides with some part of the line BA, rotates the ruler GL about the point G, because it is so joined that it always passes through the point L.

I choose a straight line such as AB to which to refer all the several parts of the curved line EC, and in this line AB [axis], I choose a point A, from which to begin the calculations [origin]. Then, selecting an arbitrary point in the curve, I draw through C the line CB parallel to GA; and since BA and BC are two variable and unknown quantities, I designate them x and y.

But to find the relation between them, I consider also the known quantities which define the curve, viz. GA which I call a, KL which I call b, and NL (parallel to GA) which I call c.

Then I say $NL : LK$ or $c : b$ as CB or $y : BK$. $\therefore$ $BK = (b/c)y$ and BL equals $(b/c)y - b$; and AL equals $x + (b/c)y - b$. Moreover, since $CB : LB$ or $y : (b/c)y - b$ as a or $AG : LA$ or $x + (b/c)y - b$, therefore, multiplying the second term by the third [means], we get $(ab/c)y - ab$ which is equal to $xy + (b/c)yy - by$, the product of the first and last [extremes]. Therefore, the required equation is $yy = cy - (c/b)xy + ay - ac$, whence we know that the graph EC is none other than a Hyperbola.

The plan of using letters from the first of the alphabet, a, b, c, etc., for constants, and from the end of the alphabet, x, y, z, etc., for variables, has been found convenient and has been in general use ever since Descartes proposed it in this article. In this passage, the square of y is written yy, but in other parts of the same treatise, it is written with an index y^2, the cube, y^3, and so forth. Descartes' sign of equality has been replaced in general by the sign $=$, originated by Dr. Robert Recorde, who was court physician to Edward VI of England. Recorde also introduced for the operations of addition and subtraction the signs $+$ and $-$.

De façon que, multipliant la feconde par la troifiefme, on produift $\frac{ab}{c}y - ab$, qui eft efgale a $xy + \frac{b}{c}yy - by$, qui fe produift en multipliant la premiere par la derniere; & ainfi l'equation qu'il falloit trouuer eft :

$$yy \approx cy - \tfrac{c}{b}xy + ay - ac,$$

de laquelle on connoift que la ligne E C eft du premier genre : comme, en effect, elle n'eft autre qu'vne Hyperbole*.

FIG. 83. A sentence from Descartes showing his sign for equality. Thus a curve can be described by an equation.

The equation which Descartes has just derived may be written $cxy + by^2 - b(a + c)y + abc = 0$. If this is compared with the general equation of the second degree and of all conic sections, viz., $Ax^2 + 2Hxy + By^2 + 2Gx + 2Fy + C = 0$, it is seen that $A = 0$, $B = -b$, and $H = \tfrac{1}{2}c$,

$\therefore AB = 0 \times (-b) = 0$ and $H^2 = \tfrac{1}{4}c^2$.

$\therefore AB - H^2 = 0 - \tfrac{1}{4}c^2$ $\therefore (AB - H^2) < 0$ (negative).

This test shows that the equation represents a hyperbola. In other words, if we strike out terms that are not in the second degree, the resulting equation has real roots ($H^2 > AB$).

In the Cartesian diagram of fig. 84a, the graph evidently represents the motion of a body B which travelled with a constant speed of 6 cm. per sec. Its speed,

$$c = \frac{distance}{time} = \frac{s}{t} = \frac{AG}{OG} = \frac{MH}{OH} = \frac{EK}{OK} = \tan \theta = \frac{18}{3} = 6 \text{ cm.p.s.}$$

Hence a graph of uniform speed in a distance-time diagram $(S\text{-}T)$ is a straight line. The speed is measured by the slope of the graph, $\tan \theta$. This simple diagram contains the germ of the differential calculus and indicates that the work of Descartes contributed to the theory of limits and prepared the way for the invention of the infinitesimal calculus.

The same facts, charted on a speed-time diagram (fig. 84b), are represented by the graph $QAME$. The equal ordinates, QO, AG, etc., indicate a uniform speed of 6 cm.p.s. Since the distance traversed

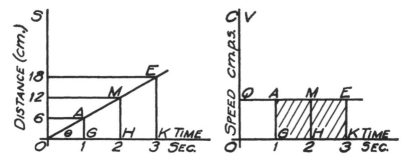

Fig. 84a. A distance-time diagram.   Fig. 84b. A velocity-time diagram.

during any time interval equals the product of the average speed and the time, or, in algebra,

Eq. 4 $s = ct,$

therefore, distance traversed in any interval, e.g. GK, is measured by the product $AG \times GK$, namely, by the *area under the graph* for that interval (AE). Hence we derive the general statement that the graph of uniform speed or velocity in a speed-time or $V\text{-}T$ diagram is a straight line parallel to the time axis OT. The distance traversed in any time interval is measured by the area under the graph for that interval.

This Cartesian diagram contains the germ of the Integral Calculus, which is one of the most useful and powerful tools employed in modern mechanics.

The Cartesian Diver or Devil (fig. 85), is an intriguing and instructive philosophical toy which is named after Descartes although it was invented half a century before his birth. It is not mentioned in his writings. It is a primitive submarine and helps to explain the action of the swim-bladder of a fish. It is an application of Archimedes' principle, Pascal's principle, and Boyle's law. It is sometimes made as an amusing parlour puzzle from a medicine bottle containing some water and an inverted pill-bottle as diver.

Aristotle recorded the fact that rock powder (dust) and gold dust can rest on the surface of water although both powders are denser than water. He said also that a needle cannot thus rest on the surface of water. Descartes was first to record the fact that a needle can rest on the surface of water. He made the slip of saying that it floats. The passage occurs in his treatise *On Meteors*, which, as it name implies, makes a study of the earth's atmosphere. "Meteor" was an old name for any atmospheric phenomenon; it persists in the word "meteorology," the study of the atmosphere, and "meteorological bureau," the weather bureau. Here are the words of Descartes:

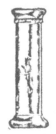

FIG. 85.
The Cartesian
D|ver.

I assume that water and all other substances which surround us are composed of numerous little particles of different shapes and sizes which are never so closely packed as not to have many spaces among them: also that these interstices are not empty but filled with that very subtle medium by the agency of which I have previously said that the action of light is transmitted. Then, in particular, I assume that the tiny particles of which water is composed are long, united and slippery[!] The surface of water is much more difficult to pierce than its interior so that we find by experiment that àll bodies sufficiently small, even though of very dense substances, such as small steel needles, can float [rest] on the surface and remain at the top as long as the surface is not pierced; but when that occurs, they sink promptly to the bottom.

After Aristotle, this is the first reference in the literature to the *surface tension* of liquids; and it happens to have a number of other interesting features. It has special interest because it introduces the term *aether*. We have seen that Aristotle considered the world a plenum whereas Democritus considered matter as composed of discrete particles, atoms, separated (or joined?) by vacuum. Descartes seized both horns of the dilemma. He accepted the corpuscular or atomic idea with modifications; the intermolecular spaces he considered as filled with a tenuous or even imponderable "substance" which he called aether, and in this way subscribed to Aristotle's idea that nature is a plenum. The source of the term aether is the Latin word for the upper heavens or topmost sky, where the air is thinnest.

This interview closes with a characteristic sentence from Descartes' greatest book, *Principiae Philosophiae:* "I hope you will not believe anything I have said unless you are persuaded by the evidence and the force of reason."

CHAPTER FOUR

The School of Galileo

MORE than one scientist has had the bitter fate of receiving evil for good, persecution and even death in return for his labours. Examples that we have met were Pythagoras, Anaxagoras, Aristarchus, Roger Bacon, and Bruno. Now we come to one of the most shameful instances. In the City of Florence, in 1638, we find Italy's greatest physicist, Galileo Galilei (1564-1642), imprisoned because he had proclaimed and defended what he believed to be the truth. He had supported the Copernican theory. Few men were better able than Galileo to compare the merits of the two rival hypotheses in celestial mechanics.[1] Newton referred to him as an intellectual giant. Students had come from all Europe to study under him. Some of his audiences had reached an enrolment of two thousand, a large audience even in these days. But ignorance and bigotry, invested with power, tyrannically crushed Galileo because of his great knowledge. As he once said, "Of all hatreds, the worst is the jealous fury of the ignorant against those whom they know to have superior knowledge."

As we see him in the famous picture (fig. 86), Galileo is a broken relic of his former self. He is nearly blind. The death of his beloved daughter has left him inconsolable. He is afflicted with arthritis and, since the Inquisition dealt with him, he suffers from hernia. He still has faithful friends, however, and he does his best to stifle heartache with ceaseless work. It was under these conditions that he completed his greatest book, *Dialogues on Two New Sciences*,[2] from which most of our quotations will be taken. The dialogue form of presentation was used by Plato and by many others since. Galileo was a very able writer in this form.

[1]Zsolt de Harsanyi's *The Star Gazer* (Putnam, 1939) is an excellent modern biography of Galileo.
[2]*Dialogue on Two New Sciences.* I am indebted to the translation by Crew and DeSalvio (Macmillan, 1914) for many excellent renderings, but I have not followed closely this or any other translation, and have made my own arrangement of the material for this exposition.

With Galileo in the picture is seen a distinguished visitor, the English poet John Milton, in whose mind the great poem *Paradise Lost* (1667) is already being planned. Milton's presence indicates that Galileo's fame was world wide, and that his admirers were numerous.

Vincenzio Galilei was an impoverished Venetian noble who had inherited a gentleman's tastes and abilities without the appropriate bank account. Though he was an adept mathematician and the best lute-player in Italy, his lot was one of grinding poverty. So he determined that his clever son Galileo would have a profession in which he

Fig. 86. Galileo entertaining Milton.

could receive decent returns for his talents and industry. He strongly advised the boy whatever befell not to become a mathematician, a musician, or least of all, a teacher. He sent the youth to study medicine under a famous professor, Dr. Cesalpino; but medicine was not Galileo's best love. One day he heard, through an open door, Professor Ricci giving a lecture algebra. That was like a match to a tinder-pile. After the class, Galileo waylaid Ricci and begged for instruction in algebra. Ricci refused on the ground that Vincenzio objected, but the boy would not take a refusal. So Ricci succumbed rather willingly but on condition that the lessons be kept secret. Vincenzio, however, was not deceived. His response was after this fashion: "I know what is going on. You are studying algebra. I can tell it by that happy gleam in your face. I tried to save you but now that you have burnt yourself, you may just sit on the burn." Galileo became an excellent mathematician, a competent musician in both performance and composition, and finally a great teacher. So much then for poor Vincenzio's practical advice, his hopes, and his attempt to guide the career of a genius.

While still a medical student, Galileo made his first invention, the pulsilogium, which was essentially a pendulum of adjustable length for reading a patient's pulse. The watch had not yet been invented. The doctor adjusted the length of the pulsilogium until its oscillations synchronized with the pulse-beats. Then the pulse-rate was read off directly from the graduations on the instrument.

When Vincenzio Galilei died he left to his son the heavy responsibility of supporting four younger brothers and sisters. It was by making and selling pulsilogia and other instruments that Galileo succeeded in this onerous duty in spite of his mean salary as a teacher.

In the pulsilogium there was already an earnest of Galileo's chief life work, the study of moving bodies. The science that deals with motion is called kinetics (Gk. *kineo*, move). An older name was dynamics (Gk. *dunamis*, force), and Galileo has been called the "Father of Dynamics." Dr. Ivor Hart remarks, however, that if Galileo is to be called the father of dynamics, then da Vinci should be called the grandfather. It is significant that the classical cycle of culture produced the fundamental treatise on statics (Archimedes') and the Western cycle of culture produced the science of kinetics. These facts are symptomatic of the difference between the pagan and western cultures.

In kinetics, the ideas of uniform motion and of uniform acceleration are fundamental, and in the following quotation these two terms are defined in a dialogue between Salviati, an Academician, and Sagredo, an intelligent amateur in science.

SALVIATI. I consider a motion steady or uniform if the distances traversed by the moving body during any equal time intervals are equal . . . I say that motion is steadily or uniformly accelerated which acquires, in any equal times, equal increments of velocity.

SAGREDO. How can I offer any rational objection to this or to any other definition which may be devised by any author whatever, since all definitions are arbitrary?

In Aristotle's writings, we found the statement that bodies fall freely with speeds that are proportional to their weights. Now we come to a famous passage which marks Galileo's correction of that old error. If a coin and a slightly smaller disc of cardboard are used to test the statement, it may be that the coin, which is heavier, will fall faster.

But if the cardboard is placed on the coin with no edge projecting, they fall together, for in this way the card is shielded from air friction.

The character Simplicio, who now enters the dialogue, brought Galileo plenty of trouble, for he is evidently an Aristotelian and the butt of Galileo's satire. Moreover, he bore an uncanny similarity to the Pope, and there were those who took the trouble to draw this co-incidence to His Holiness's attention. Even the name Simplicio is hardly complimentary, for one of its connotations is contained in our word "simpleton."

SALVIATI. We infer, therefore, that large and small bodies fall with the same speeds, provided they are of the same specific gravity.

SIMPLICIO. Your discussion is really admirable: yet I do not find it easy to believe that a bird-shot falls as swiftly as a cannon-ball.

SALVIATI. Now Simplicio, do behave; you might as well say a dust-mote as swiftly as a grindstone. But I trust you are not going to follow the example of many others who . . . fasten upon some statement of mine which comes within a hair's breadth of the truth and, behind that hair, try to hide another scientist's error which is as big as a ship's cable. Aristotle declares that bodies of different weights, in the same medium, travel, in so far as their motions depend on gravity, with velocities that are proportional to their weights. He says that an iron ball of one hundred pounds falling from a height of one hundred cubits, reaches the ground [when] a one-pound ball has fallen a single cubit. . . . I say that they arrive at the same time. You find, on doing the experiment, that the larger outstrips the smaller by two finger-breadths. Surely, surely, you would not try to hide behind those two fingers the ninety-nine cubits of Aristotle's error, nor would you, I hope, draw attention to my small error and at the same time attempt to gloze over his very large one. A gold ball, at the end of a descent of one hundred cubits, would not precede one of copper by four fingers. Having observed [at the tower of Pisa] how they fall, I came to the conclusion that all bodies would fall with equal velocities if the resistance of the air were totally removed.

SIMPLICIO. I shall never believe that, even in vacuum, if motion in such a medium were possible, a lock of wool and a bit of lead can fall with equal velocities.

SALVIATI. Since we cannot obtain such a space, we shall observe what happens in the rarest medium.

This passage is of outstanding historical importance. It is an instance of the pen doing as mighty work as ever the sword performed, for it opened a new epoch in the world. Galileo's mind was remarkably free from the trammel of tradition. No matter whose statement was under consideration, even Aristotle's, he tested it by experiment and believed the experimental result in spite of any authority whatever to the contrary. With the zeal of a missionary for proclaiming the truth, he would oppose that authority with all the power of his brilliant logic and rhetoric. He also made hilarious use of satire and ridicule in thrusting his arguments down his opponents' throats. This habit, of course, was hardly apt to make the opponents love him more.

Tradition says that to demonstrate his contention in this instance, Galileo carried two iron blocks of different weights to the top of the Leaning Tower of Pisa, his native city, and dropped them at the same moment (fig. 87), while observed by a group of peripatetics on the ground below (and a little to one side). The reverberations of those two blocks crashing to earth together have not died down yet, and may they never fade away, for Galileo thereby broke the Aristotelian shackles and helped to secure for his successors a freedom of speech and thought which we should appreciate and use, and, if necessary, defend with our lives.

When the peripatetics saw the result of the experiment, they said, "It is impossible, for Aristotle says the heavier one wins." You and I can hardly credit such slavish adherence to dogma; but it is surprising to find how many unthinking persons in our own day still hold Aristotle's opinion on this point. The peripatetics resented what they considered Galileo's attack on their faith and they wreaked on him humiliation and anguish. Sir Oliver Lodge's *Pioneers of Science* describes Galileo's examination under menace of torture when, by denying that the earth moves, he escaped death but not imprisonment. Since he knew that the In-

Fig. 87. Galileo testing his theory of falling bodies.

quisition had burned Bruno at the stake, his legendary remark, *Eppur si muovi* ("It moves just the same"), is doubtless apocryphal, for it would have meant certain death, but just as likely, it represents what he thought. Galileo's exposure of the false doctrine about falling bodies is typical of much of his work; for his "best eulogium is a recital of the fallacies he corrected."[3]

If we were seeking an appropriate symbol for the work of Galileo, we could hardly find a better one than the pendulum, for it figures in his career from first to last, from his invention of the pulsilogium to his specifications for a pendulum clock, dictated on his death-bed to his pupil Viviani. The following excerpt refers to his discovery that the frequency of a simple free pendulum is constant and independent

[3]O. Lodge, *Pioneers of Science* (Macmillan, 1904).

of the amplitude, provided the angle of swing is not greater than about 20°. It also illustrates his relish for poking fun at his opponents. The peripatetics believed that the air, which is the only body touching the pendulum, must be the agent that oscillates it. That is why Simplicio thought a body could not move in a vacuum.

> Thousands of times I have studied vibrations, particularly in churches where lamps hanging by very long cords had inadvertently been set in motion by someone or other. But the most I could gather from such observations was the improbability of the opinion of those who believe that such oscillations are maintained and continued by the medium, that is by the air. For it seems to me very clear that the air must have a marvellous judgment and must have altogether too little to do with itself, if it kills hour after hour of its good time in poking a hanging weight hither and yon with such amazing regularity.

The story of Galileo observing a swinging censer or candelabrum in church is famous. To time its oscillations, as there were no watches, he used his pulse as a stop-watch (or, one might say, as a wrist-watch). Think of the indomitability of this man, in attempting to carry on three separate trains of thought at the same time: counting the oscillations of the censer, counting his pulses, and—pondering the sermon. Throughout his book, the reader finds that in spite of the shackles and indignities his enemies have cast upon him, he is still, as in this passage, the same peppery gentleman whose insuppressible spirit cannot be confined by prison walls.

Aristotle said, but apparently without experimental basis, that the density of air is one-tenth that of water. Galileo was first to make a quantitative determination of this density. Here is his report on two methods that he used in the determination. He first shows that air has weight:

> SALVIATI. As to the other question, namely, how to determine the specific gravity of air, I have employed the following method. I secured a rather large glass bottle or flask with a narrow neck and attached to it a leather cover. . . . In the top of this cover, I fastened firmly the valve of a leather bottle. Through the latter, I forced into the glass bottle, by means of a syringe, a quantity of air. . . . With an accurate balance, I weighed this bottle of compressed air with utmost precision, adjusting the weight with fine sand (a grain at a time). Next, I opened the valve and allowed the compressed air to escape. On replacing the bottle upon the balance, I found it perceptibly lighter. From the sand, I now removed enough to restore equilibrium. There can be no doubt that the weight of the sand thus laid aside represents the weight of the air which had been forced into the bottle.
>
> In order to know the weight of the air as compared with water, however, I must measure the volume of the air that was pumped into the bottle.
>
> A second bottle with a narrow neck similar to the first one was used. Over its mouth was slipped a leather tube which was bound tightly about the

neck of the bottle. The other end of the tube embraced the valve of the first bottle.

The second bottle was provided with a hole in the bottom through which an iron rod could be placed so as to open the above-mentioned valve at will. The second bottle was filled with water. When the valve was opened, air rushed into the bottle containing water and drove water through the hole at the bottom. Evidently, the volume of the water thus displaced equals the volume of air escaped. On weighing this water, we found how many times its weight contains that of the removed sand.

The English chemist Joseph Priestley is sometimes given the credit for introducing the method of collecting gases by displacement of a liquid; but this report shows that he was anticipated by Galileo. Priestley was probably first in using mercury instead of water. It is not surprising to find that Galileo's method was more cumbersome than that of Priestley; it is just another example of the development from complex to simpler.

SALVIATI. The second method is more expeditious. Here no air is added to that which the vessel contains initially, but water is forced into it without allowing air to escape. The water compresses the air. Place it on a balance and weigh it accurately. Next, open the valve. Weigh the bottle again. The loss in weight represents the weight of a volume of air equal to the volume of the water in the vessel.

Thus we determine how many times denser water is than air. Contrary to the opinion of Aristotle, we found that this is not 10 times but more nearly 400 times.

The modern value is about 776 times. In this instance, we see the advancement from complex to simpler, from cumbersome to elegant, in the technique of two experiments performed by the same man. It is easy to see that Galileo was an adept experimenter. As a boy he was fond of making mechanical toys, and the training in manual dexterity which he thus acquired was an invaluable asset in all his scientific investigations. He was also a brilliant student; his intellectual curiosity was avid, his perception quick, and his retention tenacious. He excelled at music, painting, languages, and mathematics.

In his study of the motion of a ball falling freely through the air, Galileo found the velocity so great that it was difficult to make accurate measurements of the distances traversed and the corresponding time intervals. He therefore let the ball roll down a sloping plank, as he said, "to dilute the motion," i.e. to slow the motion or, in other words, to decrease the acceleration. The inclined plane that he employed for the purpose is still used in mechanics courses and is called the Galileo board. A common modification is the Duff inclined plane. Here is the inventor's account of such an experiment:

A piece of wooden scantling about twelve cubits long and half a cubit wide was used. On its edge, a channel was cut. . . . Having made this groove very straight, smooth, and polished, we rolled along it a hard, smooth, well-rounded bronze ball. Placing the board in a sloping position by lifting one end of it about one or two cubits above the other, we rolled the ball along the channel, noting in a manner, presently to be described, the time required to make the descent. We repeated this experiment more than once [duplicate experiments] and succeeded in measuring the time with an accuracy such that the deviation between two observations did not exceed one-tenth of a pulse-beat. . . .

We now rolled the ball only one-quarter of the length of the channel. We found the time of its descent precisely one-half of the former. . . . We always found that the spaces traversed were to each other as the squares of the times . . . for all inclinations of the plane.

Galileo's algebraic expression of his conclusions from a series of such experiments was as follows: Let s be the distance traversed; t, the time; u, the initial speed; v, the final speed; and a, the uniform acceleration. Then by definition:

Eq. 1.
$$a = \frac{v - u}{t}.$$

Eq. 2.   ∴. $v = u + at.$

Also, from *Eq.* 4, p. 93,

Eq. 3.
$$s = \left(\frac{u + v}{2}\right) t.$$

By eliminating t from *Eq.* 2 and 3, he obtained

Eq. 4. $s = ut + \frac{1}{2}at^2.$

If a body commences from rest, then $u = 0$. In that case

Eq. 5. $s = \frac{1}{2}at^2.$

In this case, since $\frac{1}{2}a$ is a constant, $s \propto t^2$, the distance varies directly as the square of the time, as Galileo has just stated.

Nowadays, the time intervals in such experiments are measured by a stop-watch, but Galileo and his students had no such convenience. Here is his description of the water-clock employed:

For the measurement of time, we used a large pail of water, placed in an elevated position. To the bottom of this vessel was soldered a pipe of small diameter giving a thin jet of water which we collected in a small beaker during the time of each descent. . . . The water thus collected was weighed, after each descent, on a very accurate balance.

A man who can contrive, from a pail of water and a balance, a stop-watch that measures time intervals to the tenth of a second, is not a man who can easily be stopped by difficulties.

As the ball rolls down the plank, it gathers speed, and if it commences from rest ($u = 0$), its speed at the bottom of the slope is,

Eq. 6. $\qquad\qquad v = at \qquad$ (by *Eq. 2*, p. 102).

Galileo found that after the ball reaches the bottom, it rolls along a smooth horizontal surface with speed that is nearly uniform. In other words, it keeps the speed v and its acceleration becomes nearly zero. As a matter of fact, it has a slight retardation or deceleration due to friction with the table.

From these facts, Galileo abstracted the following law of motion and, this time, he was in accord with Aristotle:

> Furthermore, we may note that any velocity imparted to a body is indelibly impressed upon it . . . as long as external causes of acceleration or retardation are excluded.

Later, we shall see the influence of these words in Newton's expression of the first law of motion.

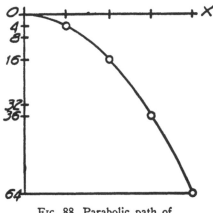

FIG. 88. Parabolic path of projectile.

Galileo showed that the paths of projectiles in general are parabolas, or would be if there were no air friction. Here is an excerpt that gives the basis of the argument:

> SALVIATI. A projectile which is impelled with a uniform horizontal motion compounded with a naturally accelerated vertical motion describes a path which is a parabola.
>
> SAGREDO. The argument rests on the hypothesis that such motions and velocities combine without altering or hindering each other.

Suppose the ball is thrown from O (fig. 88) with a horizontal speed of, say, 16 ft. per sec. By the law of the previous section, the ball will continue to move horizontally with uniform speed according to the equation:

Eq. 7. $\qquad x = ut \qquad$ (*Eq. 1*, p. 102, or *Eq. 4*, p. 102, $a = 0$).

Eq. 8. Also $\qquad\qquad t = \dfrac{x}{u}$.

Let O be, say, 64 ft. above the ground. From the instant of its projection, the ball falls with a downward vertical uniform acceleration of about 32 ft. per sec. per sec. according to the equation

Eq. 9. $\qquad y = -\tfrac{1}{2}at^2 = -\tfrac{1}{2} \times 32t^2 = -16t^2 \qquad$ (*Eq. 5*, p. 102).

At the end of 0.5 sec., it has gone $x = 16 \times \tfrac{1}{2} = 8$ ft. horizontally

(*Eq.* 7), and vertically $y = -16(\frac{1}{2})^2 = 4$ ft. downward. Its position is given by the first plot-point. Similarly for the others. Joining the plot-points gives a graph which represents the path of the projectile (its trajectory). The equation of the graph is obtained by combining the equations of the two motions (*Eq.* 7 and 9, p. 103) thus,

$$y = -\tfrac{1}{2}a\left(\frac{x}{u}\right)^2 . \qquad \therefore \quad ax^2 + 2u^2y = 0.$$

A simple test for *parabola* is this: in its equation, the terms in the second degree, x^2, xy, and y^2, form a perfect square. In this case, ax^2 is the only term in second power and it is the square of $\sqrt{a}.x$. If we wish a general test such as we used for a hyperbola (p. 92), we can proceed thus: If $Ax^2 + 2Hxy + By^2$ is a perfect square, then $H = \sqrt{AB}$ or $H^2 = AB$. In this case, $H = 0$, $A = a$, and $B = 0$. $\therefore H^2 = AB$. Therefore, the graph is a parabola.

Before Galileo's time, and in his day, there was a group of terms such as force, efficacy, virtue, action, momentum, power, energy, impulse, and several others which were used in a vague confused interchangeability without any rigour of definition. In fact, you can hear and see them pretty badly confused to this day in many conversations, in newspapers, and in books written by non-scientific authors. Galileo performed the useful service of making clear a definition of one of these terms when he defined the *momentum* of a body as the product of its weight and its velocity at any instant. The following passage bears witness to the struggle by which he clove his way through the mediaeval fog:

> Momentum is the force, efficacy or virtue with which . . . a body resists. It depends not only on the weight of the body but also on its velocity. If we consider a single striking body, the difference of momentum in its blows can depend only upon the difference of velocity.

Algebraically, this definition is expressed thus,

Eq. 10. $M = wv.$

Modern texts say:

Eq. 11. $M = mv,$

where m is the *mass* of the body. Galileo did not reach the concept of mass. That idea would not likely arise until someone moved a body from one latitude to another in the middle of an experiment. Huygens and Newton were pioneers in introducing the concept of mass into the literature. For the product of mass and velocity Newton used the term "quantity of motion"; but it has long since been replaced by Galileo's term, "momentum."

Frequently, when any of Galileo's acquaintances had equipment
that baulked or gave trouble, they brought it to him—ostensibly for
advice but really to get him to repair it and make it work. One such
instance led later to the invention of the barometer and hence indirectly to that of the extraction air-pump which furnishes experimenters
with the very kind of space that Salviati said was not available,
namely, a so-called vacuum, a space containing relatively little
matter. This may raise the question, "Doesn't
vacuum mean empty?" and the Gilbertian answer, "Well, nearly empty."

SAGREDO. I once saw a cistern that had been provided with a pump. The latter worked perfectly as
long as the water in the cistern stood above a certain
level, but below that level it failed to work. When I
first noticed this fact, I thought the machine was out
of order; but the pump-maker whom I called in to repair it told me that the defect was not in the pump.
He added that it is not possible either by a pump, or
by any other machine that works on the principle of
attraction [suction?], to lift water a hair's breadth
above eighteen cubits [about 25 feet]. Whether the
pump be large or small, this is the extreme limit of the
lift.

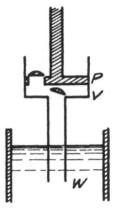

FIG. 89. The lift-
pump.

The Aristotelian explanation of the action
of the common lift pump was that 'Nature ab-
hors a vacuum." When the piston P (fig. 89) rises, if the water W in
the stem of the pump did not do something about it, there would be
a vacuum at V and therefore (?) the water rises in the tube so as not
to horrify Nature. Discussing the rebellious pump with his students
Torricelli and Viviani, Galileo remarked in his usual gay banter, that
Dame Nature's horror of the void, by some mysterious whim, seemed
to peter out suddenly at about 18 cubits. He suggested that if they
would investigate the problem, they would likely learn something important and useful. As we shall soon see, the old maestro was right
again. He could scent a good problem from afar. His blindness had
not caused any dimming of his intellectual insight. Genius can be
evinced in the selection of a problem as well as in method of attack.

Galileo was too occupied with other work and too physically weak
to undertake the research himself. During his last year on earth he
worked on several problems, among them that of constructing a
pendulum-clock. He designed and built an instrument which had a dial
graduated in minutes and a hand that was operated by a pendulum;
but it had to be given a nudge every now and then to keep it going.
What was needed was a method of keeping it going for a number of

hours at least. It is said that on his death-bed the idea he had been seeking came to him. To Viviani he said, "Quick, while there is yet breath," and gasped out the specifications of the new instrument. Thus he died as he had lived, inventing something that would be useful to his fellow scientists and to the world.

In the year of Galileo's death, 1642, there was born in England, on Christmas Day, in the household of Newton, a puny posthumous child who would be christened Isaac, and who would build on the firm foundation laid by Galileo a towering structure in physics and mathematics of great strength and beauty.

Since we shall encounter other researches of Galileo in later sections, the close of this visit is only *au revoir*. As he said in one of his dialogues, "I shall not fail to be with you at the next session hoping not only to render you service but also to enjoy your company."

THE INVENTION OF THE BAROMETER

Pursuing Galileo's suggestion, two of his disciples, Evangelista Torricelli and Vincenzo Viviani, ultimately arrived at the invention of the barometer. Here is part of a letter in which Torricelli announced the new instrument:

Florence, June 11, 1644.

To MICHELANGELO RICCI
AT ROME

Most Illustrious Sir,
I have already intimated that I do not know of any previous scientific experiments having been done on vacuum, not merely to obtain one but to prepare an instrument to indicate the changes in the atmosphere which is at one time denser and at another lighter. We live submerged at the bottom of an ocean of air which, by experiment, undoubtedly has weight, with greatest density near the surface of the earth and of about the four hundredth part of the density of water.[4]
I have made a large number of glass instruments like those in the figure [fig. 90], labelled A and B, with tubes two cubits long. Each of these was filled with quicksilver, its mouth closed with a finger, then it was inverted in a vessel C, containing quicksilver. When the finger was removed, the tube was seen to empty partly without anything entering while it emptied. The part of the tube, AD, however, remained always full to the height of a cubit, a quarter and a finger extra. To show that the tube was perfectly empty above BA, we filled the bowl beneath with water up to the brim D, and gradually raised the tube until its mouth was seen to reach the water. The mercury suddenly dropped down from the tube and, with a horrible impact, the water shot up in the tube to the mark E. This account makes it plain that the vessel AE was empty and that the mercury, though very dense, was supported in

---

[4]Galileo's value (p. 101). Quite possibly, Torricelli assisted Galileo in the determination.

the tube *AC* by some force which held it up in spite of its natural tendency to fall down.

Up to the present, it was. ascribed to the natural condition in the tube, namely, to vacuum in the extremely rarefied bulb. But I claim that the force is external. On the surface of the liquid in the bowl, a fifty-mile column of air presses down by gravity.[5] Is it any wonder if in the glass *CE*, where the mercury is at rest, with nothing to resist its entry, it rises until it balances the weight of the air outside which supports it?

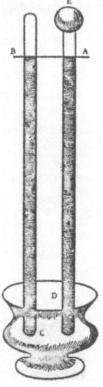

Water, then, in a similar tube but much longer, will rise to a height of eighteen cubits [as in a pump], i.e. as many times higher than mercury as mercury is denser than water, in order to counterbalance the same force which thrusts each as much as the other. My thesis is confirmed by the experiment performed at the same time with the tube *A* and with the tube *B*, in which the mercury remained constantly at the same level, marked *B*, as witness that the force was not interior since, in that case, the bulb *AE* would have had greater force . . . than the very small space at *B*.

Your devoted servant,

V. TORRICELLI

In this episode, Torricelli and Viviani show themselves as able scientists quite worthy of their great master. The space above the mercury is called a Torricellian vacuum; and it appears that, by this experiment, the "empty" space has been obtained which Salviati said was impossible to secure in his day. As to whether the space is entirely empty, we now know that it contains a *soupçon* of mercury vapour. A difficulty still remains: this vacuum is hardly available for experimentation because of its inaccessibility. We shall see that difficulty overcome by the work of von Guericke, Boyle, and Hooke.

FIG. 90.
Torricelli's
barometer.

It is interesting and not unprofitable to trace the threads of influence which run through the story of science, the relationships which bind one scientist to another. The work of Torricelli and Viviani was inspired by their old maestro, Galileo. Torricelli's letter was sent by Ricci to Père Mersenne of Paris, through whose wide correspondence the barometer became known throughout Europe. The new instrument led to new researches. It led Pascal of Paris to study equilibrium of liquids and to invent the hydraulic press. Von Guericke of Magdeburg, Germany, from his study of the Torricellian vacuum reached

[5]See p. 249 (Ptolemy and Alhazen, height of atmosphere).

the invention of the extraction air-pump. The consequences of this last invention were very far reaching indeed. Boyle and Hooke of London improved the air-pump and their studies led them to the discovery of Boyle's law. Torricelli and Viviani with some others of Galileo's old students formed a science club at Florence which carried on some excellent researches. Their report was sent to the Royal Society of London. Thus the influence of the school of Galileo spread through Europe and, finding a congenial place in London, gave rise to the English school typified by Newton. "Their echoes roll from soul to soul."

THE HYDRAULIC PRESS

In 1651, the brilliant young Frenchman Blaise Pascal (1623-1662), who was an acquaintance of Descartes, wrote a treatise, *On the Equilibrium of Liquids*.[6] He was only twenty-eight at the time but the book was not published until 1663, one year after his death. From that memoir comes the following passage which demonstrates the hydrostatic law known as Pascal's Principle:

Accordingly, if a vessel, filled with water, had only one opening, for example, one square inch in area, to which we apply a piston with a weight of one pound, that weight will exert a thrust on every surface of the vessel in general, because of the continuity[7] and fluidity of the water. To determine how much thrust each area undergoes, here is the rule. Every part of the surface of one square

FIG. 91. The Puy-de-Dôme, Auvergne.

[6]B. Pascal, *Œuvres* (Hague, 1779), IV, 228.
[7]p. 46 (Archimedes).

inch area, the same as the opening, undergoes the same thrust as if it were pressed by the one-pound weight (excluding the weight of water, of which I am not speaking here, for I am speaking only of the piston's effect), because the one-pound weight presses the piston which is at the opening. Each portion of the vessel, large or small, is pressed more or less, precisely in proportion to its area, whether that portion be opposite to the opening or to one side, whether it be far or near. The continuity and fluidity of the water render all these factors equally immaterial. Accordingly, the material of which the vessel is made must have sufficient resistance in all its parts to sustain all these thrusts.

FIG. 92. Blaise Pascal.

If its resistance is less than that at any place, it breaks. If it is greater, it furnishes what is necessary and the rest remains unused in that case. Accordingly, if we make a new opening in this container, in order to arrest the water that spouts through it, a force will be required that is equal to the resistance which that part of the wall had to have, that is to say, a force which is to that of one pound as the area of the latter opening is to that of the former.

Pascal's father was an eminent and distinguished lawyer and his mother a talented and accomplished woman. Theirs was a very cultured and happy home, in which the children, Blaise, Gilberte, and Jacqueline, were trained with great knowledge and care. But when

Blaise was only four years old, the beautiful home was irrevocably ruined by the untimely death of his mother. The neighbours said with French pessimism that such happiness could not long escape the envious blow of fate. The father did not remarry but devoted himself to the education of his motherless children.

M. Pascal must have been a remarkable teacher and unquestionably he had unusually good students, for all three became widely known for their accomplishments. Blaise won world fame in science, in mathematics, and as a writer. He was a child prodigy in geometry, and belongs to that minority of prodigies whose mature performance is in keeping with their early promise. Fearing that the boy was spending so much time on geometry that he was endangering his health, which had never been robust, the father ultimately forbade him the use of the books and blackboard in the study. Later it was found that he had resorted to the attic and there, using the floor as a blackboard, discovered for himself, at the age of eleven, a number of important theorems. Jacqueline claimed that one of these was the theorem of Pythagoras.

One important application of Pascal's Principle is the hydraulic press. Its structure is shown in fig. 93, and it is seen in action in fig. 94a, where it is bending a 10-inch sheet of cold steel! The following brief sentence from the same memoir would have been quite sufficient in itself to have made its author famous; it announces his discovery of this machine, which is unique in being the only addition made to the list of simple machines since prehistoric times.[8]

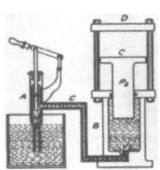

FIG. 93. Bramah's hydraulic press.

Thus it appears that a vessel full of water [or oil] is a new principle in mechanics and a new machine which can multiply force to any degree we choose.

Pascal's treatise *On the Weight of the Atmosphere*, written in 1651, contains the passage which gave the coup de grâce to the old Aristotelian "horror of the void" and established Torricelli's preferable theory of *atmospheric pressure*. At Pascal's request, M. Périer, Gilberte's husband, carried a barometer up the mountain Puy-de-Dôme, near his home in Auvergne (fig. 91). As the instrument went higher, its read-

[8]Sir William Bragg develops this topic in *Old Trades and New Knowledge* (Bell, 1926).

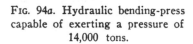

FIG. 94a. Hydraulic bending-press capable of exerting a pressure of 14,000 tons.

FIG. 94b. Hydraulic press used for pressing corrugated ends for railway cars.

ing decreased until, at a height of 3,000 feet, the mercury level had sunk about three inches.[9]

The experiments performed in the mountains have upset the universal belief that Nature abhors a vacuum, and have established the knowledge which never can perish, that Nature has no horror of the Void, that she does not create anything to abhor it and that the weight of the atmosphere is the true cause of all those effects which heretofore have been attributed to this imaginary cause.[10]

Pascal's greatest brilliance was in mathematics, but that is another story. He is best known generally for his literary and religious writings. He was a master of style in French prose. The following quotations from his *Pensées* will give a closer glimpse of the man.

The immortality of the soul is a matter which is of such importance to us and which concerns us so profoundly that one must have lost all feeling to be indifferent about knowledge of it. All our actions and all our thoughts must take such different courses according as there are or are not eternal joys to be hoped for, that it is impossible to take one step with sense and judgment

[9]E. J. Holmyard, *Physics for Beginners* (Dent, 1930), p. 82.
[10]B. Pascal, *De la Pesanteur de l'Air* (*Œuvres*, IV, iii, 325).

without determining it by our opinion on this point which should be our main consideration. Thus our prime interest and our first duty is to secure light on this subject upon which our whole conduct depends.

Every person is a combination of two contrasting parts which differ in kind, namely, body and soul. For it is impossible that the part in us which reasons should be anything other than spiritual. If anyone maintains that we are altogether corporeal, that would all the more exclude us from a knowledge of bodies, there being no idea so inconceivable as that matter should be aware of itself. It is impossible to imagine how matter could be conscious of its own existence.

Vanity is so anchored in the human heart that even a soldier's servant, a scullion, or a porter, brags and wishes to have admirers. The very philosophers themselves crave approbation. Those who write belittling glory wish to have the glory of having written well. Those who read their writings seek to have the glory of having read them. As for me, who write this, I have probably the same desire and it may be that my readers will have it too.

THE PENDULUM CLOCK

We saw how near Galileo came to inventing the pendulum clock. Fourteen years later, the Dutch scientist Christian Huygens completed the invention and took out a patent. Although his home was at The Hague in Holland, he lived for twenty years at Paris as librarian of the Royal Library. Among his friends were Descartes, Pascal and Snell with whom he frequently discussed scientific questions. In one of his Latin treatises he refers to his invention of the pendulum clock in these words:

Those who know that, for several years past, astronomers have been using pendula to measure time intervals, will easily realize that these pendula suggested to me my invention. Because of the imperfections of water-clocks, astronomers, following that pioneer and Man of Genius, Galileo Galilei, adopted the method of setting a weight swinging at the end of a light chain. Its oscillations, by counting, aggregated a number of equal intervals of time. But the amplitudes of the pendula necessarily kept diminishing unless there were assistants to keep them going. Counting all the to's and fro's, moreover, was a very tedious task. As for myself, noting the remarkable regularity of this kind of motion and considering it, among all the motions that nature displays, singularly amenable to mechanical contrivance, I sought a method of accomplishing that construction as conveniently as possible and thus remedying the double inconvenience mentioned above.

After investigating a number of schemes, I was struck by the one I am about to describe as being the simplest and easiest. After its conception, when it passes into public and private use, as has now begun to occur, everyone will derive from it the advantage that we shall establish an agreement among clocks and with the Sun itself which will be better than ever before, indeed, almost better than one could have dared to hope.[11]

Then follows a detailed description of the "works" of an early form of common kitchen pendulum clock. It was by no means Huygens' only contribution to mechanics. We shall observe later some important

[11]C. Huygens, *Complete Works*, XVII, 53.

results which he obtained by his investigation of the pendulum, but this invention comes into the story here because it developed directly from the work of the school of Galileo. Some of Huygens' most important contributions will be met in the Story of Optics.

PORES IN SOLID SILVER

Viviani and Torricelli were members of the small science club or society called the *Accademia del Cimento*, i.e. the "Academy of Experimentation," which held meetings from 1657 to 1667 under the patronage of Leopold of Tuscany, who was also one of Galileo's old

pupils. That the antagonism between the church and science had not vanished is indicated by the Pope's offer to Leopold of a cardinal's hat if he would withdraw his support from the society. They both knew that this would mean the death of the club. The temptation was too strong for Tuscany; he withdrew his patronage and the society disbanded. Its secretary, Magalotti, drew up a report of the work accomplished and sent it to the Royal Society of London, which had received its charter from Charles II in 1663. From Magalotti's report the following account is taken of an experiment to find whether water is compressible[12]:

FIG. 95. Drops of water forced through solid silver.

We had a hollow sphere cast in silver [fig. 95]. We filled it with water and screwed the cover on tightly. We then proceeded to hammer the vessel gently at several points. The denting of the silver decreased the capacity of the container yet the water experienced no compression [?] for at every stroke, water drops were seen to ooze through the pores of the metal. If the vessel were made of thicker silver, we are not sure that the water might not be compressed. But this much is certain, the water resists compression infinitely more than does air.

As was remarked previously, if experiments do not succeed in clearing up completely the question under investigation, yet the luck is unusually bad if they do not at least bring to light something of value.

This experiment was inconclusive as far as its main purpose was concerned; but as the secretary naïvely and gleefully remarked, it was useful in another way. It showed that water can be forced through a sheet of solid silver. This is poor support for Aristotle's contention that solid silver is a continuum and a plenum, devoid of internal spaces. It favours strongly the opinion of Democritus that matter consists of discrete particles separated by intermolecular spaces.

[12]The library of the University of Michigan has a copy of this rare book. See also A. Wolf: *History of Science.*

Magalotti's report contains several valuable contributions to mechanics, including studies of the flight of projectiles, of the Torricellian vacuum, and of a particular case of Newton's third law of motion, which we shall consider in the story of electricity. Among the Society's vacuum experiments was the classic one of tying the mouth of a flaccid bladder with string and placing it in vacuo. When the gas pressure around it became less than atmospheric, the bladder swelled. When the container was opened to the air again, the bladder contracted to its initial flaccid condition. Thus it was seen that the volume of the air in the bladder depended on the pressure of the gas around it. It was seen, too, that the relation between the volume of a gaseous body and its pressure is inverse, for when the pressure decreases, the volume increases, and vice versa.

Magalotti's report must have been available to Robert Boyle, who was a charter member of the Royal Society of London, and it was no doubt one of the factors which incited him, with the help of his very able assistant Hooke, to investigate quantitatively the relation between the volume and the pressure of a gaseous body. The outcome was the discovery of the principle known to some as Boyle's law and to others as Mariotte's law.

THE EXTRACTION AIR-PUMP

When the inventor of the extraction air-pump, Otto von Guericke (1602-1686), was a student at the University of Leyden, he came under the influence of Dr. Willebrord Snell. This professor of physics kindled in him a life-long love of science. Throughout von Guericke's busy and useful life, even amid the horrors of the Thirty Years War, science was the pastime for what little leisure he could snatch. He admired especially the work of Galileo, Torricelli, and Pascal, and was inspired with a desire to do some original work in pneumatics, the science of gases. In 1631, when he was twenty-nine, his city, Magdeburg, was sacked and burned to the ground. Since von Guericke was an alderman and an engineer, he was placed in charge of rebuilding the city. He was elected burgomeister or mayor in 1646. Yet by 1650 he had succeeded in inventing a primitive form of the air-pump. An account of this *Arbeit* is given in his Latin treatise, *New Experiments on Empty Space*,[18] which was published at Amsterdam but, because of lack of funds, not until 1672:

Air has the property of shrinking under pressure and expanding when given more room. Any place in the atmosphere which a body vacates never

[18]*Nova experimenta de vacuo spatio*. A prime copy of this rare old book was presented to McGill University by Sir William Macdonald. From it the accompanying quotations are translated with grateful acknowledgment to the Librarian, Dr. G. R. Lomer, for use of the book.

exhibits a vacuum, but becomes filled with air just as the space occupied by a fish in water, if it changes position, is filled with water again. Reflecting on the immensity of space and the fact that it must be everywhere present, I planned the following experiments.

Suppose a wine or beer cask is filled with water and sealed at all the cracks between staves so that air cannot enter. Below the cask let a metal tube be fastened by which the water can be withdrawn. The water, then, should descend and leave above it in the cask an air-free space [?]. Accordingly, I fitted up with piston and plunger a brass squirt-pump such as they use at fires.[14] Two leather valves were also inserted, the inner one to prevent ingress of water and the outer one, egress. Fastening the pump beneath the cask, I

 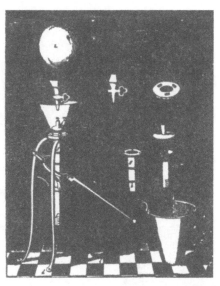

FIG. 96. Early stages of von Guericke's air-pump.

FIG. 97. Von Guericke's extraction air-pump.

tried to extract the water. In the first attempt, the iron screws gave way before the water followed the piston.

Stronger screws were used, and finally three men pulling at the plunger managed to extract water through the upper valve. But then we heard all around the cask a sound as if the water were boiling vigorously and this lasted until the cask, in place of the water removed, was filled with air. Having thus discovered the porousness of the wood, I concluded that a copper globe would suit my purpose better.

Then I proceeded to withdraw from it both water and air as before. At first, it was easy to pull the plunger, but soon this became more difficult, so that two strong men were hardly able to pull out the plunger. While they were still busy operating it and had begun to think that the air was nearly all out, suddenly, with a loud bang, startling us all, the metal globe collapsed as when

[14]See force-pump of Ctesibius, p. 59.

one crumples a cloth in the hand. When the tap was opened, air rushed in with great violence as if it wished to drag a bystander in with it.

Since it is impossible to make a cylinder and piston so perfect as to prevent all leakage of air, I built several machines, arranging the device so that the air-pump could be surrounded with water above and below. Thus I prepared the machine described below.... From the description of this instrument, it is clear that by its aid, a vacuum can be obtained and that many difficulties that previously were considered insuperable, can now be overcome.

We have heard Salviati say that, in his day, it was not possible to obtain a vacuum for experimentation. The Torricellian vacuum

FIG. 98. The Magdeburg hemispheres experiment.

was in a small and inaccessible place. The air-pump made it more convenient to place apparatus in a vessel and extract a considerable portion of the air from the vessel. Naturally, von Guericke was first to try some new experiments which the air-pump made possible. For example, indulging his fondness for spectacular and large-scale demonstrations, he performed, in 1654, his famous Magdeburg hemisphere experiment before the Emperor and the assembled Reichstag. Fig. 98 is a copy of a famous old wood-cut in his book which depicts the fact (or claim) that when the air was pumped out from between close-fitting hollow hemispheres, fourteen inches in diameter, it took a tug o' war of sixteen horses, eight on each side, to pull them apart.

Torricelli, in discussing the barometer, took the density of atmospheric air as one four-hundredth that of water. In other words, he considered that water was four hundred times denser than air. We saw the ingenious way in which Galileo obtained that value. Von Guericke went on to make use of the air-pump to measure the density of air by an experiment which is still used in science courses. He described it in these words:

> For the further study of vacuum, I utilized a glass vessel such as druggists use. A glass of this kind is brought to equilibrium on a balance. Then we pump it quite empty [?]. We then observe that after the extraction of the air it weighs 1 or 2 ounces less, according to the magnitude of its content. Free access to the atmosphere is then allowed. Not only do we hear air entering the globe with a hissing sound but also we see how the vessel gradually regains its former weight, giving us a splendid demonstration of the weight of the air.

From measurements of volume and weight in this experiment, von Guericke calculated that water is 947 times denser than air. Since his day, the density of air has been determined with many refinements of accuracy by numerous scientists, notably the master precisionist Régnault of Paris. The value which is now accepted as standard is 28.8 gm. per 22.4 litres or 1.29 gm. per L. at a pressure of one atmosphere (76 cm. mercury), and the temperature of melting ice (0°C.). Since 1 L. water weighs 1000 gm., this means that water is $1000/1.29 = 776$ times denser than air. Galileo's value was 2.5 gm. per L. which was about 93 per cent too high. Von Guericke's result was 1.05 gm. per L., which was about 18 per cent too low. Yet both these researches were excellent, the first in discovering a new constant, the density of air under fixed conditions, and the second in obtaining its value with greater accuracy.

This research was only one of many in which the air-pump added to the laurels of its inventor. If we pause to list the subsequent advances in science which would have been impossible without the air-pump, it will help us to realize that this contribution of von Guericke was of incalculable value. In the story of acoustics, for example, we shall meet the famous bell-jar experiment, which is a case in point. In every way von Guericke was a great citizen, and here is some parting advice taken from his book: "We must place more confidence in experiments than in the conjectures of that folly which is ever ready to fabricate preconceived ideas about nature."

CHAPTER FIVE

The School of Newton

SIX years after the inauguration of the Florentine Academy, Charles II of England founded the Royal Society of London, "For Improving Natural Knowledge" as Francis Bacon had recommended four decades earlier. One of the charter members was the Honourable Robert Boyle, seventh son and fourteenth child of the Earl of Cork. Boyle was also elected one of the Society's first Councillors. His chief contribution to mechanics was his discovery of the quantitative relation between the volume of any isothermal gaseous body and its pressure. Some refer to this as Boyle's law, but in France it is often named after the physicist Edmé Mariotte, who discovered the law independently at about the same time.

Because of his private fortune by inheritance Boyle was able to devote much leisure time to the study of science. Very different was the legacy of Dr. Robert Hooke, who was Boyle's laboratory assistant. Hooke was a clever scientist in both practice and theory, but his health was frail and he was very poor. Consequently Boyle was able to hire him as assistant. After several years of that service, however, Hooke was appointed curator of experiments in the Royal Society and later became its Secretary. By a series of experiments with coiled springs he discovered the law of elasticity known as Hooke's law, which we shall discuss in the next interview. It is of much broader scope than Boyle's law, for it applies to all bodies. Boyle saw that air also is elastic. He published an essay entitled "The Spring of the Air," in which he spoke of air particles behaving as if they were minute coiled springs. The essay was attacked by a peripatetic named Linus. This roused Boyle's Irish combativeness and offered him a welcome opportunity to deal a peripatetic a shrewd blow. By Hooke's help he nailed his opponent with irrefutable experimental data in a treatise entitled *Defence of the Spring of the Air*, from which we shall read. He showed that the volume of a gas varies inversely as its pressure "or thereabouts" (said Boyle).

The following quotation is condensed from Boyle's text but gives the main outline of his famous experiment:

BOYLE'S LAW[1]

We took a long glass tube, which by a dexterous hand and the help of a lamp [a blast burner], was in such a manner crook'd at the bottom [fig. 99], that the part turned up was almost parallel to the rest of the tube, and the orifice of this shorter leg of the siphon (if I may so call the whole instrument) being hermetically sealed, the length of it was divided into inches (each of which was divided into eight parts) by a streight list [i.e. a narrow strip] of paper, which, containing those divisions, was carefully pasted all along it, then putting in as much quicksilver as served to fill the arch or bended part of the siphon, that the mercury standing in a level might reach in the one leg to the bottom of the divided paper and just to the same height or horizontal line in the other; we took care, by frequently inclining the tube, so that the air might freely pass from one leg into the other by the sides of the mercury (we took, I say, care), that the air at last included in the shorter cylinder should be of the same laxity with the rest of the air about it. This done, we began to pour quicksilver into the longer leg of the siphon, which by its weight, pressing up that in the shorter leg, did by degrees streighten [compress] the included air; and continuing this pouring in of quicksilver till the air in the shorter leg was by condensation [compression] reduced to take up but one half the space it possessed (I say possessed, not filled) before.... We cast our eyes upon the longer leg of the glass, on which was likewise pasted a list of paper carefully divided into inches and parts, and we observed, not without delight and satisfaction [!] that the quicksilver in that longer part of the tube was 29 inches higher than the other.   Now that this observation does both very well agree with and confirm our hypothesis, will be easily discerned by him that takes notice what we teach, that the greater the weight is that leans in upon the air, the more forcible is its endeavour of dilation, and consequently, its power of resistance (as other springs are stronger when bent by greater weights). For this being considered, it will appear to agree rarely well with the hypothesis that as according to it, the air in that degree of density and correspondent measure of resistance to which the weight of the incumbent atmosphere had brought it, was able to counterbalance and resist the pressure of a mercurial cylinder of about 29 inches, as we are taught by the Torricellian experiment; so here the same air being brought to a degree of density about twice as great as that it had before obtains a spring twice as strong as formerly. The several observations that were thus successively made and as they were made, set down, afforded us the ensuing table:

Fig. 99.
Boyle's
J-tube.

A. The number of equal spaces in the shorter leg that contained the same parcel of air diversely extended. B. The height of the mercurial cylinder in the longer leg that compresses the air into those dimensions. C. The height of the mercurial cylinder that counter-balanced the pressure of the atmosphere. D. The aggregate of B & C. E. What that pressure should be according to the hypothesis that supposes the pressures and expansions to be in reciprocal proportions.

[1]Robert Boyle, *Complete Works*, I, 6.

A	B	C	D	E
12	00	$29\frac{2}{16}$	$29\frac{2}{16}$	$29\frac{2}{16}$
11	$02\frac{3}{16}$		$31\frac{5}{16}$	$31\frac{12}{16}$
10	$06\frac{3}{16}$		$35\frac{5}{16}$	35
$8\frac{1}{2}$	$12\frac{8}{16}$		$41\frac{10}{16}$	$41\frac{2}{17}$
5	$41\frac{9}{16}$		$70\frac{11}{16}$	70
4	$58\frac{2}{16}$		$87\frac{14}{16}$	$87\frac{3}{8}$
$3\frac{1}{2}$	$71\frac{5}{16}$		$100\frac{7}{16}$	$99\frac{6}{7}$
3	$88\frac{7}{16}$		$117\frac{9}{16}$	$116\frac{4}{8}$

The following snippet is from a book that Boyle wrote on heat, to which he gave the title, *Experiments Touching Cold*.

We have shewn, That the strengths required to compress air, are in reciprocal proportions or thereabouts, to the spaces comprehending the same portion of the Air.

In spite of its fundamental importance, Boyle's account of his experiment, only a small part of which has been given here, could have been given in far fewer words. He is so eager to make plain every detail that he becomes painfully prolix. In this regard we shall find him in sharp contrast with Newton.

Boyle often describes details of laboratory technique which have a homely touch. Here is a sample:

The tube being so tall that we could not conveniently make use of it in a chamber [room], we were fain to use it on a pair of stairs [stairway]. The lower and crook'd part of the pipe was placed in a square wooden box of a good largeness and depth, to prevent the loss of the quicksilver that might fall aside in the transfusion from the vessel into the pipe and to receive the whole quicksilver [!] in case the tube should break.

Such a confession of clumsy technique would be, in a modern paper, infra dig., for the modern chemist's technique must be, above all things, perfect in pouring.

Hooke introduced such a number of radical improvements in von Guericke's air-pump that it became practically a new instrument, but since he was merely the penniless assistant, it was called the "Boyle engine." Hooke had so many experiences of this kind with different men that he became embittered and rather cantankerous. Hooke and Boyle performed a number of experiments which the new pump made possible. Boyle made a one hundred-and-seventeen-page report of the new experiments and sent it to a nephew, the son of his oldest brother, as he said, "by way of a letter." (A correspondent who will play a trick like that is certainly a heavy responsibility.) Here is an excerpt from Uncle Robert's letter:

You may be pleased to remember that ... I told you of a book published by the Jesuit, Schottus, wherein ... he related how that ingenious gentleman,

Otto Gericke, consul of Magdeburg, had lately practised in Germany a way of emptying glass vessels by sucking out the air at the mouth of the vessel plunged into water. . . . When the engine . . . comes to be more attentively considered, . . . there appear two things to be desired in it. First, the wind-pump (as somebody not improperly calls it) is so contrived that to evacuate the vessel, there is required the continual labour of two strong men for divers hours. Next, the receiver, or glass to be emptied, is so made that things cannot be conveyed into it whereon to try experiments. Wherefore, I put Mr. G. and R. Hook

Fig. 100. Robert Boyle.

(who hath also the honour to be known to your Lordship) to contrive some air-pump . . . that might be more easily managed. The last-named person fitted me with a pump anon to be described.

Among the new experiments that Hooke and Boyle performed when "Hook had fitted them with a wind-pump" was one which has become classic as the "feather and guinea experiment." A feather and a coin are placed in a broad glass tube about a metre long. When the tube, with air in it, is suddenly turned into a vertical position with the two bodies at the top, the feather falls more slowly than the coin. But if the air is extracted the two bodies fall together. Here we have another example of the remarkable insight of Lucretius, who predicted this experiment in 55 B.C. As we recall the chain of discoveries leading to this experiment, we remember the names of Aristotle, Lucretius, Galileo, Torricelli, von Guericke, and Hooke. The work of the last three had overcome some of the practical difficulties of obtaining a vacuum to which Salviati referred.

Boyle's chief interest was chemistry, as may be seen in his book, *The Skeptical Chymist*. The term "receiver," which he was first to apply to the vessel to be exhausted by an air-pump, bears the mark of the chemist, for it is the name which the latter applies to the vessel in which he collects a distillate obtained in a distillation.

Boyle was such an ardent admirer and follower of Francis Bacon that some of his contemporaries dubbed him "Bacon II." It is not surprising, therefore, to find that, like Bacon, Descartes, Galileo, and others, he disapproved of many features of the Aristotelian schools. This was seen in the zest with which he replied to the attack made on his *Spring of the Air* by the peripatetic Linus. It was seen also in the fact that he wrote his treatises in English although he was familiar with Latin and Greek. The excerpts we have read from his works are the first in this collection whose originals are in English.

The Oxford Pulpit once declared that Boyle's researches and influence were destroying religion in the universities. In Boyle's case, therefore, as in Galileo's, there was a clash between science and the church; but that is different from saying that there was clash between science and religion.[2] Boyle had the advantage over Galileo of living in Great Britain instead of Italy; consequently his opponents could not imprison and ruin him because of his views. As a matter of fact, Boyle was so deeply religious a man that some have referred to him as "Saint and Scientist." Other indications of his religious zeal were his establishment of the Boyle Lectures in Defence of Christianity and his donation of considerable sums for the translation of the Bible into foreign languages.

HOOKE'S LAW

Reference has already been made to Dr. Robert Hooke, the curator of experiments for the Royal Society. He was professor of geometry at Gresham College and Secretary of the Royal Society for five years. In 1676, he wished to announce his discovery of the law of elasticity, but he needed more time to confirm his conclusions by additional measurements. He resorted, therefore, to a device which had been used by Galileo, Huygens, and others in similar circum-

[2]There is an exceptionally interesting discussion of this topic in Sir William Bragg's *The World of Sound* (Bell, 1921), p. 195.

stances—he published his result in the form of an anagram or cypher. Here is the passage:

> To fill the vacancy of the ensuing page, I have added a decimate of the Inventions, I intend to publish . . . (3) The true Theory of Elasticity or Springiness . . . c e i i i n o s s s t t u u . . . and (9) A new sort of Philosophical Scales of great use in Experimental Philosophy,— c d e i i n n o o p s s s t t u u .

Nowadays a caveat serves much the same purpose for inventors as the old anagram. When an inventor's device has reached what he considers a successful stage, but he does not yet wish to disclose the details, he can take out a caveat to guard his priority while he goes on with bringing the invention nearer to completion.

Two years later, in a lecture on "The Force of Restitution," which was delivered before the Royal Society, and which appears in his posthumous works under the Latin title *De Potentia Restitutiva*, Hooke deciphered these anagrams. He thereby announced his invention of the spring balance and his discovery of the law of elasticity— now known as Hooke's law. These are his words:

> The Theory of Springs . . . has hitherto not been published. I printed this Theory in an Anagram . . . viz., c e i i i n o s s s t t u u, id est, Ut tensio, sic vis; that is, The Power of any Spring is in the same proportion with the Tension thereof; That is, if one power [force] stretch or bend it one space, two will bend it two, three will bend it three and so forward.
>
> From this, it will be easie to make a Philosophical Scale to examine the weight of any body without putting in weights, which was that which I veiled under this anagram, c e d i i n n o o p s s t t u u, namely, Ut pondus, sic tensio [as the weight, so the tension]. The fabrick [structure] of which, see in the figure [fig. 101]. This Scale, I contrived in order to examine the gravitation of bodies towards the Center of the Earth, viz., to examine whether bodies at a further distance from the Center of the Earth, did not lose somewhat of their power or tendency towards it . . . and attempted the same at the top of the Tower of St. Paul's before the burning of it in the late great Fire [1666]; as also at the top and bottom in Westminster.

From Hooke's original drawing of the spring balance (fig. 101) we can see what an excellent draughtsman he was. One might wonder how he ever obtained permission to perform experiments in St. Paul's and Westminster, but one of his cronies in the Royal Society was Sir Christopher Wren. Enough said.

If a 10-lb. block B (fig. 102) is suspended by hooking it to the lower end of a wire hanging from a ceiling, the wire is somewhat stretched. Any deformation whether of shape or volume is called a *strain*. This strain of the wire is one of length and could be measured in millimetres. Where the block touches the wire, there is a *stress* consisting of two forces, the block pulling downward on the wire and the latter pulling upward on B. Each of these two forces is 10 pd. and the magnitude of the stress is also 10 pd. There is also stress throughout the wire; it is called tension. It tends to make the wire thinner.

The strict definition of stress includes the idea of area; it is force per area. Since the areas opposing each other in this case are equal, the idea of area can be disregarded. In Hooke's day and even in the writings of as meticulous a precisionist as Maxwell (1865), the idea of area was often excluded in speaking of stress.[3] A similar remark holds for tension.

If B is replaced by a 20-lb. block, the stress is doubled and, as Hooke found, the deformation or strain is also doubled. Hence Hooke's

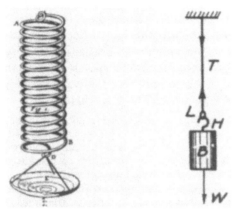

FIG. 101. Hooke's diagram   FIG. 102. Tension
of his spring-balance.   in wire.

law is generally expressed thus: Stress and strain are proportional, or, strain is proportional to stress.

If the block is removed, and if it was not too heavy, the wire returns to its original length. The force with which a body resists deformation, and by which it is restored to its initial state when the deforming force is removed, is called restoring or elastic force. This is the idea in Hooke's term "force of restitution," or *potentia restitutiva*. The property of exerting elastic force is called elasticity. All bodies have volume or bulk elasticity. It is characteristic of solids to have also shape elasticity.

The pendulum clock, invented by Huygens, and given its anchor escapement by Hooke, was quite satisfactory in stationary positions, for example, when fastened to a kitchen wall. Aboard ship, however, when rocked in the cradle of the deep, the pendulum was sadly "off tick." For such situations, what was needed was a clock that ran

[3] J. C. Maxwell, *Matter and Motion* (Sheldon, 1925), p. 26.

equally well in all positions. Unless the instrument could be supported on gimbal-rings like a compass, the pendulum would obviously have to be replaced by some other device. Hooke's special knowledge of springs enabled him to solve this problem. He replaced the pendulum by a balance-wheel with a hair-spring, as in a watch or an alarm-clock. The following passage from his posthumous works refers to his invention of the watch[4]:

> The Method I had made for my self for Mechanick Inventions quickly led me to the use of Springs instead of Gravity for the making a Body vibrate in any Posture, whereupon I did first in great and afterwards in smaller Modules, satisfy my self of the Practicableness of such an Invention and, hoping to have made great advantage thereby, I acquainted divers of my Friends and particularly Mr. Boyle, that I was possest of such an Invention, and crav'd their Assistance. . . . The discouragement I met with in the management of this Affair made me desist. . . .

Huygens also worked on this problem and took out a patent for the watch in 1675.

Hooke seems to have been born under a rather unlucky star. Poverty and ill health were part of his legacy, but, for compensation, he had an able mind and a deft hand. He seemed to be dogged by a vexatious fate which delighted in causing his discoveries to be anticipated, and the credit for his inventions given to others. Yet he had a number of staunch and very dear friends. The work of his clever mind and dexterous hand was often vitiated by his bad habit of not hoeing out his rows. His mind fairly spouted good ideas but often he pursued them only till within view of successful completion. When someone completed one of his unfinished projects and received the credit, there was friction. He has, however, a wonderful list of contributions to his account. He invented the first dial barometer and suggested the science of weather-forecasting. He invented the spirit-level, which is used by carpenters and masons and is incorporated in all instruments that must be level when used. In acoustics, he anticipated the invention of the stethoscope, Savart's wheel, and the Chladni plate. In optics, he anticipated the refractometer. He was the first to suggest the manufacture of artificial silk. He constructed a microscope for himself and by it discovered the cellular structure of plant tissues. His most famous book, *Micrographia*,[5] dedicated to Charles II, records his discoveries in biology. Its excellent drawings show what an able observer and artist he was. These instances form

[4]Copies of this book, which was published in 1705, may be seen in the library of Harvard University and in the Library of Congress, Washington, D.C. The quotation is given in R. W. Gunther, *Early Science in Oxford*, Vol. VIII, "Description of Helioscopes" (1676).
[5]The library of McGill University has a copy of this old book.

only a meagre part of his researches. It is regrettable that Hooke's works are not more easily available, for they are a veritable gold-mine of valuable ideas.

In Hooke's lecture on "The Nature of Comets," it is seen that he had a good qualitative grasp of the law of universal gravitation. This lecture was delivered before the Royal Society in 1682, five years before the publication of Newton's *Principia*, but ten years after Newton had solved the problem to his own satisfaction. Here are a few sentences from the lecture.

By Gravity, I understand such a Power as causes Bodies to be moved one towards the other till they are united. The Universality of this Principle, throughout the whole Universe . . . I shall afterwards explain . . . from the spherical surface of the Sea and from the Shadow of the Earth in Eclipses of the Moon . . . where the Shadow of it is . . . round. All the Celestial Bodies whose Shape we are able to discover, are . . . of a Globular Figure, and several of them do urn round upon their Axes. Were there not in them such a gravitating Power, all the loose Parts . . . must be shot out from them . . . or thrown away like a Stone out of a Sling.

Others have supposed various sorts of solid Orbs, Orbits, Epicycles and I know not what other wheel-work. My Attempt how trivial soever it may be supposed, hath given Mankind at least one Argument to believe somewhat better of their Mother Earth and themselves than they did before viz., that she hath Celestial Origin and that we our selves are Dwellers in the Heavens. The Principles I ground it upon are the most simple that can be. For "Nature does nothing in vain; but more is vain where less will serve."

The final quotation in this group is from the closing words of Hooke's *Micrographia:*

Wherever [the Reader] finds that I have ventur'd at any small Conjectures, at the causes of the things that I have observed, I beseech him to look upon them only as uncertain ghesses, and not as unquestionable Conclusions, or matters of unconfutable Science.

DISCOVERIES OF HUYGENS

After Huygens had invented the pendulum clock, he received from Louis XIV of France such a tempting offer of a position as librarian in the Royal Library that, although he was fairly wealthy, he could not afford to refuse. One of his duties was to assist in founding the French Royal Academy of Sciences. Since he had a great deal of leisure in his new sinecure, Huygens seized the opportunity to make an exhaustive study of the pendulum clock and related topics. We might wonder what he could find to study in a clock—had we not many times observed how much of use and interest can come from the scientific study of even the commonest and humblest of objects.

Huygens published his results in a book on pendulum clocks,

Horologium oscillatorium,[6] which is considered by many as the second greatest book in the literature of science. It consists of five parts, any one of which alone would have brought fame to its author.

Fig. 103. Christian Huygens.

In the second section of the book, Huygens followed the leadership of Galileo in making a study of projectiles. Here is a passage which deals with the motion of a pitched ball or, in general, any missile:

1. If gravity did not act and air did not resist the motion of bodies, then any given body, once set in motion, would proceed along a straight line with uniform velocity [forever].

2. Under the action of gravity, bodies move with a motion compounded from the uniform motion they have in any direction and the downward motion due to gravity.

3. Also, each of these motions can be considered separately and neither is influenced by the other.

We shall find presently that the first of these three statements is equivalent to Newton's first law of motion, which was previously

[6]The library of Princeton Observatory has a copy of this book.

stated in other words by Galileo and was even glimpsed by Aristotle. The equivalents of items 2 and 3 are to be found in Galileo's discussion of projectiles. These three statements are fundamental in considering the paths of all missiles and projectiles, spears, golf-balls, bullets, water-streams, meteors, sky-rockets, rolling-pins, etc.

We have seen that in order to describe a force or a velocity fully, we must specify its direction and sense. "Anything that has magnitude and direction is a *vector*," or directed quantity, e.g., displacement, acceleration, momentum, etc. The third of the above statements is

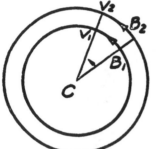

sometimes referred to as Huygens' law of the *independence of vectors*, although it occurs in the writings of Galileo.

Part V of Huygens' book deals with centripetal and centrifugal forces. The following excerpt from it may present some difficulties to beginners in mechanics; but when we propose to follow in the footsteps of giants, we must be prepared to stretch our intellectual legs.

FIG. 104. Centripetal force.

When a stone at the end of a string is swung around in a circle, like the stone in David's sling, the tightness of the string indicates a tension. One aspect of this stress is a force that pulls the stone towards the centre of motion. Accordingly, it is called *centripetal* force (Lat. *peto*, seek). The other aspect of the same stress is a force that pulls outward on the hand or centre of motion; hence Huygens called it *centrifugal* force (Lat. *fugio*, flee).

THEOREM 2. If two equal moving bodies [equal masses] are impelled with equal speeds along two unequal circumferences [fig. 104], their centrifugal forces will be in the inverse ratio of the diameters [or radii].

THEOREM 3. If two bodies, moving along equal circumferences with unequal speeds, but each with uniform speed, ... then, the centrifugal force of the swifter will be to that of the slower in the duplicate ratio of their speeds [i.e. directly as the squares].

Theorems 2 and 3 are expressed algebraically thus:

(2) $F \propto \dfrac{1}{r}$ (speed c, constant)

(3) $F \propto c^2$ (radius r, constant)

By combining these, we obtain

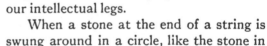

Eq. 1. $F \propto \dfrac{c^2}{r}$ or $F = H\left(\dfrac{c^2}{r}\right)$ where H is a variational constant, the letter H being chosen since it is Huygens' initial. This equation will be helpful in understanding Newton's derivation of the law of universal gravitation, for which Huygens helped to pave the way.

Consider a stone M_1 (fig. 105), whirled along a circular path at the end of a string about the centre C, with radius R, at uniform speed c cm. per sec. (cmps.) from M_1 to M_2 in 1 sec. The arc $M_1M_2 = c$ cm. and the angle ω, which equals c/R radians, is the angular speed in radians per sec. At M_1, the stone has the velocity $\mathbf{v}_1$, represented in

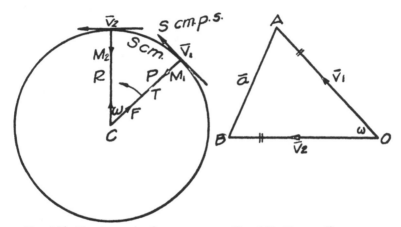

FIG. 105. Circular path of stone.   FIG. 106. Vector diagram.

fig. 106 by the vector $\mathbf{OA}$, whose length represents the speed c. At M_2, its velocity is $\mathbf{v}_2 = \mathbf{OB}$. $OB = OA$ (lengths).

In the vector triangle OAB, $\mathbf{v}_1 + \mathbf{AB} = \mathbf{v}_2$ (Stevinus);

$\therefore$ $\mathbf{AB} = \mathbf{v}_2 - \mathbf{v}_1$ = change of velocity in 1 sec.
                = $\mathbf{a}$, the acceleration, by definition.

$\because$ $\triangle AOB \;|||\; \angle M_1CM_2$, $\therefore$ $AB/OA = M'M''/R$ (Sim. $\triangle$'s)
                        $\therefore$ $a/c = c/R$.

Eq. 2 $\therefore$ $a = c^2/R$.

Assuming that Force is the only agent that can produce acceleration and that Force is proportional to acceleration, $F \alpha a$ or $F = ma$ where m is a variational constant.

Eq. 3.   Then $F = m\left(\dfrac{c^2}{R}\right)$.   (Compare *Eq.* 1, p. 128.)

Huygens had an unbounded admiration for his great English contemporary Sir Isaac Newton, and it was one of his dearest hopes that some day he might meet Newton and converse with him on mechanics and mathematics. When the Royal Society of London elected Huygens to fellowship, he seized the opportunity to go to London, for that would bring him nearer to Newton. When they met, the admiration proved to be mutual. In fact, Newton had such a high regard for Huygens as a mathematician that he did him the compliment of adopting his style in several respects. What a meeting that must have been between the authors of the greatest book in science and the second greatest. Newton tried to find a position for Huygens that would keep him in England. Huygens would gladly have stayed, but, unfortunately, Newton, in spite of his great prestige, was unable to find a single official or person of wealth who was far-seeing enough to find or make a place for Huygens.

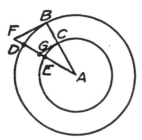

FIG. 107. Rotary motion.

When Huygens spoke to the Royal Society after the conferring of his fellowship, the topic of his address was the material of his great book. When he came to Part V, he intimated that he had, for the most part, only conclusions to offer. He had not yet found time to give the proofs in detail. These he promised to publish later. He intimated that the enunciations of his theorems would meanwhile give the boys something to shoot at. After his death in 1695, these proofs were found in his papers and were published in a posthumous treatise, *On Centrifugal Force.*

Here is Huygens' proof of the first theorem of Part V, taken from the posthumous treatise. It applies tacitly the principle of limits:

Let us consider two circumferences with radii, *AB* and *AC* [fig. 107]. Two equal bodies traverse these circumferences, revolving with the same period. Let us consider two very small corresponding arcs *BD* & *CE* [exaggerated in the diagram]. In the tangents at the points *B* & *C*, take the distances *BF* & *CG*, each equal to the corresponding arc [in the limit]. The body which revolves in the circle *BD*, has, therefore, a tendency to move away from the centre in the direction of the radius *AD*, and to traverse the path *DF* in a certain interval of time. The one which traverses the circle *CE* would, in the same time, move through the distance *EG*. Hence in the case of the greater circle the thread *AD* is stretched harder than that of the smaller in the proportion *DF* : *EG*. But *FD* : *GE* = *BF* : *CG* = *BA* : *AC* [by similar triangles]. Therefore, the centrifugal force corresponding to the greater circle will be to that of the smaller in the ratio of the circumferences or of their diameters.[7]

Before the year 1671, scientists considered the weight of a body as a constant—so much so, that if ever the weight of a body changed, it was no longer considered the same body. The only known way of changing the weight of a body was to throw some of it away or to add a new piece, and in each case the new body was not the same as the

[7]C. Huygens, *De vi centrifuga (Complete Works).*

old one. In that year, however, the French astronomer Jean Richer moved a body from one latitude to another during a series of experiments; he went to Cayenne in French Guiana near the equator (N.L.5°), to take observations of the planet Mars. He took with him a clock whose pendulum had beat astronomical seconds at Paris for years. At Cayenne, the clock ran slow. It was losing two and one-half minutes a day. To bring it into harmony with the stars, Richer had to shorten the pendulum. Since it is the force called the weight of the pendulum that pulls it toward its lowest point, Huygens was forced to conclude that the weight of the pendulum-bob had decreased. But it still looked like the same body. None of it had escaped. When the clock was returned to Paris, it ran fast, gaining about two and one-half minutes a day. It had to be adjusted to its original length to make it beat true seconds again. Instinctively, one feels that in spite of the change of weight in this body, there had been throughout the whole voyage a something in it which had remained steadily the same. To that constant property, Newton and Huygens applied the term quantity of matter. Since their day, the phrase has been replaced by the briefer term *mass*, which will come prominently into the discussion of Newton's work, for it is one of Newton's great contributions to science.

NEWTON'S *PRINCIPIA*

Several times in the story of mechanics, it has been necessary to mention beforehand the work of Sir Isaac Newton, just as one occasionally catches a glimpse of a mountain top above the trees when travelling toward it through the forest. Reference has been made to his birth on Christmas Day in 1642, the year of Galileo's death. As a boy Newton was not robust and he gave only slight indication of his genius. One meagre hint of his great life-work was afforded by the fact that, like Galileo and many other young scientists, he displayed a fondness for making mechanical toys, kites, sun-dials, and so forth. Tradition says that one day in school a big bully kicked him in the stomach and knocked out his wind. As it was not considered cricket to report such an offence, the little fellow determined to excel at lessons and to have his revenge on the bully by besting him in the classroom. At any rate it is true that at an early age Newton began to evince a passion for reading books on science and mathematics, and for thinking about their contents. When sent to the village with the hired man to do some shopping, he would often climb to the loft over the apothecary's shop where there was a pile of books, and so with the connivance of his companion get in some good solid reading until the shopping was completed. One book that made a deep impression on him was Kepler's *Optics*.

Newton ultimately became professor of mathematics at Cambridge University and for a quarter of a century he was President of the Royal Society. His name is held in highest veneration among scientists of all lands; but knowing the dangers which came from too much veneration of Imhotep and Aristotle, Newton would have been the first to urge scientists to use his work only as a basis for further progress. In 1687, he published a Latin treatise entitled *Philosophiae Naturalis Principia Mathematica*. It is generally referred to as Newton's *Principia* or "The Principia." It is looked upon as the greatest book in the literature of science, and is considered suitable reading for an honour student in physics during his third or fourth year at university. Yet Newton of his own accord would never have published it, for he greatly disliked controversy. In this, he was very different from Galileo. Had it not been for the admiration, loyal support, and insistence of one of his old students, the celebrated astronomer Dr. Edmund Halley, the *Principia* would never have appeared in print; as Newton said in the preface (dated 1686): "In the publication of this book, the most acute and universally learned Mr. Edmund Halley not only assisted me with his pains in correcting the press [proof-reading] and taking care of the schemes [diagrams], but it was to his solicitations that its becoming public is owing."

LAWS OF MOTION

One of the best known passages of the *Principia* is Newton's statement of the laws of motion, which he enunciated as axioms or fundamental assumptions. His wording of the first law runs as follows: *Corpus omne perseverare in statu suo quiescendi vel movendi uniformiter in directum nisi quatenus illud a viribus impressis cogitur statum suum mutare*, which is generally translated thus:

Every body continues in its state of rest or of uniform motion in a straight line unless compelled by external force to change that state.

We have seen the same idea expressed by Aristotle, Galileo, and Huygens, and the phrase "external force" sounds like an echo of Galileo. But for two centuries now, Newton's statement or a translation of it has been accepted as standard, although the wording "uniform motion in a straight line" contains a redundancy which is not in the original. Newton, of course, was always ready to give full credit to his predecessors and contemporaries, as when he said, "If I have been able to see a little farther than some others, it was because I stood on the shoulders of giants."

We all see countless illustrations of the first law of motion every day and apply it—when we shovel snow or coal, stamp snow from our shoes, ride a bicycle, throw a ball and so on. The law says that *the velocity of every body is constant unless changed by force.* It says that acceleration is produced by force only. Thus it assumes the existence of force and defines it as the only agent that can produce acceleration. It also implies for every body, as did Aristotle's statement, the property of offering resisting force in opposition to any body that changes or tends to change its velocity. This property, as we have seen, Kepler called *inertia.* Hence this law is sometimes called the inertia law. In all such instances, acceleration is the observed *fact* and the expression of the facts in terms of the hypothetical idea of force is a convenient way of describing the event.

Newton stated the second law of motion somewhat as follows:

Rate of change of momentum is proportional to the motive force applied and takes place in the direction in which the force acts.

We have heard the equivalent of this law enunciated by Galileo. It is a quantitative statement of the first law. It gives a method of measuring force, though not the common method, for generally a force is measured by balancing it by a known force, as in the operation of weighing a body. Consider a force f acting on a mass m for t sec. and changing its velocity from u to v. Since the momentum of a body at any instant is defined as the product of its mass and its velocity, $M = mv$ (*Eq.* 11, p. 104),

∴ the initial and final momenta in this case are mu and mv respectively and the change of momentum is $(mv - mu)$. Since this change of momentum occurs in t seconds, ∴ the rate of change of momentum is $\dfrac{mv - mu}{t}$ and the law says that

$$f \propto \frac{mv - mu}{t}, \text{ or } f = k\left(\frac{mv - mu}{t}\right) \text{ where } k \text{ is a constant.}$$

If the units are suitably chosen, $k = 1$ and

Eq. 4. $f = \dfrac{mv - mu}{t}.$ (law II)

If the body commences from rest, then $u = 0$ and

Eq. 5. $f = \dfrac{mv}{t} \text{ or } ft = mv.$

The product ft is called *impulse.* Hence a brief statement of law II is *impulse equals momentum produced.*

[8]J. B. Bélanger 1847. Cajori, op. cit. p. 59.

If $m = 1$ lb., $v = 1$ ft. per sec., and $t = 1$ sec.,

then $f = 1 \times \dfrac{1}{1} = 1$ *poundal* by definition.

If $m = 1$ gram, $v = 1$ cm. per sec., and $t = 1$ sec.,

then $f = 1 \times \dfrac{1}{1} = 1$ *dyne* by definition.

Since acceleration is defined as $a = \dfrac{v-u}{t}$ (*Eq.* 1, p. 102)

$$\therefore f = \frac{mv - mu}{t} = \frac{m(v-u)}{t} = ma \qquad \text{(law II)}$$

Eq. 6. $\therefore f = ma.$

This equation is the briefest expression of law II. It says that if we measure the mass and the acceleration, we can calculate the force by finding the product.

When a 1-lb. block falls freely, its acceleration due to gravity is about 32 ft. per sec., per sec. Hence $f = 1 \times 32$ (*Eq.* 6)

$= 32$ poundals.

But the earth's pull on a 1-lb. block is called a 1 pd. force.

$\therefore$ 1 pound (force) $= 32$ poundals.

Thus 1 poundal is about half an ounce, also 1 gram = ap. 980 dynes, and 1 dyne = ap. 1 milligram.

These are a few simple and fundamental implications of the second law. The importance of this law in mechanics could hardly be overstated.

Newton's third law of motion is as follows:

Reaction is always equal and opposite to action: or the mutual actions of two bodies are always equal and act in opposite sense.

This law is the easiest to memorize and repeat, for only the first eight words are absolutely necessary; what follows is a paraphrase. One might imagine that at the colon the author caught himself out-Newtoning Newton in brevity, and therefore added the coda to soften the brusqueness. A brief name for the third law is the reaction law. Many find it the most elusive to grasp. It is one of Newton's great contributions to science. In some instances it is even difficult to believe at first thought. It points out that forces occur only in pairs: that an apple resting on the ground cannot press down on the earth without the earth pressing upward equally on the apple. A pebble thrown at Gibraltar exerts exactly as much force on Gibraltar as the gigantic Rock does on the pebble. When a light football wing collides with a huge tackle, the smaller player has the satisfaction of knowing that he exerted as much force on the big lout as the latter did on him. The accelerations, however, were different because of the difference in the masses.

The two forces, action and reaction, together constitute a stress (again omitting the idea of area). Similarly, purchase and sale are opposite aspects of transaction. Action and reaction, though equal and opposite, cannot produce equilibrium, of course, since they act on different bodies. When father tells Willie to quit pulling the cat's tail and the boy claims that it is only the cat who is pulling, everybody knows, including Willie and the cat, how falsely he has spoken. The cat cannot pull unless Willie also pulls. Action cannot exist without reaction. They exist together or not at all.

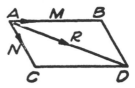

Fig. 108.
Parallelogram law.

It is noteworthy that it has taken two paragraphs to explain the third law, which itself consists of eight words. Whether one reads the original Latin of the *Principia*, or the English translation, one is struck by how very much Newton could say in a few words. He was always a diligent student of the Bible and it may well be that the simple grandeur of the Hebrew Scriptures influenced his style. Another source of his terseness was undoubtedly his training as a mathematician. In his brevity and succinctness he is a refreshing contrast to Boyle; but when he omits steps in an argument he goes to an opposite extreme.

One corollary of the laws of motion is the famous parallelogram law of vectors, which was discovered by Stevinus and, in a particular case, by Aristotle or some annotator of his writings. The *Principia* gives it the following enunciation (fig. 108):

A body by two forces conjoined will describe the diagonal of a parallelogram in the same time as it would describe the sides by these forces apart.

Newton's proof is based on laws I and II and is reminiscent of Aristotle's proof for the rectangle of vectors (displacements).

Another important corollary is derived from laws I, II, and III. It is called the law of *conservation of momentum*. When two bodies collide, they exert equal opposite forces on each other (law III), and each changes the velocity of the other (law I). Since the forces and times of contact are equal, the impulses are equal. Therefore the changes of momentum are equal (law II). But these are in opposite sense (law III). If one is plus, the other is minus. Hence their sum is zero. In any collision therefore, whatever momentum one "collidee" gains, the other loses, so that the total change in the momentum of the system is zero. No isolated system can change its own momentum.

We have seen that the remarkable insight of Lucretius gave him a glimpse of this law.

The story goes that on one occasion Newton's boyish zeal for mechanical toys and experiments led him to send up at night a lighted lantern on a kite-string. The villagers were frightened by the mysterious light in the sky and at first were seized by fear of the unknown. Then young Isaac was seized by the scruff of the neck and treated to a demonstration of impact and collision. Possibly it was during this part of the ceremony that he was struck by the idea that reaction equals action (law III).

In the third part of the *Principia*, Newton gives the following expression of the attitude of a scientist toward facts and data and the rules by which he conducts his arguments:

RULE I. We are to admit no more causes of natural things than such as are both well-grounded and sufficient to explain their appearances ... More is vain where less will serve.

RULE II. Therefore, to the same natural effects, we must as far as possible assign the same cause. As to respiration in man and in a beast; the descent of stones in Europe and in America; the light of our culinary fires and of the sun; the reflection of light in the earth and in the planets.

RULE III. The qualities of bodies which are found to belong to all bodies within the reach of our experiments are to be esteemed the universal qualities of all bodies whatsoever. We are certainly not to relinquish the evidence of experiments for the sake of dreams and vain fictions of our own devising. That abundance of bodies are hard, we learn by experience; and because the hardness of the whole arises from the hardness of the parts, we, therefore, justly infer the hardness of the undivided particles of all bodies [atomic or corpuscular hypothesis]. The bodies that we handle we find impenetrable and thence conclude impenetrability to be an universal property of all bodies whatsoever.

Rule I is an example of the "principle of economy or parsimony," of which we have seen numerous examples in the story of mechanics. As the invention and use of labour-saving devices constitute one chief goal of practical or applied science, so theoretical science seeks ever to contrive time and labour-saving methods. Hence, all simplifications are welcomed and the advance of science, as we have seen, is frequently from complex to simple by the expression of the Many as the One. All scientific laws are instances of this process. The more numerous and complex the facts which a law expresses by a single statement, and the simpler the statement, the more admired is the work which derived the law and the worker who derived it. Accordingly, the passage we next read from the *Principia* is considered the most brilliant achievement in the whole history of science, for by it and the three laws of motion, Newton expressed in a few simple sentences the infinitely numerous and complex motions of the heavenly bodies and a relation between every two particles of matter in the whole universe. The quotation is the most famous in the *Principia*, for it constitutes the theme of the book.

LAW OF UNIVERSAL GRAVITATION

There is a gravitational attraction between every two bodies in the universe, and this force is proportional to the quantity of matter [mass] of each.

The force by which the moon is retained in its orbit acts toward the earth and is reciprocally as the square of its distance from the earth's centre.

Hitherto, I have not been able to discover the cause of those properties of gravity from phaenomena and I frame no hypotheses. For whatever is not deduced from the phaenomena is to be called an hypothesis; and hypotheses whether metaphysical or physical, whether of occult qualities or mechanical, have no place in experimental science.

Newton's derivation of the law of universal gravitation is looked upon as the supreme feat of an intellect which at times seemed well-nigh superhuman. An algebraic statement of the law will be very useful in our discussion. In the original solving of the problem by Newton, algebra was crucial. The data in general were well known to Hooke and Newton and many others, and we saw how far Hooke was able to go. Newton succeeded where others failed, because of his superior mathematical power. Furthermore, the advancement possible to any physicist, or any group of physicists, is determined largely by his or their mathematical ability, so close is the relation between physics and mathematics.

If f is the gravitational attraction between masses m' and m'' and r the distance between their centres, then the law is expressed:

Eq. 7. $$f \propto m' . \frac{m''}{r^2} \text{ or } f = G\left(\frac{m' . m''}{r^2}\right).$$

where G is a constant of proportion. By a famous experiment, Cavendish found that if m' and m'' are in grams, r in cm., and f in dynes, then $G = 6.48 \times 10^{-8}$. The value of G now accepted is 6.66×10^{-8}.

The part of the law expressed in *Eq.* 7 by $\left(\frac{1}{r^2}\right)$ is called the inverse square law. The rest of the equation is fairly obvious. To simplify the argument, let us assume that the orbit of the moon or the earth is a circle. We have seen that this is very near the truth. Let us assemble our facts and express them algebraically:

(i) Huygens' centripetal law, $a = H(c^2/r)$ where c is the speed of the moon (or any satellite), a its acceleration and r its distance from the earth (its primary) (*Eq.* 2, p. 129).

(ii) Speed of moon $c = 2\pi r/p$ where p is the period of the moon.

(iii) Kepler's third law, $p^2 = Kr^3$ (*Eq. 3, p. 87*).

(iv) Newton's second law of motion, $f = ma$ (*Eq. 6, p. 134*)

$$\therefore f = ma = mH(c^2/r) = \left(\frac{mH}{r}\right)\frac{4\pi^2 r^2}{p^2} = \frac{4\pi^2 mHr}{Kr^3} = \left(\frac{4\pi^2 mH}{K}\right)\frac{1}{r^2}$$

$\because 4\pi^2 mH/K$ is a constant

Eq. 8. $\therefore f \alpha \dfrac{1}{r^2}$ or $f = N\left(\dfrac{1}{r^2}\right).$

where N is a proportionality constant. Newton's initial is used here

FIG. 109. The moon's orbit.

for the purpose. The argument is far more complex when we take the orbit as an ellipse as did Newton.

Having derived *Eq.* 8 for an elliptical orbit, Newton went on to see how it agreed with the facts in the case of the moon. Again we shall use a circular orbit. The radius of the earth is about 4000 mi. ($r = 4,000$) and the moon's distance from the earth is about 240,000 mi., which is about 60 times the earth's radius, i.e. $R =$ ap. $60r$ (fig. 109). If we take the moon's period as 4 weeks or 28 days, then

its speed c is its orbit divided by the period;

$$c = 2\pi R \div p = \frac{6.28 \times 240,000}{28 \times 24 \times 60} = \text{ap. 40 mi. per min. If } M_1 T \text{ in fig.}$$

109 represents 40 mi., then by the Pythagorean theorem

$ET = \sqrt{(240,000)^2 + (40)^2}$ mi.

$\quad = 40\sqrt{(6000)^2 + 1} = 40 \times 6000.00008$ mi.

$\quad\quad\quad\quad = 240,000.0032$ mi.

$\therefore M_2 T = 0.0032 \times 5280 = 16.6$ ft.

This is the distance the moon falls toward the earth in 1 min. The acceleration of free fall near the earth's surface, i.e. 4,000 mi. from its centre, is 32 ft. per sec. per sec. By the second law of motion $a \alpha f$; and by the inverse square law, $f \alpha \dfrac{1}{r^2}$. $\therefore a \alpha \dfrac{1}{r^2}$. $\therefore$ the moon's

acceleration in free fall would be $32(1/60)^2$ ft. per sec. per sec. = 32 ft. per min. per min. $\therefore$ in 1 min., it should fall through a distance of $s = \frac{1}{2}at^2$ (*Eq. 5, p. 102*) $= \frac{1}{2} \times 32 = 16$ ft.

The two results, 16.6 and 16, are in the same order of magnitude but with a discrepancy of 4 per cent.

A trigonometric test would be as follows. The moon describes an angle of 2° in about $28 \times 24 \times 60 \div 180 = 224$ min.

$$\therefore M_2T = \tfrac{1}{2}at^2 = R(\sec 2° - 1) = 24 \times 10^4(0.00061)5280 \text{ ft.}$$

$$\therefore a = \frac{24 \times 61 \times 528 \times 2}{224 \times 224} = 31 \text{ ft. per min. per min.}$$

which is 3 per cent less than 32 ft. per min. per min.

When Newton first made the calculations, using more accurate numbers than ours, he found a similar discrepancy of about 2 per cent for which he could not account. So he put the papers away in a trunk remarking, "Too bad." Some time later (1684), the French geodetist J. Picard made a more accurate determination of the length of the earth's radius. When Newton used Picard's new number in his calculations, the two results came out *equal—within* 0.1 per cent! They fitted like a glove. The thing was done. The law of Universal Gravitation was discovered. Newton knew very well the importance of his discovery. Tradition says that he was so overjoyed at this success that he swooned and had to be supported. After the first transports of joy had subsided, he put the papers back in the trunk, closed the lid, dusted his hands, and remarked "Well, so that's that," and went on with other work.

Some time later, Newton's former student Dr. Edmund Halley, Dr. Robert Hooke, and Sir Christopher Wren, were arguing about the moon's motions at an "after-meeting" of the Royal Society. When the three failed to reach an agreement, Halley said, "Well, I'm going to see if Newton can give some help." When he proposed the difficulty to his old professor, he was answered "right off the bat." The conversation probably ran somewhat like this: "How do you know all that?" asked Halley. "Oh, I worked it out some time ago" replied Newton, "and the papers are in that trunk." "Please, may I have a look?" "Certainly." As Halley read the papers his eyes grew large. He was learned and astute enough to realize that there was no book in science as great as this manuscript. "Of course, you will publish this?" "Of course, nothing of the kind; anything but." "You owe it to the world," argued Halley. Newton replied, "I was so persecuted with discussions arising out of my theory of light that I blamed my imprudence for parting with so substantial a blessing as my quiet, to run after a shadow." One by one Halley overcame Newton's objections; he carried copy to the printer, read the proofs, did the draughting, and even financed the whole undertaking. Finally, in 1687, the *Principia* was published. The story ends happily for, contrary to the gloomy forebodings of the Royal Society, the first edition sold out like hot-cakes; the "most acute and universally learned" Dr. E. Halley had backed a winner and made a handsome and well-earned profit.

THE INFINITESIMAL CALCULUS

Even a brief sketch of Newton's work would be incomplete if it did not mention his development of the Binomial Theorem and of the Infinitesimal Calculus. Both of these are helpful in understanding the work of Napier.

In the study of Descartes' work, we saw that to each value of t, there is a corresponding value of s (see fig. 84a). Then s is said to be a function of t or, in symbols, $s = f(t)$. Also, we saw that the speed of the moving body is measured by the slope of the graph.

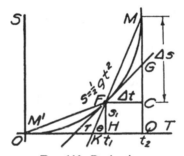

FIG. 110. Derivative.

Suppose the body is falling freely from rest, then the distance it travels in any interval beginning with the initial instant O, is represented by $s = \frac{1}{2}gt^2$ (Eq. 5, p. 102) where $g = 32$ ft. per sec. per sec. $\therefore s = 16\ t^2$. This is the equation of the graph OFM, a parabola. The continual steepening of the graph indicates the continual increase in the speed of the moving body B. Since B's speed is never the same for two instants, it would be very helpful to have a single *general expression to represent its speed at any and every instant of its motion.* Newton succeeded in deriving such an expression. It is called a *derivative.* His procedure was somewhat as follows:

Let the distance traversed in any interval commencing at O, e.g. OH ($= t_1$), be $HF = s_1$ ft. F is any point on the graph. Let the time increase. Represent its growth by the symbol Δt ($= HQ$). Then the corresponding growth in s_1 is $CM = \Delta s$. Δt and Δs are called increments. Complete the diagram as in fig. 110.

$\Delta s/\Delta t = \tan \theta = $ slope of the secant FM. Think of FM as a stiff straight wire and rotate it about the fixed point F clockwise.

Then the intersection M moves along the graph and *approaches* F or, in symbols, $M \rightarrow F$. As $M \rightarrow F$, the slope of FM decreases and approaches that of the graph at F. At the same time Δs and Δt decrease and approach zero: $\Delta t \rightarrow 0$ and $\Delta s \rightarrow 0$. If M shoots past F, the slope of FM' to the left of F is less than that of the graph at F. Therefore, at the instant M passes F, the slope of the wire equals that of the graph at F. This limiting position of the straight wire with the same slope as the graph is called the *tangent* to the graph at F.

As $\Delta t \rightarrow 0$, the slope of the secant $\Delta s / \Delta t$ $(= \tan \theta) \rightarrow$ slope of the tangent at F, i.e. the slope of the secant $\rightarrow$ the slope of the graph at F. Therefore a general expression for the slope of the tangent at any point on the graph is $\underset{\Delta t \rightarrow 0}{\text{Limit}} \left(\dfrac{\Delta s}{\Delta t} \right)$. But the slope of the graph at any point measures the speed at that instant. Hence a general expression for the slope of the graph at any point or for the speed of the moving body at that instant is the derivative

$$\underset{\Delta t \rightarrow 0}{\text{Lim.}} \left(\frac{\Delta s}{\Delta t} \right).$$

Let us calculate the value of the derivative for the graph of fig. 110.

Its equation is $\qquad s = 16\,t^2, \qquad$ (*Eq.* 5, p. 102)

$$s + \Delta s = 16(t + \Delta t)^2$$
$$= 16(t^2 + 2t.\Delta t + \Delta t^2).$$

By subtraction, $\qquad \Delta s = 16(2t.\Delta t + \Delta t^2).$

Divide by Δt. $\qquad \dfrac{\Delta s}{\Delta t} = 16(2t + \Delta t)$

$\therefore \underset{\Delta t \rightarrow 0}{\text{Lim.}} \left(\dfrac{\Delta s}{\Delta t} \right) \qquad = 16(2t + 0) = 32\,t.$

This derivative says that the slope of the graph at any point $F = 32t$. E.g. at the end of 1 sec., slope of graph $= 32 \times 1 = 32$; but the slope represents the speed which is known to be at the end of 1 sec., 32 ft. per sec. Similarly, at the end of 3 sec., the slope of the graph $= 32t = 32 \times 3 = 96$; and the speed at the end of 3 sec., $v = gt$ (*Eq.* 6, p. 103) $= 32 \times 3 = 96$ f.p.s.

The derivative is fundamental in the infinitesimal calculus. In this case, it expresses how s changes as t changes and is called the derivative of s with respect to t. It tells the slope of the graph at any and every point and in this case, therefore, measures the speed of the body at any and every instant of its motion. The calculus is sometimes called the velocity of functions and its birthplace was mechanics.

There are several ways of writing the derivative, invented by different mathematicians, each with its own advantages, e.g. the derivative

$$\text{Lim.}_{\Delta t \to 0}\left(\frac{\Delta s}{\Delta t}\right) = D_t s = \frac{ds}{dt} = f'(t) = \dot{s}$$

(Cauchy)   (Leibnitz)   (Fermat Lagrange)   (Newton)

Newton's notation, as one might expect, is by far the briefest. It is limited to cases in which the derivative is with respect to the time t, and is specially useful in mechanics. The notation invented by Leibnitz is probably the one of most general utility.

The process of finding a derivative is called *differentiation* and the branch of infinitesimal calculus which deals with differentiation is called *differential calculus*. The invention of this calculus placed in the hands of mathematicians an instrument of great power by which the solutions of many problems became simpler and more elegant and some problems, hitherto insoluble, became soluble.

The French mathematician Lagrange once remarked that "among great mathematicians, Newton shines as a great comet among stars." Newton disclosed at least part of the secret of his power when he said, "Whatever I have been able to accomplish, I attribute to my habit of attention." Nowadays that faculty is called concentration. Associated with his intense concentration, was his habit of abstraction or absentmindedness, which gave rise to some well-known episodes. When Newton fastened on a problem, he wrestled with it day and night without cessation. When dressing in the morning, perhaps when he had just put one leg in his trousers, he might become absorbed in his problem. Hours later, the housekeeper, coming in to tidy the room, would find him sitting at the side of the bed with his dressing no further advanced. In urging students to emulate Newton, it may be well to make clear that his concentration is meant, rather than his dressing habits.

In the story of optics, we shall meet Newton's study of the spectrum, which led him to investigate the colours of soap-bubbles. It is said that an old lady who once visited in a house near Newton's wrote a letter to a young lady friend in this wise, "There is a peculiar gentleman in a house near by who sits in the garden and plays with soap-bubbles. I fear things are not all that one could wish, if you know what I mean, dearie."

During Newton's first year at Cambridge, he managed, by dint of much scrimping, to purchase a second-hand copy of Euclid. He read the whole book that same night! Shortly after, he told his tutor, Dr. Isaac Barrow, that he did not think the book was any great shakes as the whole thing seemed quite obvious. "Then come tomorrow

FIG. 111. Isaac Newton.

at ten," said the professor, "and I shall examine you on Euclid."
In the examination, the Doctor bore down heavily on the candidate
and humbled him to the dust. "Remember this, Newton," said Dr.
Barrow, "there is plenty in Euclid for you and for me for many a
day." The young man took his lesson to heart and the influence of
Euclid is plainly visible in the style of the *Principia*.

If any man could have been excused for having a touch of arrogance,
Newton was probably the man; but he was, on the contrary, a very
affable, kindly, humble-minded, and strongly religious man. As he
was lecturing one day, he and his class saw through the window a
prisoner taken to the scaffold. Newton became absorbed, then remarked, "There goes, but for the grace of God, Isaac Newton."
The story, though possibly apocryphal, is in keeping with Newton's
temperament.

The following selection from the *Principia* will serve to illustrate
Newton's attitude. He has just described and explained the wonders
and beauty of the solar system as no man before him had done;
then he continues:

> This most elegant system composed of the sun, the planets and the comets,
> could not originate except by the counsel and fiat of an intelligent and all-
> powerful Being.... The Supreme God is a Being eternal and infinite and
> absolutely perfect.... He is omnipotent and omniscient. He endures from
> eternity to eternity and is present from infinity to infinity. He governs all
> things that are or can be done. Blind metaphysical Necessity which is certainly
> the same always and everywhere could originate no variety of creatures. All
> that diversity of natural objects which we find adapted to different times and
> places, could spring only from the mind and will of a Being necessarily existing...
> So much then, concerning God of whose nature as manifested in phenomena
> it certainly pertains to Natural Philosophy to discourse.

SURVEY

No matter how far the archaeologist has helped us to press back
into prehistoric times, we have not found a time when early man
did not observe natural phenomena, speculate on their causes in his
primitive way, and apply his acquired knowledge in devising and
using weapons, tools, and machines. Thus were laid the practical
foundations of mechanics. The astronomy of Babylon, Egypt and
Alexandria marked great advances in accurate observation, calculation,
speculation, and ability to predict. The experiments of Archimedes
and of the Alexandrines founded the department of statics and were
prophetic of the new tempo of progress which began with the epochal
work of Galileo, the founder of dynamics or kinetics. Since then the
pace has accelerated: now the theory and practice of mechanics are
woven into the very fabric of our civilization. In this development

the use of language has been vitally important and especially the language of mathematics. The spirit of science has been, in general, international, like that of music; and the growth in facilities for free communication of ideas and results among scientists the world over has been one of the chief contributory causes of the rapid advancement of modern science. One scientist who did much to promote such interchange was Joseph Henry, the first secretary of the Smithsonian Institution, Washington, D.C.

In the story of mechanics, we have seen several empires wax and wane—Babylon, Egypt, Greece, Rome, and Arabia. Culture and civilization spread westward from the Indus and Euphrates; and there is good reason to believe that there was a similar trend eastward to China. In the histories of those empires, one can see a recurring series of phases, from birth to death, so that it seems appropriate to describe them as "cycles" of culture and civilization, as if each cycle were an organism with a life history like that of a plant or animal. Every cycle is different, but all have a series of phases in common[9], as different years have the same seasons. Our visits with Roger Bacon and Leonardo da Vinci brought us into the Western cycle, which continues to-day.

The early stage of each cycle is, in general, agricultural and religious. As the cycle progresses, its potential vitality manifests itself in phases which can be epitomized by the following series: architecture, sculpture, painting, music, literature, mechanics and mathematics, other sciences, applied science, empire and wealth. Then come luxury, decadence, decline and over-throw. In the Western cycle, one can mark some such phases or peaks by referring to Gothic architecture, Michelangelo, da Vinci, Shakespeare, Bach, Newton. Each of these names, of course, bespeaks a group of workers.

In some of these developments, such as architecture, sculpture, and painting, much of the achievement of a cycle may perish with the cycle. In its downfall, some skills and techniques are lost or forgotten and may even be contemned by the artists of a later cycle. This has occurred to a lesser degree in science and mathematics than in most other departments. The genius of science has been happy in that, to a considerable extent, the scientists of one cycle, as we have seen, have retained the proven results of their predecessors and have forged ahead. This accumulation[10] from cycle to cycle has fostered the development of science and the speed of its advancement. In social and political matters, however, progress has not kept pace with that of science, and from such unequal development abuses and troubles have arisen. The applications of science tend to produce wealth and

[9]A. J. Toynbee, op. cit. p. 8.
[10]G. G. Sarton, op. cit.

leisure, but social and political conditions have not yet been organized so as to make just distribution of the benefits of the machine, nor to prevent unscrupulous individuals or groups from applying the machine to selfish and even to criminal purposes. Some have placed the blame on the machine, declaring that it has brought more woe than weal. It is to be hoped that the attitude and method of the scientist, which have brought him success in his own field, may prove applicable in the organization of society, so that the benefits of the machine may be justly distributed. It is a good omen that more and more scientists are coming outside their laboratories and doing their best to lend a helping hand in the solution of the infinitely complex and harassing problems of social organization. Many of the scientists whom we meet in the story of science were teachers and leaders of great ability and high ethical principles—von Guericke, Franklin, Kelvin, Young. It will be a grand outcome of science if some of her votaries are able to help and guide the people out of the wilderness into the promised land of a true democracy.

To say that mechanics reached a peak in Newton's day does not mean that it declined after 1700. Beyond that peak have risen whole mountain ranges of progress. Newton's work was an incalculably great stimulus to new endeavour and brought tremendous results, but where he was held in slavish reiterative veneration progress was deterred, just as it was by uncritical devotion to Aristotle and to Imhotep.

The liaison between mechanics and mathematics has enhanced the status of both sciences. Here we can mention only a few of the many new attainments in mechanics during the past two hundred years. Daniel Bernoulli studied hydraulics and streamlines, and discovered the theorem in fluid pressure which bears his name. He also did pioneer work in reaching the concept of energy and its conservation, which was indeed tacitly implied in a scholium of the *Principia*. The law was established by the concerted labours of Rumford, Kelvin, Joule, Helmholtz, Clausius, and many others. The blind scientist Plateau discovered laws of surface tension. Lavoisier's name will typify those who helped to derive the law of conservation of matter which we found suggested in the words of Lucretius. Laplace, "the Newton of France," published the famous book, *Celestial Mechanics*. A virile group of mathematicians such as Leibnitz, Lagrange, Poisson, and Cauchy, developed the infinitesimal calculus and other types of calculus. Lagrange, probably the greatest mathematician of his century, published his monumental book, *Mécanique Analytique*, in which he used the celebrated Lagrange equations[11] and unified the science of mechanics under the broad principle of Least Action, which includes the law of conservation of energy. This law was first hinted at in the work of Hero of Alexandria, when he said that reflection of light occurs along the shortest path. The great Dublin mathematician, Sir William R. Hamilton, did pioneer work in the same field. Hamilton, Maxwell, Professor J. Willard Gibbs of Yale University, and others developed the science of vectors and of mechanics in general far beyond Newton. The fundamental laws, however, were still Newton's laws of motion and his law of gravitation.

By the close of the nineteenth century the territory of mechanics had been so thoroughly subjugated that the physicists were inclined to imitate Alexander in sitting on the bank of the Ganges, lamenting that there were no new problems to conquer. It appeared that about all a researcher could hope to do was to repeat old experiments with new accuracy and push the results to one more decimal place.

Then came some new experimental results which were difficult to explain, or even impossible, on the basis of established principles. This led to a questioning and criticism of old foundations by Professors Max Planck and Albert Einstein of Berlin and by a cohort of investigators. Some old concepts were modified and some new ones introduced. It was stipulated, for example, that space and time are not absolute but merely relative to the observer, and are such that, for all observers, whatever their motions, light travels with the same constant velocity. As a consequence of these revisions and innovations, it became possible for the scientist to explain some events otherwise inexplicable, to make some calculations and predictions with greater accuracy than heretofore and to improve the organization of his subject. Thus what had appeared an ending was really a new beginning, the end of the classical period and the commencement of the modern. Said Confucius, "A river, like truth, flows forever, and will have no ending." On this grand vista of modern mechanics, nevertheless, we must now turn our backs, however loth, and hie us away to a new field of enquiry.

[11]Leigh Page, *Introduction to Theoretical Physics* (Macmillan, 1930), p. 168.

THE STORY OF
Acoustics

CHAPTER SIX

The Acoustics of
the Ancients

HUMAN beings, in common with many animals, have the important power of producing sounds with the voice and perceiving them with the ear. This they can do to a marvellously complex degree in speech and song. Any body of which we can become aware by hearing it may be called a *sounding body* or a sonorous body. A fog-horn is an example of a sounding body of whose existence we may become aware through hearing, when no other sense perceives it. In the story of acoustics, the science that deals with sounding bodies, the ability of human beings to devise and use instruments was of fundamental importance from early times. This is not surprising when we recall from the story of mechanics the importance of man's ability to invent machines. As practice preceded theory in mechanics, so also in the progress of acoustics there was, in prehistoric times, a preliminary stage during which musical instruments and other sonorous bodies were invented and used with little understanding of any principles of acoustics. Centuries later, with the development of speculation and theory, the science of acoustics arose and it, in turn, led to numerous improvements and inventions in applied acoustics such as the tuning-fork, metronome, megaphone, stethoscope, phonograph, dictaphone, and dozens of others. Practice can help theory and theory can assist practice.

PRIMITIVE MUSICAL INSTRUMENTS

The most primitive method of producing sound by means of an implement, one that was doubtless used in prehistoric times, was by percussion: for example, by pounding a log with a stick. To this category belong the gavel, drum, sistrum (fig. 112) door-knocker, tambourine, cymbals, gong, bell, tympani, temple-blocks, silver triangle, xylophone, piano. Neolithic man used earthenware pots

151

over whose mouths skins were stretched to form drums (fig. 113). Palaeolithic man probably used hollow tree-trunks for the same purpose. One of a young boy's palaeolithic symptoms is his delight in a drum as a present from Santa Claus; another is his barbaric

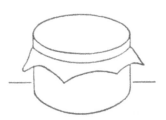

FIG. 112. The sistrum.   FIG. 113. Neolithic drum.

fondness for drumming on the table with his fingers, or fork, or lead-pencil. The sistrum made a sibilant rattling *noise* somewhat like the sound of a rattle-snake's warning, and was used in Mesopotamia and Egypt by the priests for frightening away evil spirits. A mediaeval vestige of the sistrum was the jester's bauble, and modern examples are a baby's rattle and certain festive noise-makers.

From the twanging of the archer's bow-string to the invention of stringed musical instruments was an inevitable sequence, which was suggested in Greek mythology by making the silver bow and the lyre the symbols of the same god. In fact, Apollo, the sun-god, was accredited by some with the invention of the lyre. Similarly, his twin sister, the moon-goddess Diana, was portrayed as a huntress and an expert with both bow and lyre.

FIG. 114. Sumerian lyre and sistrum.

The large family of string instruments includes the cithara, zither, guitar, harp, lute, spinet, pianoforte, violin, and many others. Fig. 114 shows part of a plaque of shell inlay dating about 3500 B.C., which was excavated in 1928 from the royal cemeteries at Ur. It shows a group of Sumerian actors dressed in animal costumes; the jackal is shaking a sistrum and the donkey playing a lyre. A harp belonging to the same period was found in the tomb of Shubad, Queen of Ur, and with it the skeleton of her harpist, who was sacrificed at the queen's funeral so that she might accompany her mistress in the hereafter. That the harp was used also by Egyptian musicians is indicated by fig. 115, which is taken from a tomb inscription at Beni Hassan in the Nile valley. It dates about 2500 B.C. Most string instruments are sounded with the bow or by plucking; the Aeolian harp, however, is sounded by the wind.

The invention of the lyre was attributed by some to Hermes, for instance, in the account given in the Homeric hymn, *To Hermes:*

In Cyllene, Hermes found a tortoise and won endless delight, for lo, it was he who first made of the tortoise a minstrel. The creature met him at the outer

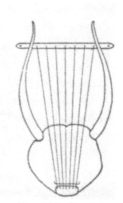

FIG. 115. Egyptian harp.   FIG. 116. Greek lyre.

door of the cave, waddling along, at sight whereof the son of Zeus laughed and straightway spoke saying, Lo, a lucky omen for me, not by me to be mocked! and raising in both hands the tortoise, went back within the dwelling bearing the glad treasure. Then he choked the creature and, with a gouge of grey iron, he scooped out the marrow of the hill tortoise. He cut to measure stalks of reed and fixed them in through holes, bored in the stony shell of the tortoise, and cunningly stretched round it the hide of an ox, and put in the horns of the lyre, and to both he fitted the bridge and stretched seven harmonious cords of sheep-gut. Then took he his treasure when he had fashioned it and touched the strings in turn with the plectrum and wondrously it sounded under his hand, and fair sang the God to the accompaniment of its tones. Then sped Hermes and Apollo to sandy Pylos. Taking his lyre in his left hand, Hermes sounded it with the plectrum. Thereat, Apollo laughed and spake to him winged words. What art is this, what charm against the stress of cares? Then answered Hermes, Since then, thy heart bids thee play, harp thou and sing and let joys be thy care, taking this gift from me.[1]

[1]*Homeric Hymns*, trans. Andrew Lang (Allen & Unwin, 1899).

Fig. 116 is from a vase in the British Museum. It shows the body of the lyre made of a tortoise shell, also the horns, bridge, and seven strings mentioned in the Homeric hymn. The reference to the stretched ox-hide suggests a relation in construction between the lyre and the drum, such as may still be seen in the banjo. This hymn was composed about 900 B.C. by the Greek poet Homer, who lived at Smyrna in Asia Minor.

Musical instruments of a third type are classified as wind instruments: those of wood in general as wood-winds and those in brass as brass winds. In whistling, the mouth is used as a wind instrument.

FIG. 117. Egyptian trumpet and drum.

The wind instruments in general have a mouthpiece or embouchure; the player's mouth is used in sounding the instrument, frequently by a process resembling whistling. Examples of wind instruments are the flute, piccolo, fife, bugle, trumpet, horn, mouth-organ, clarinet, trombone, and pipe-organ. Archaeologists have discovered palaeolithic whistles made of bone which resemble the fife. One modern vestige of that old instrument is the willow whistle that a boy makes in the springtime. Its origin may even antedate the Stone Age and hark back to the wooden age. The mouthpieces of some wind instruments contain reeds, e.g. the clarinet, saxophone, bassoon, and oboe.

The invention of wind instruments is symbolized in Greek mythology by the story of the great god Pan, the inventor of the Pandean pipes or syrinx, and of the trumpet. Figs. 117 and 118 show an ancient "full orchestra": (1) percussion—drum, taboret, and cymbals; (2) winds—Egyptian trumpet; and (3) strings—Babylonian lute and harp.

Select any key of the pianoforte and strike it first softly (*piano*), then strike the same key again but with greater force (*forte*). The two sounds differ in *intensity*, or loudness. Strike any key of the piano, then any other key, with the same intensity. The two sounds differ in *pitch*. One important feature which string and wind instruments have in common is their ability to produce sounds that have definitely recognizable pitch—if they are expertly sounded. Such sounds are called *tones*. Other sounds are *noises*. Some percussion instruments produce noises; some can produce no better than poor or fair tones; but tympani drums, piano, and some other percussionals can produce good tones. The word *note* is frequently used for the term *tone*, and the spellings of the two words are remarkably alike, but literally, a note is a mark on paper to represent a tone.

ANCIENT HEBREW ACOUSTICS

Of the many references to sounding bodies in the Hebrew scriptures, only a few will be quoted here, as the reader can readily find others for himself. The following reference is dated about 4000 B.C. by the Concordance, and mentions the harp and pipe as being already invented and their techniques developed:

FIG. 118. Babylonian lute, harp, and tabor.

And Jabal's brother's name was Jubal; he was the father of all such as play the harp and pipes.—Genesis iv. 21.

From Jubal's name, or the Hebrew word *jobel*, meaning the blast of a trumpet, come our words "jubilee" and "jubilation."

The next quotation is taken from 1 Samuel x. 5. It mentions the three types of instruments in use about 1100 B.C., percussional, wind, and string:

And it shall come to pass, when thou art come thither to the city, that thou shalt meet a company of prophets coming down from the high place with a psaltery, a tabret, and a pipe, and a harp before them.

The psaltery was a development of the lyre. "Tabret" is an archaic form for tabor or taboret (shown in fig. 118). It was a small hand drum and accordingly percussional. Its name suggests its relation to the tambourine, which is a tabor with brass jingles—these, in turn, are related to the beads of a sistrum.

In the *Book of Job*, there is also a passage which names three types of instruments:

> Suffer me that I may speak; and after that I have spoken, mock on. Wherefore do the wicked live, become old, yea, are mighty in power? Their seed is established in their sight with them. They take the timbrel and the harp and rejoice at the sound of the pipes.—Job xxi. 12. (About 1500 B.C.)

We have referred to the intensity of sounds as their degree of loudness or faintness. This feature of sounds is sometimes called "volume," a term borrowed perhaps not wisely nor too well, for loudness could hardly be measured in quarts or litres and there is little relation between the ideas of intensity and volume. The latter term has the advantage of being shorter by two syllables. Here is a passage that mentions a very faint sound:

> But if ye will not hearken unto me, and will not do all these commandments ... I will set my face against you.... They that hate you shall reign over you; ye shall flee when none pursueth you ... and upon them that are left alive of you, I will send a faintness and the sound of a shaken leaf shall chase them.— Leviticus xxvi. 36. (About 1500 B.C.)

The trembling of a leaf, just mentioned, may be taken as symbolic of the fundamental fact, known to many ancient musicians, that *every sounding body is in a condition of vibration* or relatively rapid to-and-fro motion. The following account of the fall of Jericho describes a sound that was sufficiently intense to collapse a wall:

> And seven priests shall bear before the ark seven trumpets of rams' horns ... and the priests shall blow with the trumpets. So the people shouted when the priests blew with the trumpets: and it came to pass, when the people heard the sound of the trumpet, and the people shouted with a great shout, that the wall fell down flat ... and they took the city.—Joshua vi. 4 and 20.

It is interesting to note that the noun "wave," which figures very prominently in acoustics, entered our language via the Hebrew scriptures.[2] Early uses of the term, therefore, deserve mention, for example, the following from the Psalms:

> Thou rulest the raging of the sea: when the waves thereof arise, thou stillest them.—Psalms lxxxix. 9. (About 1000 B.C.)

In the old Sumerian map (p. 16), we saw waves represented by dashes. Waves are also shown, in greater detail, in the old Babylonian raft and boat pictures of pp. 21 and 22.

THE MONOCHORD

Previous to the time of Pythagoras, musicians, typified by Jubal, Apollo, Pan, and others, had gained much practical knowledge about the *intervals* between tones, that is, their pitch relations, for instance, the intervals known as the octave, unison, the third, and the fifth.

[2]*Encyclopaedia Britannica* (11th ed.) "Waves."

Such names bespeak a musical scale, a series of tones graded as to pitch, for example, the diatonic scale or the scale of Guido, *do, re, mi, fa, so, la, te, do,* the names in which were taken from an old Latin hymn

Ut queant laxis, *re*sonare fibris
*Mi*ra gestorum, *fa*muli tuorum
*So*lve polluti, *la*bii reatum
Sanc*te* Joannes.

Ut was later replaced by *do*.

Pythagoras, however, is called the "Founder of Acoustics," because he investigated the subject experimentally and discovered the values of some common intervals by using the monochord, which consists of a tight string and a bridge. It is the oldest piece of scientific apparatus known.

Since there are no writings of Pythagoras extant, we are obliged again to rely on quotations from other authors. Heraclitus, the physicist, said of him, "He practised inquiry beyond all other men"; and Diogenes Laërtius, "Pythagoras discovered the musical intervals on the monochord."

When the monochord has a hollow base filled with air, it is called a *sonometer*. The hollow box intensifies the tones of the strings as it does in the violin and other string instruments. Musicians knew by practical experience that a thicker string gave a lower tone than a thinner one, and that if two strings were equally thick, the shorter one emitted a higher tone. Pythagoras by his experiments found that halving the length raised the pitch an octave. He found also that to obtain the third or, in Guido's scale, *do-mi*, the length must be reduced to four-fifths its initial value. Similarly, to obtain the interval of the fifth, or *do-so*, the lengths must be in the ratio 3:2. From these and other discoveries, Pythagoras came to the general opinion that the meanings of all phenomena are to be sought in the properties of numbers or, as he put it, "All things are numbers."

"THE TINY TRUMPETING GNAT"

A passage from Aristophanes' play, *The Clouds*, is quoted here because it refers to an acoustical topic. Of course, it was not written from a scientific viewpoint but from that of a dramatic poet and a satirist. In the play, poor Strepsiades is being ruined by the extravagance of his family. In desperation, he decides to study logic under the great master Socrates, in the hope of learning how to argue so as to make wrong seem right, and then, by use of his new forensic powers in the law-courts, to repudiate his debts. He meets one of Socrates' pupils:

STREPSIADES. O Zeus and king, what subtle intellects!

STUDENT. How would you like to hear one of our Master's own sayings?

STREP. O come, do tell me that.

STUDENT. Why, Chaerophon was asking him which theory did he sanction; that the gnats hum through their mouth, or backward, through the tail?

STREP. Aye, and what said your Master of the gnat?

STUDENT. He answered thus: the entrail of the gnat is small: and through this narrow pipe, the wind rushes with violence straight toward the tail: there, close against the pipe, the hollow rump receives the wind and whistles to the blast.

STREP. So, then, the rump is trumpet to the gnats! O happy, happy in your entrail-learning; full surely need he fear nor debts nor duns, who knows about the entrails of the gnats.[3]

It may seem rather far afield to seek scientific sources in the works of a satirist from whose burlesque nothing is exempt, neither science nor Socrates; but even a ribald lampoon may serve incidentally to indicate something about the state of science at the time it was written. In this case, the reference to the "hollow rump" indicates a practical knowledge of the principle of *consonance* which is applied in the hollow sound-box of the sonometer, the violin and other instruments. The principle of consonance is also employed when a tuning-fork is struck and its tone intensified by placing its foot on a table.

The work of Socrates has, of course, made his position so impregnable that he needs no defence even from an attack by Aristophanes. Socrates studied under Anaxagoras and other physicists. As a sort of corrective to Aristophanes' misrepresentation, let us hear one of Socrates' sayings:

All reasoning depends on original premises: all premises consist of words. Never attempt to reason until premises and words have both been scrupulously examined, or the reasoning will infallibly be spurious.

Responding to this excellent advice, we might well look at the word *sound*, which we have been using rather freely. Not only has it several meanings, but also a number of different acoustical meanings.

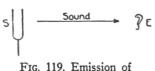

FIG. 119. Emission of sound.

One of these is exemplified by the statement, "The sound of the tuning-fork reaches the ear E" (fig. 119). This meaning of sound involves an ancient tacit assumption. Whatever body lies between a sounding body and the ear is called a *medium* (Lat. *medius*, between). The fork, though at a distance from the ear, has an effect upon it, and this without anything going on in the medium as far as one can see. Such an event or process is referred to as an "action at a distance." From expe-

[3]*The Clouds*, trans. B. Rogers (Bell, 1916).

rience, however, or instinctively, many people dislike granting that there can be an action at a distance, i.e. without any process occurring in the medium; although a body falling freely seems to be a pretty clear example. They hold that a body cannot produce an effect or an action where it is not. This view is an abstraction from many experiences. That was why Simplicio thought that it must be the air that keeps a pendulum swinging. We have seen a magician mesmerize a chair and beckon to it; and we have seen it waddle obediently to its master as if there were an action at a distance: but we were able to discern the intermediary black thread in spite of the black velvet curtain. To explain the effect of the tuning-fork on the ear or to "explain away" this apparent action at a distance, an assumption was tacitly adopted centuries ago. It may be stated thus: Let us assume that when a sonorous body B is heard by an ear, E, or by the possessor of the ear, the body B sends to E an invisible something called sound. This does not say what sound is but merely assumes its existence. The fact is that the person hears the fork: the hypothetical and convenient way of expressing the fact is to say that the fork sends a sound to the ear. Having made the assumption, it is for us to assign to sound such properties as we find necessary or probable in view of our experiences and experiments. The word *sound* is taken as a name for the science of acoustics and it is the name more commonly used. A sonorous body, accordingly, is called a *source of sound*. In most instances, the medium through which sound travels is air. The air may be said to conduct sound or to be a conductor of sound. We might even say that air is transparent to sound, although the expression is not customary.

The word sound, which implies the above-mentioned assumption, is found in ancient writings, e.g. in the Hebrew scriptures (p. 156). There the word is *kol*, meaning a voice or a sound. The pronunciation of this Hebrew word resembles that of the English word "call."

Among Aristotle's writings, there is a small treatise *On Sounds*. Some authorities deny, however, that it was written by Aristotle and ascribe it to Strato. The treatise is, in general, vague and its chief value is to show how great was the need of an adequate theory of sound. It contains, however, two passages worth quoting. The first of these by free translation may be taken as stating an important relation in acoustics which musicians of such wind instruments as the flute and oboe certainly knew by experience before Aristotle's day:

If we lengthen the air column of an oboe by closing the holes, the pitch of the tone becomes lower.

The second quotation makes a noteworthy attempt at describing the manner in which sound travels through air:

All sounds are produced by contact of the air with some body. The air is set in motion either by compression or by expansion. When the strings of a musical instrument strike the air nearest them, it is driven away, thrusting forward in like manner the adjoining air. Thus sound travels as far as the disturbance manages to reach. Sounds, if they are clear, travel far and fill all the space around them.

Evidently the writer is aware of the vibration of a sonorous body, and of the fact that the vibration is transmitted in all directions by its source, and he definitely suggests pulses of compression and of rarefaction.

TRANSMISSION OF SOUND

Lucretius' poem, *Concerning the Nature of Things*, contains several passages that are of interest in acoustics:

Sounds pass through walls and fly through the enclosed parts of houses: but were there no empty spaces through which bodies can pass, you would not find this occurring by any means.

Subscribing to the theories of Democritus, Lucretius assumed that sound emitted by a sounding body consists of small particles or corpuscles and that these pass through air, and even through solid bodies, by navigating the vacuum channels between the molecules. Such a theory is therefore called corpuscular.

Furthermore, a single word often strikes the ears of everyone in an assembly when it is shouted from the mouth of the announcer. . . .
Part of the sound that strikes a solid surface is reflected and a rock returns the sound and sometimes mocks us with the echo of a word.

In acoustics, the corpuscular theory has not proved very helpful or satisfactory. (A more helpful assumption was advanced about half a century after Lucretius by one of his fellow-countrymen, Vitruvius.[4]) When sound which is travelling, say, through the air, strikes a solid obstacle, a door or a wall, part of it is bounced or hurled back into the same medium. The process is called *reflection* and the obstacle, a reflector. Note the carefulness of this observer. Many would say that the sound is reflected, but not Lucretius. He says that part of it is reflected and he is right, for part of it enters the new medium, namely, the obstacle. Thus part of the sound striking a door passes into and even through the door. In these three quotations Lucretius has described three fundamental properties of sound.

In early times especially, there were some who offered personal or subjective explanations of natural phenomena. Such explanations

[4]p. 163.

stand in contrast with those developed by scientists, which are characteristically impersonal or objective. How differently, for example, did the poets deal with the reflection of sound and light in the myth of Echo and Narcissus.[5] When the nymph Echo was young she was gay and fickle, now sad, now merry, sometimes cruel, often mocking, according to her mood and what she had heard. Then she fell under the wrath of Juno who put on her this curse, "To the tongues of others shall thine be slave. Thou shalt repeat only the last words falling on thine ears." A shackled nymph was poor Echo. When she fell in love with the beautiful lad Narcissus, she haunted him like his shadow. Once in the forest he cried, "Who's here?" "Here," answered Echo. "Why do you shun me?" asked Narcissus "Do you shun me?" repeated Echo. "Let us join one another," requested the lad. "Join one another," said Echo as she rushed to him with open arms. The boy shrank back. Poor Echo sought the solitudes of mountains, cliffs, and caves; but finally grief killed her body leaving only her voice.

By and by, the youth Narcissus, stooping down to drink from a pool, saw a beautiful face smiling up at him and was instantly in love with the nymph of the pool. When he smiled, she returned the smile. When he stretched his arms toward her, the nymph welcomed him with open arms; but as he stooped to kiss her, she was snatched away and only the chilly water met his lips. When the pool was quieted, there she was again. He pleaded and she did not flee; but ever in vain he tried to clasp to his breast the beautiful mirrored image.

Thus where the scientists, by an objective impersonal attitude, have found the useful laws of reflection and their marvellous applications, some poets, with a personal and subjective viewpoint, have found yearning, frustration, and heartache. Yet who could say that those have found truth and these have not?

The accounts of some prehistoric inventions have come down to us through the long ages in the form of myths, for example, that which ascribes the invention of the first musical wind instrument to the god Pan. Here is part of the story as told by the poet Ovid, who lived at Rome about a generation after Lucretius. Argus, the hundred-eyed sentinel, asked Mercury how the reed-pipe came to be invented:

Then said the god, "On Arcadia's cool mountain slopes, among the wood-nymphs, there was one much sought by suitors. Her sister nymphs called her Syrinx. One day Pan saw her. The nymph, spurning his prayers, fled until she came to Ladon's stream flowing peacefully between his sandy banks. Here when the water checked her further flight, she besought her sisters of the stream to change her form; and Pan, when he thought he had caught Syrinx, held in

[5]Based on the retelling by Jean Lang in *Book of Myths* (Nelson; Putnam).

his arms, instead of her, naught but some marsh reeds. While he sighed in disappointment, the soft air stirring in the reeds gave forth a soft complaining sound. Touched by this wonder and charmed by the sweet tones, the god exclaimed, "This union, at least, will I have with thee." And so, the pipes, made of unequal reeds, fitted together by a joining of wax, took and kept the name of the maiden, Syrinx."[6]

The syrinx is also referred to as the Pandean Pipes. One modern representative of the syrinx is the harmonicon or mouth-organ, but the sounds of the latter come from vibrating reeds, whereas those from the syrinx are produced by air columns. The voice-box of a bird is called the syrinx.

ALEXANDRINE SCHOOL

The first pipe-organ was invented by Ctesibius of Alexandria, about 50 B.C. Fig. 120 is taken from the book, *Pneumatica*, which, as

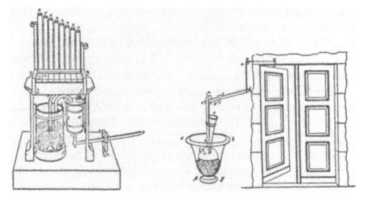

Fig. 120. Pipe-organ of
Ctesibius.

Fig. 121. Hero's door-alarm.

we noted in the story of mechanics, was published about A.D. 50 by Hero, a son and disciple of Ctesibius. The diagram indicates how the instrument operated. Its resemblance to the Pandean Pipes is quite patent and its origin harks back also to the palaeolithic whistles mentioned on p. 154. An old Scottish name for the pipe-organ was "chest of whistles" ("Kist o'wussles").

In the same book, Hero describes a door-alarm that he himself invented (fig. 121). After describing its structure, he gives an explanation of its action:

Now when the door opens, the result is that the string tightens and pulls the end of the lever. Consequently the lever is depressed beneath the hook.

[6]*Metamorphoses*, trans. F. J. Miller (Loeb Library).

As soon as the air vessel V descends, it sinks into the water and causes the trumpet to sound, for the air which V contains is forced out through the mouthpiece and the bell of the trumpet.

Here again one sees Hero's fondness for a device that is mystifying to the beholder. This elaborate scheme seems to have a slight touch of Rube Goldberg.

SOUND WAVES

In the reign of Augustus Caesar, and contemporaneously with Ovid, there lived at Rome a master architect named Marcus P. Vitruvius. He wrote a famous treatise, *On Architecture*, which is of special interest to us because, in discussing the acoustics of amphitheatres, it introduced the assumption that sound travels in the form of a wave-train, much in the manner of pond-waves, although invisible:

Now the sound of the voice is transmitted by a progressive motion in the air which can be perceived by the sense of hearing. It is propagated in the form of a countless series of concentric circles as when a stone is thrown into standing water, innumerable circular waves form and grow as they recede from the centre of disturbance as far as they can until the boundaries of the place or else some obstacle prevents the wave formation from reaching the shores. Accordingly, when opposed by obstructions, the first waves recoil and interfere with the form of succeeding waves. In water, the circles spread out only horizontally over the surface of the pond whereas the voice, on the contrary, travels not only laterally but also mounts upward wave after wave. Therefore, when no obstacle confronts the first wave, the second wave and succeeding waves also are not deflected from their courses. Hence, they all reach the ears of those in the top seats ["the gods"!], as well as of those in the lowest seats without reverberations. Accordingly, the ancient architects constructed the tiers of steps in theatres in harmony with the clues they found in nature by studying the facts about the voice.[7]

We have seen that most of the advancement in acoustics before the time of Vitruvius was chiefly of a practical or descriptive kind, without much profitable speculation or theory. It is true that Pythagoras and Lucretius did develop some theory, and Vitruvius says that the old architects of amphitheatres had developed some valid theory of acoustics, possibly a wave theory. With the advent of the *wave theory*, however, the science of acoustics had arrived, and, in the sequel, that theory became very important, not only in acoustics but also in other departments of physics.

[7]Vitruvius *De architectura*, lib. 5, cap. iii.

CHAPTER VII

Acoustics in the Western Cycle of Culture

THE work of Vitruvius in acoustics was the last which was of importance in the classical cycle of culture. During the Middle Ages, this science made little progress. In the story of mechanics the work of Leonardo da Vinci was the sign of an approaching peak in the development of that science. His manuscripts also herald a development in acoustics which derived from the growth in mechanics.

LAW OF REFLECTION

A stone flung into the water becomes the centre and cause of a number of circles.[1] In all cases, the movement of water has great similarity to that of air. Although there is shown an appearance of motion, yet the water itself is not transported from its place, because the opening made by the pebble closes itself again and causes a certain disturbance to be set up which may be spoken of as a vibration rather than a translation.

I say that the sound of the echo is cast back to the ear after it has struck [a surface], just as the images of objects strike a mirror and are there reflected to the eye. And in the same way as these images are transmitted from the object to the mirror and from the mirror to the eye at equal angles, so the sound of an echo will strike and rebound to the ear at equal angles.[2]

This passage is probably the earliest reference to the fundamental fact that when a wave-train traverses a medium, it does not transport the medium. We see here an interesting transition: although da Vinci is using the wave idea in his argument, nevertheless the influence of the corpuscular theory of Democritus and Lucretius is plainly seen in his reference to images striking a mirror and bouncing into the eye; and it is by this concept that he derives the law of reflection of sound. As far as reflection is concerned, neither the corpuscular nor the undular explanation has any advantage over the other, since the law would be the same by either assumption. This passage is the earliest to assume light waves (p. 252).

[1]E. McCurdy, *Science Note-Books of Leonardo da Vinci* (Cape, 1938).
[2]Quoted in Ivor Hart, *Great Physicists* (Methuen, 1927).

Da Vinci's note-books contain some very significant and sage advice concerning the relation between theory and practice:

> Those who become enamoured of practice without theory are like a sailor who enters a ship without helm or compass and who never can be certain whither he is going.[3] Theory is the captain; practice, the soldiers. Study first the theory and then follow the practice derived from that theory.[4]

Coming from a genius of such superlative performance as this man, the statement commands attention and should not be ignored. That Leonardo da Vinci picks up the thread of development where Vitruvius left off, is illustrative of the happy genius of science in accumulating from cycle to cycle of culture.

PERIOD OF VIBRATION

The oscillating pendulum shown in fig. 122 typifies an important part of Galileo Galilei's contribution to science and the world. He might well have chosen this design for his crest (even though the emblem of three gold spheres was already preempted in 1300 by a famous gentleman often referred to as "Uncle"), for a great deal of Galileo's work had to do with the study and application of vibration. Nor could he have chosen a more significant symbol from the whole realm of science and nature, for vibration and oscillation are everywhere, from the periodic motions of the planets and satellites down to the motions of electrons and all the particles of which matter is supposed to consist. The pendulum has furnished a mechanical model which has been widely used in the scientist's attempts to understand his environment.

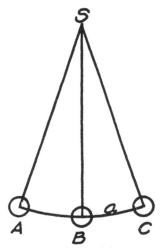

Fig. 122. Pendulum.

For his studies in acoustics, Galileo had a special zest. As we found in the story of mechanics, he became, in spite of his father's earnest advice, a good musician. In the following excerpt from his *Dialogues on Two New Sciences*, he takes for granted that every sounding body is in a condition of vibration or, briefly, that the cause of all sound is vibration. When the pendulum-bob *B* makes one complete "round-trip," it is said to make one complete (or double)

[3]J. P. Richter, *Manuscripts of Leonardo da Vinci* (Kensington).
[4]F. Martini, *Scritti*, p. 47.

oscillation or two single oscillations. The distance *BC* or *BA* from its rest-position to either extreme of its path is called its *amplitude*. The time which *B* requires to make one oscillation is called its *period* and, to the number of oscillations it makes per second, Galileo gave the name *frequency*. French authors generally speak of single vibrations or oscillations but when an English or German author uses the term vibration he means a double vibration unless the contrary is stated. In the following dialogue, the term *period* is introduced in the form "time of vibration."

SALVIATI. A string which has been struck begins to vibrate and continues that motion as long as one hears the sound. These vibrations cause the air immediately surrounding the string to vibrate and quiver: then these ripples in the air spread out into space and strike the strings of neighbouring instruments. . . .

One must observe that each pendulum has its own time of vibration [period] so definite and determinate that it is not possible to make it oscillate with any period other than its own natural one. . . . Now it is possible to confer motion upon even a heavy pendulum which is at rest, merely by blowing at it. . . . Suppose that by the first puff we have displaced the pendulum from the vertical, say, by half an inch; then if, after the pendulum has returned and is about to begin the second oscillation, we give it a second impulse, we shall impart additional motion: and so on with other puffs, provided they are applied at the right instants and not when the pendulum is coming towards us, for, in that case, the impulse would impede rather than aid the motion.

SAGREDO. Even as a boy, I observed that one man alone, by giving impulses at the right instants, was able to ring a bell so large that when six men seized the rope to stop it, they were lifted from the ground, all of them together being unable to overcome the momentum which one man had given it by properly-timed pulls.

SALVIATI. Your illustration is quite well fitted to explain the fact that a vibrating string can set another string in vibration and cause it to sound.[5]

Continuing the discussion, Salviati announces in clear and simple terms the very far-reaching principle of *sympathetic vibration* or *resonance*. In so doing, incidentally, he introduced the useful term *frequency*, to stand for the number of vibrations a body makes in a second.

Since that string which is tuned to unison with one plucked, is capable of vibrating with the same frequency [It. *frequenzia*], it acquires at the first impulse, a slight amplitude; after receiving two, three, and up to twenty or more impulses, delivered at proper instants, it finally accumulates a vibratory motion equal to that of the plucked string. Accordingly, if we attach to the side of a spinet small pieces of bristle, we shall observe that when the instrument is sounded, only those bristles respond that have the same period as the string which has been struck. The remaining pieces do not vibrate in response to this string; nor do the former pieces respond to any other string.

[5]G. Galilei *Opere*, xiii, 100; or, in translation, Crew and De Salvio, *Dialogues on Two New Sciences*, p. 97.

The spinet used in this experiment was one of the ancestors or forerunners of the modern pianoforte. The experiment with Frahm's "resonance-top" (fig. 123) is a modern form of Galileo's experiment with bristles on a spinet. The modern cyclotron applies the principle of "impulses exerted at proper instants."

FIG. 123. Frahm's resonance-top.

Galileo performed an experiment in resonance from which he was shrewd enough to discover that the value of the interval of an octave is 2:1 or 2. In other words, if the interval between two tones is an octave, as judged, say, by the ear, then the frequency of the higher tone is double that of the lower one. From this big step, he could determine the value of any other interval.

A glass of water may be made to emit a tone by the friction of a [moistened] finger, run along the rim of the glass.... If now we bow the low string of a viola vigorously and bring near it a goblet of fine smooth glass with the same tone as the string, the goblet will vibrate and emit the same tone audibly.... In the water is produced a series of regular waves.... I have observed that when, as sometimes happens, the tone of the glass jumps an octave higher, each of the aforesaid waves divides into two, a phenomenon which shows clearly that the ratio involved in the octave is 2:1.

In this clever experiment, Galileo was observing what received in 1825 the name stationary or standing waves. The sound from a goblet produced, as Galileo describes, with a moistened finger, is one which Sir William Bragg classified as a "stick-and-slip" sound. It will be discussed more fully when we come to consider the work of Sir William.

Galileo carried out a series of sonometer experiments with greater completeness and mastery than Pythagoras:

If a string stretched upon a sonometer be sounded, first the whole string, and then a bridge be placed at the middle and half of the string sounded, the octave [higher] is heard. If the bridge be placed at one-third of the string, on plucking the whole string and then the two-thirds we produce the interval called a fifth [do-so]. . . .

There are three methods of sharping [raising the pitch of] the tone of a string. One is by shortening it, another is by increasing its tension, i.e. stretching it, and the third is by decreasing its diameter. Keeping the length and diameter constant, if we wish to raise the pitch an octave by increasing the tension, it is not enough to double the tension for it must be quadrupled.

He adds a fourth method:

If two spinets are strung, one with gold wire and the other with brass, and if the corresponding wires have the same length, tension and diameter, the instrument with gold wires will have a pitch about a fifth lower than the other, for gold has a density almost double that of brass.

I say that an interval is dependent directly upon the ratio between the frequencies of the air-waves that strike the tympanum of the ear causing it to vibrate with the same frequency.[6]

An indication of the power of this great investigator may be seen in the fact that even the few contributions which have been selected for this chapter constitute more advancement in acoustics than the combined labours of all his predecessors since the Stone Age.

From Galileo's statement about density, it is seen that the relation between the density or specific gravity (s) of a string and its frequency is inverse or indirect, for the denser wire had the smaller frequency (f). To find the relation precisely we have the interval, fifth, $f_2 : f_1 = 3:2$ and the inverse ratio of the densities $s_1 : s_2 = 19.3 : 8.6 = 2.25 : 1 = 9 : 4 = 3^2 : 2^2$

$$\therefore \ \left(\frac{f_2}{f_1}\right)^2 = \frac{s_1}{s_2} \ \therefore \ \frac{f_2}{f_1} = \frac{\sqrt{s_1}}{\sqrt{s_2}} \ \text{ or } \ f \, \alpha \, \frac{1}{\sqrt{s}}.$$

Thus the frequency of a string varies inversely as the square root of its density or specific gravity.

The four sonometer laws can be expressed most briefly and conveniently in algebra thus

$$f \, \alpha \, \frac{\sqrt{t}}{Ld\sqrt{s}}$$

where d is the diameter of the string, s its density, t its tension, and L its length.

PÈRE MERSENNE

Père Marin Mersenne of Paris (1588-1648), a contemporary of Galileo, was also an ardent student of acoustics. He was an old schoolmate of Descartes. In 1636 he published a French treatise entitled *L'Harmonie Universelle* which includes his contributions to acoustics.[7]

Mersenne discovered that "An open string when plucked, emits several tones at the same time."[8] He found further that many other sounding bodies similarly can emit two or more tones at the same time. Such tones are described as *complex*. Galileo called the lowest

[6]*Dialogues on Two New Sciences.*
[7]The Library of Congress, Washington, D.C., has a copy of this rare book, which is too precious to be sent out on the Interlibrary Exchange. Those who wish to consult it must go to Washington, or else make use of that library's film service, which furnishes a small negative of any desired page at a nominal cost of a few cents. Such negatives can be read with a lens or by means of a projector. Large libraries are installing projectors for the purpose. The films are shipped in metal cans. To the long list of canned goods, therefore, we may now add "canned books."
[8]H. Ludwig, *M. Mersenne und seine Musiklehre* (Waisenhaus, Berlin, 1935).

tone of a series of tones the first tone. To the lowest tone of a complex tone, Mersenne applied the term "principal tone" (*la principale ou la plus basse voix*). Eighty years later, we shall find it called by Sauveur (1701) the "fundamental tone" (Lat. *fundamentum*, foundation), and this name has now long since become standard. All tones of a complex tone higher than the fundamental are called *overtones* or partials. When comparison of the tones of two strings is made, as in the sonometer laws, the reference is to their fundamental tones.

Mersenne was first to determine the frequency of a sounding body. To find the frequency of a sonometer wire of length 2 ft. and tension 28 pd., which, of course, made far too many vibrations per second to be counted by eye, he used an ingenious device. From the same spool, he obtained a wire of length 17 ft. and slacked it until at a certain tension (7 pd.) he could count the vibrations. His reading was eight single vibrations per pulse-beat, although he must have had to count them as double vibrations. The tone of the longer wire (if any) was of too low pitch to be heard. By the sonometer laws, he calculated the frequency of the short wire thus: $f = 8 \times (17/2) \times \sqrt{(27/8)}$ = 136 single vibrations per second, or 68 double vibrations per second. Hence, by his knowledge of intervals and by further use of the sonometer laws, he was able to determine the frequency of any tone.

Mersenne also determined (1620) the speed of sound in air by using an echo. Later, he employed a direct method, without echo, using a shotgun for source of sound and a pendulum for measuring time in seconds or, as they were called in those days, "second-minutes." The hour was divided into 60 minute parts, called, therefore, "minutes." Each minute was divided into 60 equal parts of a second degree of minuteness. Hence these were called "second-minutes." Later, this name was shortened to "seconds." In Mersenne's day, the change to the shorter term was taking place. In his treatise, he usually uses the formal term, second-minute, but occasionally unbends enough to use the colloquialism "second."

We can determine the speed of sound (in air) from experiments with echoes. Now a person readily pronounces two syllables in succession while the pulse beats once, i.e. in a second-minute.[9]

As a priest and a student of the classics, Mersenne had much to do with recitation of prayers and poetry; and he had great skill in reciting so that each syllable received the same time as all others. He next refers to a certain place where an echo was obtainable.

Now since this particular echo repeats the seven syllables, "Benedicam Dominum" and since the last syllable, "num," returns in a second-minute,

[9]M. Mersenne, *L'Harmonie Universelle*.

it must travel 485 Royal feet both going and again in returning, during the lapse of one second, that is to say, about 162 toises [1000 ft.].

The value now accepted for the speed of sound in air at 0°C., the temperature of melting ice, is 1088 (or 1089) ft. per sec., or 332 metres per sec. Hence Mersenne's result was about 9 per cent low; but his achievement was very creditable, when we consider the apparatus at his disposal. At room temperature (20°C.), the speed of sound in air is 1128 ft.p.s. or 344 m.p.s. about a mile in 5 sec., or 720 m.p.h.

The acoustical phenomenon called "beating" or beats, was first described and named by Mersenne, but he failed to reach a satisfactory explanation.

It is difficult to find out why two organ pipes that differ in pitch by not more than a quarter or half a tone produce a throbbing or thrumming sound similar to that of beating a tambourine when they are sounded together; for one can perceive ten beats, clearly audible in the interval of a second minute. One might expect a greater number of beats when the pipes differ by a semitone than when the difference is a tone; but the facts are quite the reverse, for the greater the interval, the greater is the number of beats per second.

Here the word *tone* is used, unfortunately, with a new meaning, namely, the interval between two tones of a scale, for example between C and D on the piano. A half-tone would be, for instance, that between C and C sharp. Mersenne also suggested the use of beats in tuning two sonorous bodies to unison. This of course has long been included in the technique of piano-tuning.

It is therefore, appropriate to remark in this connection that organ-builders should see to it when tuning organs that none of the pipes differ slightly from unison.

Mersenne was personally acquainted and maintained an active correspondence with a considerable proportion of the scientists of his day. In this way, his letters served, to a limited extent, the same purpose in his times as the science journals do in ours. It was through a letter to him from Ricci at Rome that the invention of the barometer by Viviani and Torricelli was announced to the scientists of Europe. The free exchange of information among scientists has been of fundamental importance in the advancement of science. The increase in the tempo of advancement has been partly owing to improvements in the facilities for publishing results and in the accessibility of results.

The records of the Florentine Academy[10] give some account of acoustical experiments, for instance, on a direct method of finding the speed of sound in air. Because of its quaint style, the translation by

[10]*Saggi di Naturali Esperienze fatte nell' Accademia del Cimento* (1667), p. 241. A copy is in the library of the University of Michigan.

R. Waller, Secretary of the Royal Society of London in 1684, is given here:

Sound, that Noble Accident of the Air, keeps so unchangeable a Tenour in its Motions, that a greater or lesser impetus wherewith the Sonorous Body produces it, is unable to alter it. Gassendus affirms that All Sounds, whether great or small [i.e., loud or faint], pass the same space in the same time, which we found undoubtedly true. We made this Experiment in the Night, with three several sorts of Pieces [cannon], planted at Three Miles distance from the place of Observation where we could well discern the flash of the powder in firing the pieces. From this flash then, we always counted an equal Number of Vibrations of the Pendulum Clock. In repeating this Experiment, we found the sound travelling the known space of a Mile in five seconds.

The French scientist Descartes, who was a close friend of Mersenne, stated in 1629 that the speed of sound is independent of its pitch. One exception to the Florentine statement is that the initial speed of the sound of a cannon, i.e. near the cannon's mouth, is somewhat above the normal value of 1128 ft.p.s. or 344 m.p.s.

CONDUCTION OF SOUND

A great deal of the advance in science has depended, of course, on the invention and improvement of instruments and apparatus. When Otto von Guericke had invented the extraction air-pump, accordingly, he naturally proceeded to per- form, by the aid of the new machine, some experiments that were impossible without it. Outstanding among these was the *bell-jar experiment*. Fig. 124 shows a modern form of the apparatus and here is the account given in von Guericke's book, *De vacuo spatio*:

In an evacuated container, bells ring exceedingly faintly. To demonstrate this experimentally, I suspended a loud-tick- ing clock-work in a glass receiver by a thread, fastened to its mouth. I had previously arranged the clockwork so that, by the striking of a small hammer, the bell emitted a clear tone at fixed intervals during half an hour. This done, I closed the receiver and proceeded to extract the air from it and ob- served that when the air was partly pumped out, the sound emitted was weaker, and when the air was completely [?] ex- tracted, no sound reached me. If, however, I brought my ear near the glass, a dull thudding caused by the stroke of the hammer was heard as if someone were holding the whole bell in his hand and striking it with the hammer, so that a dull thud or noise was heard at each blow but not a tone. On the other hand, as soon as I gave the air free access into the receiver, and it rushed in, immediately the sound returned to the original tinkling tone.[11]

FIG. 124.
Bell-jar
apparatus.

[11]O. von Guericke, *De vacuo spatio* (1672). III, xv. A copy may be seen in the library of McGill University. (Presented by Sir William Macdonald.)

By this experiment, von Guericke showed that the transmission of sound requires a material medium, or that vacuum cannot conduct sound. Modifications of the bell-jar experiment show that all gases, liquids, and solids, in fact all bodies, conduct sound, some substances better than others. The bad conductors are useful as insulators, rubber being one of the best bad-conductors or insulators. The bell in this

FIG. 125. Otto von Guericke.

experiment is always suspended or supported by an insulator. In other departments of physics, we meet analogues of the bell-jar experiment which are also quite important.

In this experiment, if we could, we should like to have the bell surrounded on all sides with vacuum only, as are the heavenly bodies, and thus cut it off utterly from the rest of the world. The extreme difficulties we encounter in the attempt bring home to us the realization that this material world, which it is the scientist's privilege and joy to study, is so marvellously constituted that in spite of the myriads of motions and transformations continually occurring in it, nevertheless, there is, in general, at all times a material chain joining every body in the world to every other part of the planet. To sever that chain is in general impossible. In the Feather and Coin experiment (p. 121), however, during the fraction of a second when the two bodies are falling they are surrounded completely by vacuum. But even in that case, there is still a modicum of air in the tube. Furthermore, it would be very difficult to employ this plan in the bell-jar experiment. No wonder Aristotle considered the world a plenum.

The *Philosophic Transactions* of the Royal Society of London for 1677 contains a letter from Dr. John Wallis which records the first use of a paper indicator in the study of the vibrations of strings:

Oxford, March 14, 1677.

Sir,

I have thought fit to give you notice of a discovery that hath been made here which I suppose may not be unacceptable to those of the Royal Society who are Musical and Mathematical. Supposing *AC* be an upper Octave of *EF* and, therefore, an Unison to each half of it, stopped at *B*: Now if while *EF* is open, *AC* be struck: the two halves of this other, that is, *EB* and *BF*, will both tremble: but not the middle point at *B*. Which will easily be observed if a little bit of paper be lightly wrapped about the string *EF* and removed successively from one end of the string to the other. This was first of all (that I know of) discovered by Mr. William Noble, a Master of Arts of Merton-Colledge and after him by Mr. Thomas Pigot, a Batchelour of Arts and Fellow of Wadham-Colledge.[12]

In the work of Sauveur (1701) we shall see these new indicators modified to paper riders which straddle the string like loose clothespins.

THE WAVE FORMULA

The relation known as the *wave formula* is fundamental in acoustics and in all sciences that have to do with waves. Here is Newton's derivation of the equation as given in the *Principia* (1687):

PROPOSITION xlvi. To find the velocity of waves.

Let *ABCDEF* [fig. 126] represent the surface of stagnant water ascending and descending in successive waves. In truth that ascent and descent is performed rather in a circle (*C,D*). That which I call the breadth of the waves [wavelength, symbol *L* or λ] is the transverse measure lying between the deepest parts of the troughs (*B,D*) or the tops of the crests (*A,C*, etc.). In media of uniform density and elastic force, all the pulses are equally swift. Let the number of vibrations of the body by whose tremors the pulses are produced, be found to any given time [frequency, *f*]. By that number, divide the space which a pulse can traverse in the same time [i.e. speed or velocity, *v*], and the part found will be the breadth of one pulse [*L*].

[12]*Philosophic Transactions of the Royal Society of London* (1676), p. 839. May be seen in the library of the University of Pennsylvania.

Expressed in algebra this is

Eq. 1. $v/f = L$ or $v = fL$ (or $f\lambda$).   (the wave formula)

A set of corks floating on the water in fig. 126, indicates that, when a wave-train is being transmitted through a medium, the particles of the medium vibrate with the same frequency as the disturbing vibrating source of the waves. Pond waves are *transverse* because the water particles vibrate across the direction of wave transmission.

When a particle of a medium is vibrating (fig. 126), each stage along its orbit is called a *phase*. Similarly, we speak of the phases of the moon, four of which have common names. In the case of the

FIG. 126. Pond waves.

water particles, however, only two phases have common names, viz. crest (*c*) and trough(*T*). Any two phases which are at opposite ends of a diameter are called *opposite phases*. They occur half a wave-length apart, whereas *like* phases occur one wave-length apart. Crests and troughs are the characteristic phases of transverse waves. The question arises, "To what extent are sound waves like pond waves?"; and in particular, "Are they transverse waves?" A knowledge of phases in waves will be of service in subsequent arguments.

HARMONICS

The astronomer Joseph Sauveur, who was professor of mathematics at the College Royale of Paris in 1701, published a treatise on *Sound Intervals* which is rich in sources of acoustical terms and ideas. He christened the science:

I thought, therefore, that there should be a science which I have named Acoustics and which takes Sound in general for its field whereas Music deals with sound insofar as it brings pleasure to our sense of hearing.[13]

He next introduces the term *fundamental* for the lowest tone a sonorous body can produce. Then he defines the important term *harmonic*:

I apply the term, harmonic sound of a fundamental sound, to one which makes a whole number of vibrations while the fundamental makes only one.

Continuing, he describes the *harmonic scale*, which he found to

[13]*Histoire de l'Académie Royale des Sciences* (1701), p. 297. A copy is in the library of McGill University.

be always and necessarily present in all instrumental and vocal music, regardless of the scale in which the music is written.

Similarly, if in a wind instrument, one blows harder and harder, the tone continues to rise but only according to the series of harmonic tones.

The frequencies of the tones of a harmonic scale are to one another as the natural numbers, 1, 2, 3, 4, 5, etc.

Sauveur introduced the term *node*, which he borrowed from astronomy and used to advantage in discussing his demonstration of a string vibrating in *segments*.

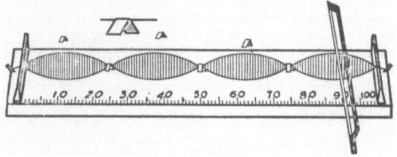

FIG. 127. Sauveur's experiment.

Divide the string of a monochord in equal parts, for example, in 5. Pluck this open string and it gives a sound which I call the fundamental of this string. Now place on one of these divisions a light obstacle such as the end of a feather, D, so that the motion of the string is communicated to the parts on both sides of the obstacle. It will emit the fifth harmonic. I shall call these points A,D,E,F,G,B, the *nodes* of these vibrations and the intervening parts of the string will be called 'bellies' [English, ventral segments]. If we place some little pieces of black paper at the divisions E,F,G etc., and some white paper riders on the middle points of the segments [now called loops or antinodes], upon vibrating AC, we shall see that the white riders which are at the loops will jump off and that the black ones at the nodes will remain on the string.

This experiment is known as Sauveur's experiment. From Mersenne's observation that a string can emit two or more tones at the same time, it follows that a string must be able to vibrate in more than one set of segments simultaneously. It is then said to vibrate complexly.

Sauveur was first to reach a satisfactory explanation of beats, which we found were first described and named by Mersenne in 1620. Sauveur interpreted them as due to *interference* or superposition of sound waves, i.e. by the combined action of two wave-trains. When they strike the ear *in the same phase*, their combined effect is a sum and produces maximum intensity, but when they arrive *in opposite phase*, their combined effect is a difference and produces minimum intensity by partial or total neutralization. Thus in all cases, the net effect is the algebraic sum of the two effects. Hence the throbbing or alternate increase and decrease of loudness. He also concluded that the number of beats per second is the difference of the two frequencies, or, in algebra,

Eq. 2. $b = f'' - f'.$ (Sauveur's beat formula)

In the following sentence, Sauveur urged the desirability of an accurate standard of pitch:

A standard tone, e.g. one of 100 vibrations per second, would be of such great utility in acoustics and even in music that we should take steps to investigate every means of determining it accurately in order to compare all others with it.

Pursuing this idea, he obtained such a standard of pitch in the following ingenious manner. Having a pair of pipes which gave 4 beats per second, he compared each of them with a monochord and found the interval between their tones to be 25/24. Hence by calculation he knew that the frequency of the higher pipe was 100 v.p.s. (vibrations per second).

$$ i = \frac{f''}{f'} = \frac{25}{24} \quad \therefore \quad f' = \frac{24}{25} f''. $$

Also the number of beats per sec. $b = f'' - f' = 4.$

$\therefore \qquad f'' - (24/25)f'' = 4$

$\therefore \qquad\qquad\qquad f'' = 4 \times 25 = 100$ v.p.s.

and $\qquad f' = \qquad\qquad 96$ v.p.s.

From the excellence of Sauveur's work in acoustics, one might not suspect that he laboured under a serious handicap. As a matter of fact, however, he had such a poor ear for tones and intervals that he was obliged to rely on the testimony of a musical assistant in recognizing unison, an octave, or other intervals.

Here it might be mentioned that it was an English contemporary of Sauveur, namely, John Shore, Handel's first trumpet, who invented the tuning-fork in 1711. Thus Shore could tune his instrument without depending on the oboe and could even check on the latter when he "sounded his A." The tuning-fork, of course, has been of tremendous importance in acoustics. "Thou knowest not which shall prosper this or that."

CHAPTER VIII

The Nineteenth Century
and After

FROM about 1750 to 1850, there flourished in France a galaxy of physicists and mathematicians of a brilliance and vitality which has seldom if ever been equalled. It is graced by such names as Coulomb, Charles, Lavoisier, and dozens of others. Strangely enough, however, acoustics received comparatively little attention from these great luminaries. She seemed to be the Cinderella of the sciences. Yet she, too, had her devotees. Mersenne, Descartes, Sauveur, and others had made fine contributions, and now comes the work of Dr. Ernst Chladni (1756-1827). The writings of the Swiss scientist, Daniel Bernoulli, made a deep impression on young Chladni, and helped to kindle in him a zeal for science that became a life interest. In 1802, Chladni published a German treatise, *Die Akustik*. Later he came to Paris, where he became a great favourite of Emperor Napoleon I, of whom this good can be said that he admired able scientists and mathematicians and gave them much encouragement and assistance. In 1807, Napoleon offered Chladni six thousand francs to translate his book into French. Chladni responded to his patron's generosity by writing him a new treatise, entitled *Traité d'Acoustique*, which was published in 1809.

CHLADNI FIGURES

As an amateur in music, the rudiments of which I had begun to study at the rather late age of nineteen, I noted that the theory of Sound was more neglected than several other branches of physics, an observation which kindled in me a desire to reduce this deficiency and to be of service to this department of physics by making some discoveries. . . . In 1785, I had observed that a plate of glass or metal, if clamped at the centre, gave different sounds when struck at different places: but I found nowhere any account of these different modes of vibration. The journals had given at that time some notices about a musical instrument made in Italy by Abbé Mazzochi, consisting of bells to which he applied violin bows. . . . When I applied a bow to a round plate of brass fixed at its centre . . . it emitted different tones. . . . But the nature of the motions to which these sounds corresponded was still unknown to me. . . . The experiments on electric figures that form on a resin plate, dusted with powder, discovered and published by Lichtenburg,[1] led me to surmise that the different vibratory motions of a sonorous plate might also present a different appearance if I sprinkled on the surface a little sand.

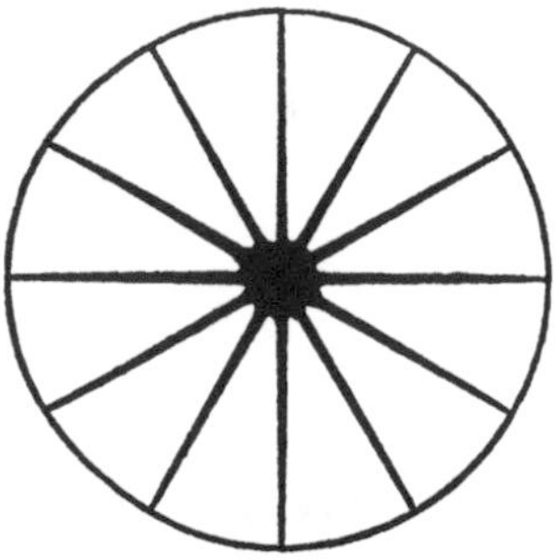

Upon employing this device, the first figure to present itself to my eyes on the round plate resembled a star with 10 or 12 rays much like fig. [128]. Just imagine my astonishment and delight upon beholding this sight which no one had ever seen before!

In order to produce each kind of vibratory motion in a plate, and to render visible the *nodal lines*, it is necessary to hold one (or more than one) part motionless [damping] and put some movable part in motion by means of a violin-bow after having sprinkled on its surface a little sand which is buffetted by the vibrating parts [loops] and consequently accumulates upon the nodal lines.

Figures produced in the manner just described are called Chladni figures.

FIG. 128. Chladni figure.

FIG. 129. Chladni figure on Chladni-plate.

[1] *Mémoires de la Société Royale*, Göttingen.

One hates to discount Chladni's priority, but the fact remains that that remarkable man Galileo had anticipated Chladni in the matter of sand-figures. He had seen a type of Chladni figure and had recorded it in these words:

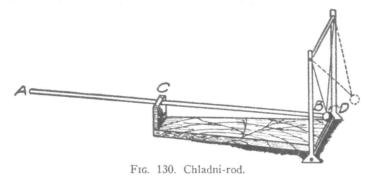

FIG. 130. Chladni-rod.

As I was scraping a brass plate with a sharp chisel to remove some spots, I heard the plate emit a clear whistling tone. I noticed a long row of streaks [of small particles] parallel to each other. When the tone was higher, the streaks were closer together.

Figure 129 shows a Chladni figure obtained on a square Chladni plate in the manner described by the inventor. Chladni's book contains many useful and interesting items: for instance, there is the invention of the Chladni rod (fig. 130), and a neat discussion of elasticity, showing that without that property there would be few if any sonorous bodies, for it is elasticity that keeps a tuning-fork vibrating. The same principle holds for most sonorous bodies. The Chladni rod is clamped at its centre C. When properly stroked with a rosined thumb and finger of a gloved hand, it emits a tone. This is another example of a stick-and-slip sound. Whenever the sounding rod touches the ivory ball D, the latter is shot away like a billiard ball from a cue.

FIG. 131. Ernst Chladni.

Hence the vibrations of the rod must be parallel to its long axis and are therefore longitudinal whereas those of a tuning-folk are transverse.

In another passage, Chladni made a calculation of the height of the atmosphere based on the fact that a shooting star had been heard to explode overhead ten minutes after the explosion was seen. At a speed of one mile in 5 sec., (p. 170), this indicates that the meteorite was at a height of $\dfrac{10 \times 60}{5} = 120$ miles. The atmosphere must reach still higher. Furthermore, the sound of the explosion, to be heard at such a distance, must have been of terrific intensity.

That Chladni richly achieved his ambition to be of service to acoustics is indicated by the fact that John Tyndall, in his famous lectures on sound (1873), referred to him as the "Father of Modern Acoustics." The researches of Chladni undoubtedly influenced the work of the four members of the French school whose contributions we shall next consider: Laënnec, La Tour, Savart, and Colladon.

THE STETHOSCOPE

As early as 400 B.C., Hippocrates, the great Greek physician who became the friend of Democritus, used the method of diagnosis by direct *auscultation*, that is, by applying the ear to the patient's body and listening to the internal sounds. In 1665, Robert Hooke predicted the invention of an instrument for the purpose, and described the sounds produced by the "works" of his watch as he heard them by applying the blade of a screwdriver to the watch and the wooden handle to his ear; but it was not until 1816 that the prophecy was fulfilled. At that time the brilliant young doctor René Laënnec of Paris was studying diseases of the chest. One patient sent to his clinic was a poor girl who was so fat that no sounds could reach his ear through the wall of adipose tissue. In the following excerpt from Laënnec's book on auscultation, he tells how this difficulty led him to invent the stethoscope (Gk. *stethos*, chest):[2]

I happened to recollect a simple and well-known fact in acoustics . . . the distinctness with which a person hears the scratch of a pin at one end of a stick of wood if he applies his ear to the other end of the stick. I rolled a quire of paper into a cylinder . . . and applied one end of it to the region of the patient's heart and the other end to my ear and was not a little surprised and pleased to find that I could thereby perceive the action of the heart in a manner more clear and distinct than I had ever been able to do by the immediate application of the ear. . . . The longitudinal aperture which is always left in the centre of paper thus rolled, led accidentally in my hands to an important discovery. . . . I now employ a cylinder of wood, an inch and a half in diameter and a foot

[2]Quoted in C. N. Camac, *Epoch-Making Contributions to Medicine*, trans. J. Forbes (Saunders, 1909), p. 162.

long, perforated longitudinally by a bore three lines wide, and hollowed out into a funnel-shape to the depth of an inch and a half at one of its extremities. . . . This instrument, I have denominated the stethoscope.

A garage-man sometimes diagnoses quite astutely a gas-engine's symptoms by using Hooke's trick with a screw-driver; and inspectors of the water-works department use a metal rod running through a shell like that of a telephone receiver or ear-piece to locate breaks in underground pipes without excavation.

THE SIREN

The French scientist Baron Cagniard de la Tour invented the *siren* in 1819 and gave it its name. It has become more important both in theory and in practice than its own inventor ever hoped or imagined. Here is part of his description of the instrument:

If sound, produced by instruments, is due, as physicists believe, to a regular series of reiterated impulses which they give to atmospheric air by their vibrations, it seems natural to think that by means of a mechanism which would be devised to strike the air with uniform period, one could cause the production of sound. Such is indeed the result that I have obtained ... by making the wind from a bellows issue through a small orifice before which is placed a circular plate turning about its centre. . . . The plate is pierced obliquely by a certain number of apertures, arranged in a circle. . . . By the motion of the plate, these openings will successively come opposite the orifice. . . . The [intermittent] current [of air] gives the exterior air a regular series of puffs which produce a sound that is higher or lower according as the current turns the plate with greater or less speed.

We see that the aim of this apparatus was to produce sound ... by a rotary motion which is very easy to measure by means of gears. . . . That is what I have done in the model I possess. It consists of a circular brass box about four inches in diameter. The top of the box is pierced obliquely with a hundred holes. . . . Its centre carries an axis which serves as centre of rotation for the plate which surmounts the apertures. This disc is also drilled with a hundred holes corresponding to those in the box and likewise oblique but in the opposite sense. The obliquity of the openings serves merely to give the current the proper direction to spin the disc. . . .

If we send water through this siren instead of air, it produces sound equally well, even when it is entirely immersed in that fluid. It is because of this property of singing in the water that I thought I might give it the name by which it is known.[3]

The siren, of course, has many common applications, such as fog-signals at light-houses or harbour-gaps, on ambulances, fire-reels, bicycles, and policemen, at circuses, and even in pop-corn bags if the luck is good. In the hands of later scientists, notably Helmholtz, the

[3] *Annales de Chimie et de Physique* (1819), XII, 167.

siren became the fundamental instrument in acoustics and a powerful help in research. La Tour foresaw some of this development when he

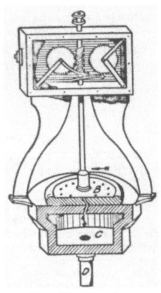

said "a sound . . . which is very easy to *measure*." In other words, this instrument enabled the science of acoustics to advance in a quantitative manner.

By slowing a siren more and more and thus lowering the pitch through a continuous series of tones, it is found that the lowest tone audible to human ears has a frequency of about 20 d.v.p.s. (double vibrations per second). Below that value we hear the individual puffs of air. If we speed the instrument, the puffs fuse at about 16 to 20 d.v.p.s. to form a tone. This agrees with Mersenne's experience (p. 169) that 4 d.v.p.s. is below the lower limit of audible tones for human ears. Chladni's experiments with strings led him to put the limit at about 30 single vibrations per second or 15 d.v.p.s. The value varies from individual to individual and even with

FIG. 132. The siren.

the same person according to age and physical condition. The currently accepted value is about 20 d.v.p.s.

SAVART'S TOOTHED WHEEL

A contemporary and compatriot of La Tour who made valuable additions to knowledge in acoustics was Dr. Felix Savart (1791-1841). He was at first a physics teacher but was later emancipated and became the conservator of the physical cabinet at the Collège de France at Paris. The toothed wheel which he devised and used in his researches is named after him. He wished to determine the upper limit of tones for human ears. A circular saw in a carpenter-shop gives the clue which was fundamental in this research, for the pitch of its sound varies with the speed. Although Savart did not succeed in finding the upper limit of tones, yet his work proved new and valuable and led to other researches.

Several distinguished physicists have sought to determine the limits beyond which sounds both high and low are no longer perceptible to the human ear. There is fairly general accord in setting the limit for low sounds around thirty single vibrations per second. Chladni admits that the ear can still hear sounds which reach 12,000 single vibrations per second. . . . It seemed to me that the question could be cleared up . . . by means of a wheel rotating with sufficient speed and having its circumference equipped with a convenient

number of teeth which would strike . . . against some thin body such as a card. . . .
My first experiments were made with a brass wheel, 24 centimetres in diameter,
whose circumference bore 360 teeth. When the speed was doubled, the sound
rose an octave. . . . I set up another apparatus whose toothed wheel . . . had a
diameter of 82 cm. and its circumference was cut into 720 teeth. Then the sounds
were again perceptible even when the number of collisions was 24,000 per
second . . . but up to the present, I have been unable to carry these researches
further because of the cost of the machines that would have to be made.[4]

The state of Savart's finances as a teacher is indicated by his admission
that the research came to an end because he could not afford the
necessary apparatus. The first scientist to sug-
gest this use of a toothed wheel was Robert
Hooke (1660) but he did not carry the idea
to completion.

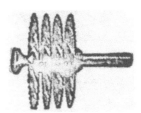

Any sounding body that produces a tone
may be called a tonal body. When the teeth
of the wheel are all equal in size and the speed
of the wheel uniform, the card's frequency
is constant. It is said to have *regular vibration*
and Savart found that *every tonal body has*

Fig. 133. Four
Savart toothed
wheels.

regular vibration. If a wheel has teeth of unequal sizes, the card's vibration becomes irregular and Savart found that the sound produced in
that case is a noise. The wheel in these experiments is itself a sonorous body being set in vibration by the blows from the card.

Savart's wheel can be used instead of a siren to determine the
frequency of a tone, although the technique is not as accurate as with
the siren.

If the wheel has t teeth and makes r rotations in s seconds, when
in unison with the tone under examination, i.e. when it has the same
pitch, then the frequency

Eq. 1.
$$f = \frac{tr}{s} \text{ d.v.p.s.}$$

The same relation holds for the siren.

In another research, Savart located the nodes in a number of
sounding organ-pipes by lowering into the pipe a small tambourine
on which some sand was sprinkled. Does it not seem that there was
some of Chladni's influence hovering near this research?

By means of a Savart wheel, it is found that the pitch of middle-C
is 256 d.v.p.s. (philosophic standard of pitch). Hence the range of
tones in a piano is as follows:

A_3	C_3	C_2	C_1	C	C'	C''	C'''	C^{IV}
				(middle)				
26	32	64	128	256	512	1024	2048	4096	d.v.p.s.

[4]*Annal. de Chim. et de Phys.* (1830), I.XIV, 337.

Galileo had shown that the value of the interval octave is 2:1 but the proof which Savart gave in the quotation we have just read is better than that of the Italian.

SPEED OF SOUND IN WATER

The speed of sound in water was determined in 1826 by two scientists who were chums and frequently worked together on a problem. They were Daniel Colladon, professor of mathematics at

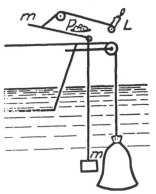

Geneva, Switzerland, and Jacob Sturm (1803-55), who left Geneva in 1830 to become professor of mathematics at Paris. Colladon announced their findings from the Lake Geneva experiments in a letter to M. F. Arago, secretary of the Royal Academy of Paris, from which letter this excerpt is taken. (Fig. 134 is copied from their report.)

Fig. 134. Colladon and Sturm's experiment.

The bell we used for these experiments ... was suspended by a beam beside a boat and kept under water at a depth of a metre. To this beam .. was fixed a bent lever, *mm*, whose upper end was in the boat while the other was immersed in the water and served as a hammer for striking the bell. All the experiments were made at night not only to be wholly undisturbed by interfering sounds but above all for the precision of the signals which were given by burning powder *P*. We took a cylindrical tube of thin sheet iron three metres long and about two centimetres wide. We had it closed at the end which was to dip into the liquid. . . . The open upper end, to which I applied my ear, projected only 5 or 6 decimetres above the surface. . . . In this way, we found 1435 metres per second as the . . . speed of sound in water at 8° C.[5]

One modern application of the Lake Geneva experiments on speed of sound in water is the instrument by which a ship's captain can know at all times the depth of water beneath his vessel. The instrument is called a fathometer because it "sounds" the ocean by the echo method, and, by the reading of its dial, gives the sounding in fathoms.

STANDING WAVES

Our itinerary takes us now to Germany where a school of acoustics developed under the leadership of Dr. Hermann von Helmholtz of Bonn and Berlin. We shall consider some of the results obtained by the Weber brothers of Leipzig, by Helmholtz, by Melde of Marburg, and by König of Paris, who often agreed with Helmholtz's ideas and sometimes opposed him.

[5] *Annal. de Chim. et de Phys.* (1826), XXXVI, 240.

One of the greatest books ever written about waves was published in 1825 at Leipzig and dedicated to Chladni by two brothers, Ernst and Wilhelm Weber, who were both eminent physicists. It contains the following discussion of what are called *standing waves* or stationary waves. The term is very important as we shall see, although rather paradoxical if we accept the dictionary definition that a wave is a progressive form, for in that case a wave could not stand still.

Equilibrium is that condition of a body in which the actions of two or more forces mutually cancel each other and thereby produce a condition of rest. A body is in vibratory motion when its parts under the tendency toward equilibrium, alternately approach and recede from the position in which equilibrium can occur [e.g. pendulum]. Vibratory motion is of two sorts, (1) a progressive mode called *oscillatio progressiva* and (2) a stationary kind named *oscillatio fixa*. Progressive vibration is synonymous with wave motion or *motus undulatorius*.

In the winter of 1821-2, one of us, while purifying some mercury by pouring it through a funnel from one flask to another, observed that at the surface of the mercury in the second flask the stream . . . caused an exceedingly complicated but regular pattern. . . .which was seen to persist . . . unaltered as long as the inpouring mercury struck the surface at the same place and with the same velocity. But if these conditions varied, the pattern also changed. He recognized this figure as resulting from waves continually intercrossing each other at the same place. . . . These results are best comprehended as due to the second mode, namely, stationary oscillations. These actually occur more frequently than progressive waves but so far, they have been almost entirely overlooked. It is to standing waves that we wish to direct the attention of physicists and mathematicians. . . .

The essential cause of this second type of vibration is as follows: two wave-trains of the same wave-length and intensity but in opposite sense . . . encounter each other and by their mutual influence, transform the progressive vibration of both trains to standing waves.[6]

The regular pattern produced by the Weber brothers as just described was known to Galileo (p. 179) and can readily be seen where a smooth stream of water descends from a tap into an open sink, if light from a near-by lamp shines across the surface of the water. The characteristic features of standing waves are nodes and loops or antinodes. Sauveur observed these in a string and introduced the term node. The lines of a Chladni figure are called nodal lines. Hence standing waves were present in the experiments of Galileo, Sauveur, and Chladni and will be further discussed in connection with the work of Melde. Indeed standing waves were present, though not recognized, whenever the first tone was produced in prehistoric times, for investigation has shown that *every tonal body vibrates in standing waves* and this is one reason for the importance of standing waves in

[6]E. H. and W. Weber, *Die Wellenlehre.* This book may be seen in the Library of Congress, Washington, D.C.

acoustics. A knowledge of stationary waves has been of tremendous assistance in the advancement of acoustics and other branches of physics; for example, it was by use of this knowledge that Hertz, one of Helmholtz's students, succeeded in discovering the electric waves which we call radio waves.

QUALITY OR TIMBRE

One of the greatest books in acoustics was published in 1862 by Dr. Hermann von Helmholtz of Bonn (1821-94), a scientist of

FIG. 135. Hermann von Helmholtz.

tremendous power and scope. One of his early teachers recognized his unusual ability and, knowing that he was too poor to afford university training, managed to have him admitted under army auspices so that the costs were defrayed by the state. Helmholtz began as a physiologist and by his study of the ear and eye rose to preeminence. His work necessitated a knowledge of physics and he became one of the foremost physicists of his century. Realizing that the mastery of physics was impossible without mathematics, he studied mathematics—and to such purpose that he became one of the most accomplished mathematicians of his day. He was a fascinating and inspiring lecturer, a great teacher, a leader, and a noble citizen.

Helmholtz's book on sensations of tone, *Die Lehre von den Tonempfindungen*, has been called the Principia of physiological acoustics. Much of it is too advanced for this book. It discusses the role of elasticity in acoustics, and here is one of its sentences on that topic:

Physical acoustics is essentially nothing more than a section in the theory of the motions of elastic bodies.

Chladni had pointed out the fundamental significance of elasticity in acoustics, showing that the vibrations of sonorous bodies are maintained by this property, that the transmission of sound through a medium is an elastic phenomenon, and that the reception of sound by the ear would be impossible without elasticity. Without elasticity, no sound. Chladni confessed in his book of 1809 that the physical cause of *timbre* was unknown to scientists, but he gave a very shrewd guess at the riddle. Finally the problem was solved by two men, Helmholtz and König, each working by his own method but both reaching the conclusion, which Helmholtz expressed in one of his famous lectures, delivered in 1857 at Bonn.

A sound of a certain frequency is always of the same pitch no matter from what instrument it originates. Now the feature which still distinguishes the tone *A* of the piano from the tone *A* of the violin, flute, clarinet, etc., we call timbre. The wave-length determines the pitch of a tone; to this I add that the amplitude of the wave-crests or, referring to the air, the degree of the varying compression and rarefaction, determines the loudness or intensity of the tone. But waves of equal amplitude [and wave-length] can still have a difference of form. The peaks of their crests, for example, can be rounded or sharp. Corresponding differences may also be present in sound waves of the same pitch and intensity; and indeed, it is the timbre that corresponds to the form of the water waves.[7]

The work of Mersenne and Sauveur had shown that the tone of a string is frequently complex, consisting of a fundamental and one or more overtones or harmonics which Helmholtz called partials. Both Helmholtz and König came to the conclusion that the timbre of a tone depends on its complexity or in other words, on its admixture of overtones.

The method by which Helmholtz discovered the cause of timbre was that of analysing sounds by means of an invention of his, called Helmholtz resonators. He describes the device:

Stretched membranes . . . are very convenient for . . . experiments on the partials of complex tones. They have the great advantage of being independent of the ear, but they are not very sensitive to the fainter simple tones. Their

[7]H. von Helmholtz, *Vorträge und Reden*, I, 122.

sensitiveness is far inferior to that of the resonators[8] that I have introduced. These are hollow spheres of glass or metal, or tubes with two openings as shown in the figure [fig. 136]. One opening, *a*, has sharp edges, the other, *b*, is funnel-shaped and adapted for insertion into the ear. This smaller end I usually coat with melted sealing-wax and when the wax has cooled enough not to hurt the finger on being touched, but is still soft, I press the opening into the entrance of the ear. The sealing-wax thus moulds itself into the shape of the inner surface of this aperture, and when I subsequently use the resonator, it fits easily and is air-tight. Such an instrument is very like a resonance bottle, already described, in which the observer's own tympanic membrane has been made to replace the former artificial one.[9]

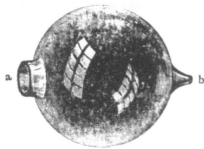

FIG. 136. Helmholtz resonator.

To analyse the sound of a string, for example, middle C (256 v.p.s.), set it vibrating and listen to it with a 256 resonator attached to one ear. If the two are in unison the tone of the string will be intensified by the resonator. Comparison can be made with the other ear. Then try other resonators, e.g. 512, 768, etc. By such a series of trials, one can find which partials or overtones are combined with the fundamental and their relative intensities. This is the analysis of a complex tone.

A MINIATURE HARP OF 3,000 STRINGS

Helmholtz describes how the ear recognizes the pitch of a tone:

If several tones are sounded simultaneously near a piano, each individual wire can vibrate in sympathy only if among them its own pitch is present. What happens in the same case in our ear is probably similar to the process in the piano, just described. In the interior of the hard bone within which our inner ear is situated, we find a special organ, the cochlea[10], so named because ... it resembles closely the inner cavity of the shell of the common vineyard snail. ... Now the canal of the cochlea is divided throughout its length, at the middle of its height, by two stretched membranes, into three compartments, an upper one, a middle, and a lower. In the middle chamber, the Marquis of Corti discovered some very remarkable structures, innumerable tiny microscopic rods [the rods of Corti] which, like the wires of a piano, lie near one another in regular arrangement. At one end, each is attached to the fibres of the auditory nerve and its opposite end is suspended from the stretched membrane. ... That these rods are set in sympathetic vibration by the sound vibrations carried into the ear, their anatomical arrangement leaves hardly any doubt. If, further, we consider the opinion that each minute hair, like a wire of the piano, is limited

[8]An interesting discussion of resonators is given in Poynting and Thomson, *Text-Book of Sound* (Griffin, 1909), p. 68.
[9]Ibid., *Werke*, "Tonempfindungen" (Vieweg Braunschweig), p. 73. A translation by O. Ellis, *Tone Sensations*, has been published by Longmans.
[10]See Sir James Jeans, *Science of Music* (Cambridge, 1937), p. 2.

to one tone, then we see that only if this tone sounds can the corresponding rod vibrate and the nerve ending belonging to it be stimulated to sensation; and also that the presence of each individual tone in a complex tone must always be detected by the corresponding sensation.

These few excerpts have only touched the fringe of the contents of Helmholtz's monumental book.

Dr. Franz Melde, professor of physics at Marburg in 1859, introduced an experiment which has been named after him. The fol owing excerpt from his treatise on acoustics affords a pleasing example of an inventor admiring improvements in his own invention made by another scientist:

The method of vibrating thin threads in standing waves by means of a tuning-fork, I published in 1859. . . . When I stretched a silk thread over the edge of a bell along a diameter and stroked the bell with a violin-bow, 90° from the end of the thread, . . . if the thread had the right tension, I saw that it

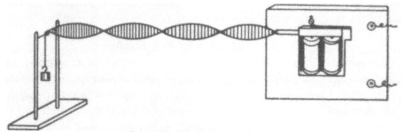

FIG. 137. Melde's experiment.

displayed the most beautiful stationary waves. Now we shall turn our attention to another apparatus that I constructed as shown in the figure [fig. 137]. It is diagrammed in all the texts. Tyndall with his characteristic insight examined my method, . . . then used it to carry out some brilliant demonstrations . . . in a series of lectures at the Royal Institution of London. . . . He used a thin platinum wire, one end of which was fastened to the tuning-fork and the other to a copper rod. If we connect the fork and rod to an electric battery, the wire can be brought to a red glow, uniform throughout its length. If then the fork is stroked with a violin-bow and the wire given such a tension that the vibration in segments appears, the nodes appear in gleaming light which decreases in intensity towards the loops, indeed even entirely disappears.[11]

Instead of a tuning-fork, the hammer of an electric bell is sometimes used in Melde's experiment. From fig. 138 a very useful relation can be derived, which may be referred to as the standing-wave formula. Let λ be the wave-length of the two similar wave-trains whose interference or superposition produces the standing waves. Then

Eq. 2. $\lambda = N_1 N_3 = 2N_1 N_2 = 2L_1 L_2 = 4N_1 L_1$.

The importance of this equation can be better realized if we recall

[11]F. Melde, *Akustik* (Brockhaus, Leipzig, 1883), p. 74.

that every tonal body vibrates in standing waves. To realize its power, we should see it in action. Suppose we are asked to calculate V, the speed of sound in brass, from an experiment with a Chladni rod whose tone, produced by longitudinal vibrations, is found to have a frequency of 1810 v.p.s. The rod's length is 100 cm. Since the rod is

FIG. 138. Standing wave.

a tonal body, it vibrates in standing waves with a node where it is clamped at the centre, and a loop or antinode at each end where there is maximum freedom of motion and hence maximum amplitude.

∴   wave-length $\lambda = 2L'L'' = 200$ cm. $= 2$ metres.   (Eq. 2)
∴ $V = f\lambda$ (Eq. 1, p. 174) $= 1810 \times 2 = 3620$ metres per sec.

This speed is about ten times that of sound in air and two and one-half times that of sound in water.

KÖNIG'S MANOMETRIC FLAME

Rudolph König (1832-1901) was a celebrated instrument-maker of Paris whose solving of the problem of timbre has been mentioned. His treatise, entitled *Experiments in Acoustics*, contains an account of his method.

Early in the year 1862, I had thought of a new method of observation, designed to render sound waves capable of being examined by the eye. The

FIG. 139. König's manometric capsule.

little device on the use of which my method ...is based and which I call a "manometric capsule,"[12] consists of a cavity formed in a block of wood [fig. 139] and enclosed by a thin membrane MB. Two tubes enter it, one of which carries illuminating gas and the other, ending in a nipple, gives egress to the gas and allows it to burn at F. Now suppose the air contracts or expands suddenly near the membrane; in the first case, the membrane, driven toward the interior of the capsule, compresses the gas in it and consequently the flame lengthens; in the second case, the membrane is drawn outward, the cavity enlarges and by the rarefaction of the gas, the flame being drawn in, must shorten. ... If we make the image of these flames form in a rotating mirror, it depicts at the same time all the phases of their movements, and one can not only determine the absolute number of the oscillations of different sounds and their ratios but also can observe the images produced by the combinations of a number of sounds.[13]

[12]A manometer is an indicator of fluid pressure.
[13]R. König, "Quelques Expériences d'Acoustique," *Poggendorf Annalen* (1872).

The rotating mirror which König used was invented in 1833 by Sir Charles Wheatstone of London for another purpose.[14] From his optical analysis of tones, König reached the same conclusion as Helmholtz, that the timbre of a sound depends on its admixture of overtones, i.e. on which overtones are present and on their respective intensities.

Another passage in the same book describes apparatus for demonstrating that the combined effect of two sounds may be silence, namely, when two wave-trains of the same frequency and intensity continue meeting at a point in opposite phase and thereby neutralizing each other. The phenomenon is, therefore, a case of superposition or interference.

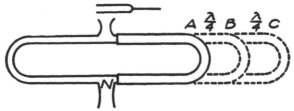

Fig. 140. Herschel's divided tube.

Herschel . . . produced interference by causing sound waves . . . from the same source to traverse two channels that differed from each other by half a wave-length and finally to unite again. I have constructed on the same principle an instrument designed to study the phenomenon of interference. . . . If, now, we lengthen one of the two branches of the tube [fig. 140] until it exceeds the other in length . . . by half a wave-length of the sound of the tuning-fork, the waves that arrive by the two channels neutralize each other at N, the far end of the tube.

Three men share the credit for the invention of this *divided tube*: Sir William Herschel, König, and Quincke, a German professor of physics. As a master mechanician, König was a genius. He is famous as the maker of the best tuning-forks in the world. He did more than any other person to carry out Sauveur's advice about obtaining a standard of pitch. Though hampered by poverty, he laboured during two or three decades at the making of his tuning-forks and the result is a monument to his skill and perseverance. The University of Toronto, the Case School of Applied Science, Cleveland, Ohio, and the Military Academy at West Point, are fortunate possessors of rare sets of König tuning-forks, which were brought to the Pan-American Exposition in 1901, the year of König's death, and left here under tragic circumstances. Their ideally pure tones are so pleasant to hear that an audience takes pleasure in listening to the

[14]*Phil. Trans. Roy. Soc. Lond.* (1833).

sound of even a single fork as long as it continues to "sing" and, *a fortiori*, to a diatonic chord from four such forks. The pleasure of this experience is akin to that of beholding the pure colours of a rainbow which makes "the heart leap up."

The fame of Sir William Herschel is so great as an astronomer that it comes as a surprise to some that he should make an acoustical

FIG. 141. Sir William Herschel.

invention; but in truth acoustics preceded astronomy in his life. When he came to England from his home in Hanover, Germany, he came as a musician, and he was first an organist and music teacher. The desire to know something of the science that underlay his art led him to the study of acoustics. In those days, however, the library shelf for acoustics would likely have plenty of spare room, enough indeed to accommodate some books on astronomy. Thus we see the fate of an individual determined by the accident that the names of two sciences have the same initial. Caroline Herschel said of her brother, "William used to retire to bed with a bason of milk or a glass of water, with Smith's *Harmonics* and Ferguson's *Astronomy*, and so went to sleep buried under his favourite authors." That Sir William's interest in acoustics was shared and maintained by his great son, Sir John Herschel, is indicated by the fact that Sir John's article on sound in the *Encyclopaedia Metropolitana* (1845) is considered one of the best outlines ever written.[15] The Herschels we may look upon as a link between the German school of acoustics and the British, to which we shall now turn.

THE BRITISH SCHOOL OF ACOUSTICS

Most of the scientists whom we have met thus far belong to the explorers. They are the researchers, discoverers, and inventors. But a devotee of science may serve her effectively in other capacities. For example, she has need of biographers and historians and of men who have abilities as speakers, writers, and teachers, to clarify and explain scientific knowledge and to bring a knowledge of science to people generally. A case in point is that of Dr. John Tyndall, professor of natural philosophy at the Royal Institution of London in 1854, who was a collaborator and biographer of Faraday. He took his doctorate under Bunsen at Marburg and thus formed another link between the German and British schools of acoustics. He was a successful researcher and engineer but probably the most valuable part of his work came from his power of clear and vivid explanation and demonstration.

He made a study of fog-horns, sirens, and lighthouses on the British coast which led to improvement in the coastal service of Great Britain and the saving of much life and property. In 1867 he published a text, *On Sound*, which became standard throughout the English-speaking scientific world for three decades. Nor is it certain that its successors have been improvements in every way. Its demonstrations and arguments are clear and often brilliant. It has a very human viewpoint, and such an appealing biographical and historical approach that it still makes interesting reading even to the layman. It possesses a freshness and clarity of language which is not always characteristic of text-books.

Tyndall's unrivalled demonstration of the famous bell-jar experiment includes the following:

> The motion of sound . . . is enfeebled by its transference from a light body to a heavy one. I remove the receiver which has hitherto covered our bell; you hear how much more loudly it rings in the open air. When the bell was

covered, the aerial vibrations were first communicated to the heavy glass jar, and, afterwards by the jar to the air outside; a great diminution of intensity being the consequence.

The action of hydrogen gas upon the voice is an illustration of the same kind. The voice is formed by urging air from the lungs through an organ called the larynx. In its passage it is thrown into vibration by the vocal chords which

FIG. 142. John Tyndall.

thus generate sound. But when I fill my lungs with hydrogen, and endeavour to speak, the vocal chords impart their motion to the hydrogen, which transfers it to the outer air. By this transference from a light gas to a heavy one, the sound is weakened to a remarkable degree. The consequence is very curious. You have already formed a notion of the strength and quality of my voice. I now empty my lungs of air and inflate them with hydrogen[!] from this gas-holder. I try to speak vigorously, but my voice has lost wonderfully in power, and changed wonderfully in quality. You hear it, hollow, harsh and unearthly: I cannot otherwise describe it.

The same book gives a remarkably vivid demonstration of the reflection of sound:

> Every experiment on the reflection of light has its analogue in the reflection of sound. On yonder gallery you see a electric lamp, placed close to the clock of this lecture room. An assistant in the gallery ignites the lamp, and directs its powerful beam upon a mirror, . . . behind the lecture table. By the act of reflection the divergent beam is converted into this splendid luminous cone. I mark the point of convergence; and the lamp being extinguished, I place my ear at that point. Here every sound-wave sent forth by the clock, and reflected by the mirror, is gathered up, and I now hear the ticks as if they came, not from the clock, but from the mirror. I will stop the clock, and have a watch held . . . at the place occupied a moment ago by the electric light. At this great distance I distinctly hear the ticking of the watch. My hearing is much aided by introducing the end of a glass funnel into my ear, the funnel here acting the part of an ear-trumpet.
>
> We know, moreover, that in optics the position of a body and of its [real] image are reversible. I place a candle at this lower focus; you see its image in the gallery above, and I have only to turn the mirror on its stand, to make the image fall on any one of the row of persons who occupy the front seat of the gallery. Removing the candle, and putting the watch in its place, the person on whom the image of the candle fell, distinctly hears the ticking. [*Assent and applause.*]

When sound travels through tubes, it is reflected from the walls and prevented from spreading out in all directions. The effects produced are quite remarkable and useful in application.

> The celebrated French philosopher Biot observed the transmission of sound through the empty waterpipes of Paris, and found that he could hold a conversation in a low voice through an iron tube 3,120 feet in length. The lowest possible whisper, indeed, could be heard at this distance, while the firing of a pistol into one end of the tube quenched a lighted candle at the other.

It was Tyndall who introduced the term "overtone" in British acoustics:

> All bodies and instruments, then, employed for producing musical sounds, emit, besides their fundamental tones, tones due to higher orders of vibration. The Germans embrace all such sounds under the general term *Obertöne*. I think it will be an advantage if we, in England, adopt the term *overtones* as the equivalent of the term employed in Germany.

There are good reasons for thinking of the human ear, especially the ear of a talented musician, as the most wonderful "instrument" in the world. In one passage Tyndall attempts to describe the infinitely complex action of the ear in listening to the sounds from an orchestra:

> In the music of an orchestra; not only have we the fundamental tones of every pipe and of every string, but we have the overtones of each, sometimes audible as far as the sixteenth in the series. We have also resultant tones; . . .

all trembling through the same air, all knocking at the self-same tympanic membrane. We have fundamental tone interfering with fundamental tone; we have overtone interfering with overtone; we have resultant tone interfering with resultant tone. And besides this, we have the members of each class interfering with members of every other class. The imagination retires baffled from any attempt to realize the physical condition of the atmosphere through which these sounds are passing. And as we shall learn in our next lecture, the aim of music, through the centuries during which it has ministered to the pleasure of man, has been to arrange matters empirically, so that the ear shall not suffer from the discordances produced by the multitudinous interference. The musicians engaged in this work knew nothing of the physical facts and principles involved in their efforts; they knew no more about it than the inventors of gunpowder knew about the law of atomic proportions. They tried and tried till they obtained a satisfactory result, and now, when the scientific mind is brought to bear upon the subject, order is seen rising through the confusion, and the results of pure empiricism are found to be in harmony with natural law.

J. C. Doppler, an Austrian scientist, had shown in 1842 that the pitch of a tone is sharped if the source is approaching the ear, i.e. while the distance between them decreases, and is flatted while the source recedes from the ear. The phenomenon is called the Doppler effect. Tyndall's reference to this effect is worth noting:

An extremely instructive effect may be observed at any railway station on the passage of a rapid train. During its approach the sonorous waves emitted by the whistle are virtually shortened, a greater number of them being crowded into the ear in a given time. During its retreat, we have a virtual lengthening of the sonorous waves. The consequence is, that when approaching, the whistle sounds a higher note and when retreating it sounds a lower note than if the train were still. A fall of pitch, therefore, is perceived as the train passes the station. This is the basis of Doppler's theory of the coloured stars.

THE GALTON WHISTLE

Savart failed to produce, by means of a toothed wheel, a sound of such high pitch that the ear could not perceive it, but Dr. Francis Galton of London, a cousin of Charles Darwin, achieved that result by means of a very short pipe (fig. 143), known as a Galton whistle. In his fascinating autobiography he refers to the instrument, and his words rouse in one the suspicion that he had a strong vein of whimsy in his nature:

Among other instruments that I contrived ... were small whistles with a screw plug for determining the highest audible tone, the limit of which varies much in different persons and at different ages. A parcel of schoolboys might interchange very shrill and loud whistles quite inaudible to an elderly master. I found them to produce marked effects on cats, and made my experiments at a house where I often stayed, in which my bed-room window overlooked a garden much fre-

FIG. 143.
Galton
whistle.

quented by them. My plan was to watch near the open window, and when a cat appeared and had become quite unsuspicious and absorbed, to sound one of these notes inaudible to most elderly persons. The cat was round in a minute. I noticed the quickness and precision with which these animals direct their eyes to the source of sound. It is not so with dogs. I contrived a hollow cane . . . having a removable whistle at its lower end, with an exposed india-rubber tube under its curved handle. Whenever I squeezed the tube against the handle, air was pushed through the whistle. I tried it at nearly all cages in the Zoological Gardens but with little result of interest except that it certainly annoyed some of the lions . . . who turned away and angrily rubbed their ears with their paws.[16]

The same topic is discussed in Galton's *Inquiries into Human Faculty*.[17] His experiments inaugurated a new department of acoustics called suprasonics, and ultrasonics, the name given to sounds above the upper human limit of pitch (about 30,000 d.v.p.s.). Incidentally, Galton was also a pioneer in the science of eugenics and in the technique of identification by finger-prints.

RAYLEIGH'S RÉSUMÉ

In the acoustical world the greatest event of the year 1877 was the publication of a book entitled *The Theory of Sound*, by John W. Strutt, Baron Rayleigh, of Trinity College, Cambridge. Indeed, it was one of the greatest events in the whole history of acoustics, for there are those who consider this book so great (if one dare peer into the future) that it might ultimately be judged the third greatest book in the literature of science, approaching the eminence of Newton's *Principia* and Huygens' *Horologium Oscillatorium*. All its chapters but the first are couched in the language of mathematics, however, so that the motto of the second chapter might be taken with slight modification from the portals of Plato's Academy, "Only mathematicians may enter here." The first chapter of Lord Rayleigh's book is a résumé of acoustics, and our selections from it will serve as a summary of the story of acoustics:

Directly or indirectly, all questions connected with this subject must come for decision to the ear, as the organ of hearing; and from it there can be no appeal. . . . When once we have discovered the physical phenomena which constitute the foundation of sound, our explorations are in great measure transferred to another field lying within the dominion of the principles of Mechanics.

Very cursory observation often suffices to shew that sounding bodies are in a state of vibration, and that the phenomena of sound and vibration are closely connected. . . . But, in order to affect the sense of hearing, it is not enough to have a vibrating instrument; there must also be an uninterrupted communication between the instrument and the ear. . . .

[16]F. Galton, *Memories of My Life* (Methuen, 1908), p. 247.
[17]Available in Everyman's Library (Dent), No. 263.

The passage of sound is not instantaneous. . . . The first accurate experiments were made by some members of the French Academy, in 1738. . . . The general result has been to give . . . a value for the velocity of sound [in air]—about 332 metres per second at 0°C.

It is a direct consequence of observation that, within wide limits, the velocity of sound is independent . . . of its intensity, and also of its pitch. . . . But when the disturbances are very violent and abrupt, so that the alterations of density concerned are comparable with the whole density of the air, the simplicity of this law may be departed from.

Although, in practice, air is usually the vehicle of sound, other gases, liquids, and solids are equally capable of conveying it. . . . In the year 1826, Colladon and Sturm investigated the propagation of sound in the Lake of Geneva. . . . At a temperature of 8°C., the velocity in water was thus found to be 1435 metres per second.

In an open space the intensity of sound falls off with great rapidity as the distance from the source increases. . . . Anything that confines the sound will tend to diminish the falling off of intensity . . . ; the most effective [method] of all is a tube-like enclosure which prevents spreading altogether. The use of speaking-tubes . . . is well known.

Before proceeding further we must consider a distinction, which is of great importance, though not free from difficulty. Sounds may be classed as musical or unmusical: the former for convenience may be called *notes* [tones] and the latter *noises*. The extreme case will raise no dispute; every one recognizes the difference between the note of a pianoforte and the creaking of a shoe.

Musical sounds arrange themselves naturally in a certain order according to *pitch*—a quality which all can appreciate to some extent . . . .

Many contrivances may be proposed to illustrate the generation of a musical note. One of the simplest is a revolving wheel whose milled edge is pressed against a card [Savart]. Each projection as it strikes the card gives a slight tap whose regular occurrence, as the wheel turns, produces a note of definite pitch, rising in the scale, as the velocity of rotation increases. But the most appropriate instrument for the fundamental experiments on notes is undoubtedly the Siren, invented by Cagniard de la Tour.

One of the most important facts in the whole science is exemplified by the Siren—namely, that the pitch of a note depends upon the period of its vibration. . . . In passing from any note to its octave, the frequency of vibration is doubled.

The French pitch makes $a' = 435$. In Handel's time the pitch was much lower. If c' were taken at 256 or 2^8, all the c's would have frequencies represented by powers of 2. This pitch is usually adopted by physicists and acoustical instrument makers, and has the advantage of simplicity.

The determination *ab initio* of the frequency of a given note is an operation requiring some care. The simplest method in principle is by means of the Siren. . . .

Let us consider the disturbance due to a simultaneous sounding of a note and any or all of its harmonics. By definition, the complex whole forms a note having the same period (and therefore pitch) as its gravest element. . . . In point of fact, it has long been known to musicians that under certain circumstances the harmonics of a note may be heard along with it, even when the note is due to a single source, such as a vibrating string [Sauveur]: but the significance of the fact was not understood. . . . It has been proved (mainly by the labours of Ohm and Helmholtz) that almost all musical notes are highly compound, consisting in fact of the notes of a harmonic scale. . . .

The effect of the harmonic overtones is then to modify the *quality* or *character* of the note independently of pitch. [*Footnote*: German, *Klangfarbe*— French, *timbre*.]. . . . Musical notes may thus be classified as variable in three ways: First, *pitch*. Secondly, *character*, . . . : and thirdly, *loudness*.

The knowledge of external things which we derive from the indications of our senses, is for the most part the result of inference. . . . The whole life of each one of us is a continued lesson in interpreting the signs presented to us, and in drawing conclusions as to the actualities outside.[18]

THE AMERICAN SCHOOL OF ACOUSTICS

As we trace the development of the famous schools of scientists, we notice that such schools do not "just grow." There are several factors which may be conducive to such development: patronage, sufficient wealth in the society to allow men to devote time and talents to research, or the influence of an able and progressive leader. In the Florentine Academy, for example, the fostering factors were the patronage of Leopold and the influence of Galileo and Torricelli.

Until about 1750 North America made few contributions to science, partly because her people were still battling with the new environment to secure the means of existence. Since the middle of the nineteenth century, however, an important school of acoustics has developed in North America, and as usual it began not with theory but with a practical invention, namely, Edison's phonograph. Important results of this school in acoustics were obtained by Edison, Mayer, Sabine, Miller, and Knipp.

It is said that of all Edison's inventions, the one that gave him the greatest personal satisfaction was the phonograph, because of the entertainment, culture, and happiness it can bring to home life. Some details of the instrument are given in his patent papers:

December 24th, 1877.

Patent no. 200,521. To all whom it may concern, Be it known that I, Thomas Alva Edison, of Menlo Park . . . have invented an Improvement in Phonograph or Speaking Machines, of which the following is a specification. The object of this invention is to record in permanent characters the human voice and other sounds, from which characters, such sounds may be reproduced and rendered audible again at a future time. The invention consists in arranging a plate, diaphragm (*G*), or other flexible body capable of being vibrated by the human voice or other sounds, in conjunction with a material capable of registering the movements of such vibrating body by embossing or indenting or altering such material in such a manner that such register-marks will be sufficient to cause a second vibrating plate or body to be set in motion by them, and thus reproduce the motions of the first vibrating body. . . . *A* is a cylinder having a

18Lord Rayleigh, *Theory of Sound*, I, i.

helical indenting groove, cut from end to end, say, ten grooves to the inch. Upon this is placed the material to be indented, preferably metallic foil. *B* is the speaking-tube or mouth-piece. The clock-work is set running and words spoken in the tube *B* will cause the diaphragm to take up every [?] vibration, and these movements will be recorded with surprising accuracy by indentations in the foil.... *C* is a tube similar to *B*.... When the cylinder is allowed to rotate the spring *D* is set in motion by each indentation corresponding to its depth and length. This motion is conveyed by the diaphragm *F*...and the voice of the speaker is reproduced exactly and clearly.... It is obvious that many forms of mechanism may be used.... For instance, a revolving plate may have a volute spiral cut on both its upper and lower surface....

This instrument is an example of a comparatively simple contrivance producing astonishingly complex results, namely, the reproduction of speech and song. Even its own inventor was so pleasantly

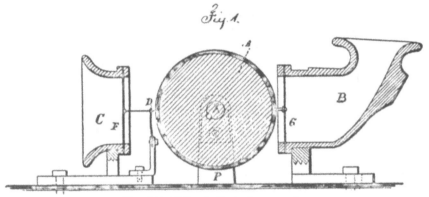

FIG. 144. Edison's first phonograph.

"surprised" at his own success as to display emotion in a patent paper! Another instance of the same sort was the first telephone. That is why a great physicist once said of Graham Bell that if he had known any physics, he would never have invented the telephone. The answer to the riddle in both cases is the property of matter called elasticity. It is interesting to note that although the first Edison phonograph used cylinders as records, yet Edison foresaw the use of discs.

An elementary text, entitled *Sound*, was published in 1881 by Professor Alfred Mayer of Philadelphia. It is remarkable for the simplicity, inexpensiveness, and effectiveness of its home-made apparatus. One of its experiments is a modification of a famous one performed first in 1876 by August Kundt of Würzburg, Germany, and named after him.[19]

[19]*Poggendorf Annalen* (1876), CLVII, 353.

Experiment 33. The chimneys of student-lamps have a fashion of breaking just at the thin narrow part near the bottom. Such a broken chimney is very useful in our experiments. At A in the figure [fig. 145] is such a broken chimney, closed at the broken end with wax, W. A cork is fitted to the other end of the chimney, and a hole bored through its centre. In this hole is inserted part of a common wooden whistle. Inside the tube is a small quantity of very fine precipitated silica [or cork dust]. Hold the tube in a horizontal position and blow the whistle. The silica powder springs up into groups of thin vertical plates, separated by spots of powder at rest, as in the figure. This is a very beautiful and striking experiment.[20]

Mayer's last remark in this quotation is decidedly no overstatement. The observer beholds particles actually lifted up in mid-air and held there "without visible means of support," yet vibrating ecstatically, parallel to the long axis of the tube. This observation shows that sound waves are not transverse like pond-waves but longitudinal, for as the cork particles indicate, the air particles vibrate in a direction

FIG. 145. Mayer's experiment with a cork-filing tube.

parallel to that in which the sound waves are travelling through the tube. The characteristic phases of longitudinal waves are *pulses of compression* and *pulses of rarefaction*.

When the sound of the fife ceases, the cork particles drop down in a pattern of windrows indicated in Mayer's diagram (fig. 145). These seem to indicate nodes and loops and, therefore, the existence of standing waves in the air, produced by the superposition or interference of two longitudinal wave-trains in opposite sense as in Melde's experiment. A set of larger nodes are found to be about 6 cm. apart. Hence the wave-length of the sound producing them, L, $= 2\times6 = 12$ cm., by the standing wave formula (*Eq.* 2, p. 189). Therefore,

the pitch of the tone $f = \dfrac{v}{L}$ by Newton's wave formula (*Eq.* 1, p. 174)

$$= \frac{34400}{12} \text{ (p. 170)} = 2866 \text{ d.v.p.s.}$$

A tone with this frequency is in the top octave of the piano scale and is in this case the fundamental of the fife's tone.

The fine lines are 2 mm. apart. If they are loops (a disputed point), the wave-length of the tone producing them can be determined thus:

[20]A. Mayer, *Sound* (Macmillan, 1881).

$L'' = 2 \times 2 = 4$ mm.; and the pitch of such a tone

$$f'' = \frac{v}{L''} = \frac{344000.}{4} = 86,000 \text{ d.v.p.s.}$$

Such a tone would be a suprasonic or ultrasonic. In that case, the eye sees an effect produced by a tone which the ear cannot hear. All this and more, Mayer's experiment obtains from a few cents' worth of very simple apparatus. Whatever the explanation, Mayer was right in declaring that the experiment is well worth seeing.

Mayer's simplification of Wheatstone's rotating mirror (p. 191) was quite characteristic. He placed two plane mirrors back to back with a vertical rod between them and twirled the system by hand, moving his open palms past opposite sides of the rod in opposite sense.

ARCHITECTURAL ACOUSTICS

The following selection is from a lecture on *Architectural Acoustics* delivered in 1915 by Professor Wallace C. Sabine of Harvard University. Almost without exception, the scientists we have met thus far were impelled to their work, even in spite of great difficulties and discouragement, by their enthusiasm and the urge of that intellectual appetite called curiosity. Of course, there were sometimes other goals also, such as the application of their findings to human affairs and the justifiable expectation of adequate remuneration for services performed. But in this instance we find a scientist who undertook a research mostly as an unavoidable job of work. It was discovered that the acoustics of the Harvard auditorium was atrociously bad. Something had to be done. On whom should the task be imposed if not the professor of physics? As the work progressed, however, Sabine became absorbed in the problem, then enthusiastic, and finally emerged as one of the most eminent authorities on architectural acoustics of his day.

The conditions surrounding the transmission of speech in an enclosed auditorium are complicated, it is true, but are only. such as will yield an exact solution in the light of adequate data. It is, in other words, a rational engineering problem. . . . The solution of the problem should be quantitative, not merely qualitative. . . . It should be such that its application can precede not merely follow the construction of the building. . . . Scientifically, the problem involves three factors, reverberation, interference, and resonance.

Sound, being energy, once produced in a confined space, will continue until it is either transmitted by the boundary walls or is transformed into some other kind of energy, generally heat. This process of decay is called absorption. . . In the lecture-room of Harvard, . . . the rate of absorption was so small that a word spoken in an ordinary tone of voice was audible for five and a half seconds. . . . With an audience filling the room, the conditions were not so bad but still not tolerable. This phenomenon will be called reverberation, including as a special case, the echo. . . . Broadly, . . . there are only two variables in a room, . . . shape (including size) and materials. . . .

The first work was to determine the relative absorbing power of various substances. . . . On bringing into the lecture-room a number of cushions having a total length of 8.2 metres, the duration of audibility fell to 5.33 seconds. . . . When all the seats (436) were covered, the sound was audible for 2.03 seconds. The coefficients of absorption were determined as follows—open window . . . 1.000; hard pine . . . 0.061; plaster on lath . . . 0.034; glass, single thickness . . . 0.027; plaster on tile . . . 0.025. . . .

The formation and propagation of echoes may be admirably studied . . . by the construction of a model of the auditorium. . . . As the sound is passing through it, it is illuminated instantaneously by the light from a very fine electric spark. . . .

It is entirely possible to calculate in advance of construction whether or not an auditorium will be good, and, if not, to determine the factors contributing to its poor acoustics and a method for their correction.[21]

Sabine derived a working rule, called Sabine's law, for determining the time of reverberation for any auditorium, namely,

$$t = 0.05 \left(\frac{V}{a_1s_1 + a_2s_2 + a_3s_3 + etc.} \right) \quad \text{where } V \text{ cu. ft. is}$$

the volume of the room, a_1, a_2, a_3, etc., are the absorption coefficients of the materials exposed, and s_1, s_2, s_3, etc., their respective areas in square feet.

Another prominent member of the American school of acoustics is Dr. D. C. Miller of the Case School of Applied Science, Cleveland, Ohio, who published in 1916 a fascinating book on *The Science of Musical Sounds*,[22] from which the following excerpts are quoted:

Noise and tone are merely terms of contrast, in extreme cases clearly distinct, but in other instances, blending: the difference between noise and tone is one of degree. The drawing of a cork from a bottle expands the contained air; when the cork is wholly withdrawn, the air, because of its elasticity, vibrates with the frequency dependent on the size of the bottle. The resulting sound is of short duration and is thought of only as a popping noise while it is in reality a musical tone. The musical characteristic is made evident by drawing the corks from several cylindrical bottles . . . the tones of which are in the relations of the common chord, do, mi, so, do.

The simplest possible type of vibration which a particle of elastic matter can have . . . is called simple harmonic motion; it takes place in a line . . . the middle of which is the position of rest of the particle; when the particle is displaced from this position, elasticity develops a force tending to restore it, which force is directly proportional to the amount of the displacement [Hooke's law]. If the displaced particle is now released, it will vibrate to and fro with simple harmonic motion. The name originated in the fact that musical sounds in

[21]Reprinted by permission of the publishers from W. C. Sabine, *Collected Papers on Acoustics* (Harvard, University Press, 1922).
[22]D. C. Miller, *The Science of Musical Sounds* (Macmillan, 1916).

general are produced by complex vibrations which can be resolved into component motions of this type.

In order to analyse complex tones with greater precision than was possible by König's manometric capsule and to record the facts

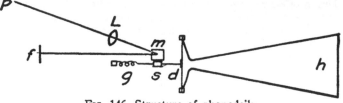

FIG. 146. Structure of phonodeik.

photographically, Dr. Miller designed and constructed the *phonodeik* (fig. 146) which his book describes:

> The sensitive receiver . . . is a diaphragm *d* of thin glass placed at the end of a resonator horn *h*; behind the diaphragm is a minute spindle mounted in jewelled bearings, to which is attached a tiny mirror *m*; one part of the spindle is fashioned into a small pulley; a string of silk fibres . . . is attached to the centre of the diaphragm and, being wrapped once around the pulley, is fastened to a spring *g* . . . ; light from a pinhole *p* is focused by a lens *L* and reflected by the mirror to a moving film *f* in a special camera. . . . The phonodeik responds to 10,000 complete vibrations per second.

KNIPP TUBES

For vivid demonstration of beats no instrument can compare with the sonorous tubes invented in 1918 by Dr. Charles T. Knipp of the University of Illinois. Here is part of his report:

> The new source of sound that was described and exhibited for the first time by the writer before the American Physical Society presents many interesting features. The figure [fig. 147] pictures one type of the tubes used. It is a

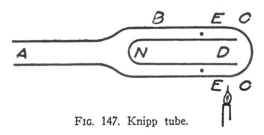

FIG. 147. Knipp tube.

resonator with a loop at *A* and a node at *N*, so that the distance *ABCN* constitutes approximately one fourth of the wave-length of the sound given out by the tube when operating. The air surges back and forth at *A* with the greatest velocity and displacement. From this point, the to and fro motion of the air grows gradually less until it becomes zero at *N*. . . . If the motion is to be maintained, energy must be supplied to the vibrating system.... This energy is furnished at CC by a gas flame.... The motion is maintained first by the heating effect [expansion] at D and a half period later, by the cooling effect [contraction] at EE.... Demonstrations that can be made with a Knipp tube are: 1. conversion of heat to sound; 2. production of sound by vibrating an air column; 3. relation between pitch and length of air column; 4. with a pair of tubes, the phenomenon of beating. The sound produced is very penetrating and when beats are produced by the proper tuning of the shorter tube, the intensity of the beats is so great as to be almost painful.[23]

THE WORLD OF SOUND

Our study of acoustics could not be concluded more acceptably than by a visit with Sir William Bragg of London, who was President of the Royal Society in 1938. In 1921 he published a little book entitled, *The World of Sound*, from which our quotations will be taken.[24] Books of popular science, fortunately, are increasing in numbers. The leaven of science cannot too soon leaven the whole lump. Such books, of course, are not all of equal merit; they run the risk of pandering so far to popularity as to become a shade weak in their science. One of the happiest combinations of entertaining treatment and irreproachable science is *The World of Sound*. Those who have not read it are defrauding themselves, for it is one of the most instructive and charming books ever written.

How shall we detect the submarine?.... Unless we can detect at a mile or so with fair certainty, we are practically helpless. The position would be like that of a very short-sighted golfer who had not seen which way his ball went, the ball being probably able to see him and, in any case, anxious to keep out of his road.... There is one very simple way of transferring the sound which can readily be tried; it was used long ago by Colladon when he measured the velocity of sound on the Lake of Geneva. It is enough to lower into the water a hollow body, such as a tin can or a rubber ball or even a rubber tube, plugged at one end, and connect the air inside with the ear by a sufficient length of tubing or pipe: a stethoscope makes a convenient finish, but is not absolutely necessary.... If we take one of the instruments we have devised and try to use it on board ship, we find that the water around the ship ... is full of noise which may "drown" the sounds we want to hear.... Under such conditions it is as difficult to hear a submarine as it would be for the occupant of a motor car to hear a sparrow moving in a hedge as he went by. And in war it is generally too dangerous to stop for the purpose of listening.... The modern submarine can go very quietly if it does not go too fast. Most of the noise of a moving ship comes from the screw.... Through it all, we hear the beat of the engines or the whirr of the turbines or other special noises.... If the screw goes slowly, ... there is very little noise.... We see, therefore, that the transference of sound from the sea is a matter of many difficulties. In practice, it is found that ways of avoiding them can be devised— But I may not follow the subject further in that direction.

[23]*The Physical Review*, xv, 155.
[24]Sir William Bragg, *The World of Sound* (Bell, 1921; in U.S., Dutton).

In the last sentence Sir William seemed to clamp down the lid suddenly just when our curiosity was at its height. If you wonder why, it may be helpful to know that he was a consulting scientist to the Admiralty Board during the First World War.

There are those who consider the ear the most wonderful instrument in the world. Sir William is not the only scientist who has attempted to express his astonishment at the marvels it performs.

FIG. 148. Sir William Bragg discussing science with a juvenile audience in the Royal Institution, London.

We say that we hear a sound, which means that somewhere or other an air quiver has been started and has reached our ears. As the life and processes of the world go on, the actions which take place are accompanied by these tremors, and we live in this world of sound. We can interpret what we hear because all the tremors are different and we have learned to know them all. We can tell the sort of tremor that is made by the rustle of the leaves from the sort that is made by thunder or the call of an animal. In fact, it seems quite absurd to think that there is anything wonderful in it, because the sounds seem so different. But, of course, that is just where the wonder lies; only air tremors in every case, and yet the ear has such marvellous powers that it can sort them all out from each other, can tell one person's voice from another, can tell one word from another, can even tell by the minutely differing shades of inflection, the spirit that lies behind the word. The more one thinks about it, the more wonderful one finds it to be.

In one passage of the book, Sir William very neatly labels a certain group of sounds "stick and slip" sounds because of the method by which they are produced.

Sometimes when we walk on the sands of the seashore we find it giving out curious sounds as it shifts under the pressure of our feet. They are squeals more nearly than anything else. . . .

There is little doubt that the sound belongs to the "stick and slip" noises. A pencil squeaking on a slate is a familiar sound which is caused by the pencil sticking and slipping quite regularly as it is pushed across the slate. In fact, if it is held lightly as it is pushed we can see a regular succession of dots . . . . Little toys which imitate the singing of a bird are made of a piece of soft metal turning in a wooden seat. These work in the same way. A bearing squeaks when it wants oil; and the oil turns the jumping motion into a steady one. . . . When the bow is drawn across the string of a violin, there is an action of a similar kind. . . .

SCIENCE AND RELIGION

As an illustration of the statement that the ken of many scientists extends well beyond the confines of their laboratories and of their own special sciences, we could hardly do better than quote the closing paragraph of Sir William Bragg's *World of Sound*:

There is just one point more. In what I have been saying I have had in mind more than the applications to times of war. It is the fact that in our lives, in all that we work at and strive for, it is of first importance to know as much as we can about what we are doing, to learn from the experience of others, and, not stopping at that, to find out more for ourselves, so that our work may be the best of which we are capable. That is what science stands for. It is only half the battle, I know. There is also the great driving force which we know under the name of religion. From religion comes a man's purpose; from science, his power to achieve it. Sometimes people ask if religion and science are not opposed to one another. They are: in the sense that the thumb and fingers of my hand are opposed to one another. It is an opposition by means of which anything can be grasped. It is right, therefore, with all our heart to learn what will help us in the work we want to do, so that when the call comes we can say, "I am here and ready; I want to play my part, and I have tried to fit myself to play it well."

CO-OPERATIVE RESEARCH

Before turning to the Story of Optics let us glance at two trends in modern science which are well illustrated by a group of researches conducted in the Bell Telephone Research Laboratories. One of these trends we have met in passages which contrast theory and practice, for example in the writings of da Vinci. It might be called the theoretical approach to practical problems. The second is more novel in science though not new in some departments: it might be called collaborative research. The two trends are not unrelated and can be mutually helpful.

In a direct utilitarian way, a telephone company would be interested in improving its system of telephones. Such a system may be divided into three categories, transmitters, intercommunicating conductors, and receivers. The knowledge of electric currents and of vibrating bodies had been brought to a considerable degree of perfection about four decades ago. The need for research concerned, therefore, chiefly transmitters and receivers. Instead of proceeding with a number of isolated problems, the Bell Telephone Co. set out upon a project of wide scope—an investigation of speech and hearing—under the guidance of Dr. H. D. Arnold and Dr. Harvey Fletcher. An account of valuable progress achieved by the research staff is given in Fletcher's book, *Speech and Hearing*.[25]

At first glance such a choice of objective, involving physiological and psychological difficulties, might seem unnecessarily broad, rather theoretical than practical, even visionary. The results show, however, that the choice was decidedly fruitful and sagacious and has justified the heavy cost of the undertaking. Similarly Archimedes showed King Hieron that the best way to solve the practical problem of launching a huge trireme was to know the theory of pulleys and to apply that knowledge in practice.

In order to investigate speech and hearing it became necessary to improve existing instruments and to devise new ones with new functions and new degrees of accuracy and precision. Many of these new developments have since been employed to great advantage not only in telephone systems and in other acoustical problems but also in the improvement of the phonograph, in radio, in aid for the deaf, in sound-movies, in an instrument which enables patients to speak whose vocal cords do not function, in the solution of some war problems, in the introduction of a useful unit of intensity level, the *decibel*, named after Graham Bell, in a more accurate determination of the limits of audibility for human ears at 20-20,000 cycles per sec. These will serve to give a hint of the results which have spouted forth, as from an oil-gusher, in the Bell Telephone programme of research. Dr. Fletcher's confidence in the power of this method in science is seen in his remark, "no such persistent and thorough-going study can be carried through without [making] large additions to the philosophy of the subject."

It is easy to see that few of the great scientists whom we have interviewed would have had adequate funds to conduct a research of such magnitude, or time to achieve such a large quota of results. We can see how tragically the work of some investigators has been

[25]Van Nostrand, 1929.

hampered by lack of funds, for instance that of the Curies. Success in the Bell Telephone researches has been gained by the concerted labours of a large group of researchers backed in their efforts by the financial support of a wealthy organization. The production of the atomic bomb is another illustration of collaborative research.

Science has here given democracy an object lesson. If all members of a community could apply the same concerted effort in peace as in war, there would be simply no limit to the magnitude and wonder of the benefits that could be gained for all. If instead of proceeding selfishly by the jungle law of tooth and claw, we could collaborate like an army or the artisans of a building, we could make war on such difficulties as have to do with, say, noxious animals, plants and viruses, control of weather and climate, forest fires and reforestation, poverty and the just distribution of wealth, monopolies and cartels, educational problems, crime reduction, and on a host of enemies which could be controlled or eradicated. In solving such problems we would gain tremendous returns for our investments of labour, time, and money, whereas in war the chief return for our heavy investment is bitter ceaseless anguish, the destruction of life and property, and the loss of many priceless values. Such collaboration would certainly help to "make the world a better place in which to live."

THE STORY
of Optics

CHAPTER NINE

Through Ancient Eyes

W E all have a considerable fund of practical knowledge concerning optics, the science that deals with visible bodies.
Some of this knowledge is as old as the human race, but much of it
must be learned anew through experience by each individual for himself. In the stories of mechanics and of acoustics we found in prehistoric
times a preliminary practical period, which was characterized by the
invention of machines and instruments. Studying the same period
from the standpoint of optics, we shall look for the ancient gathering
of optical facts and for the invention of optical devices.

If there is one visible body which is, in general, more important in
human affairs than any other, it is the sun. Naturally, the sun became
in prehistoric times a prime object of worship, as indicated by fig. 11,
p. 9—a Sumerian pictograph signifying "day" and also "worship."
The Egyptian sun-emblem of fig. 149 also suggests the veneration
and awe in which the sun-god, Ra, was held in Egypt (fig. 151).

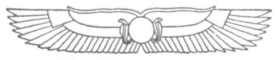

FIG. 149. Egyptian sun-emblem.

A familiar sketch of the dawn which often adorns the exercise-
books of artistic children (fig. 150) is partly factual and partly
imaginative. The sun, trees, horizon, and the three black crows are
visible bodies, but the sun's features and the strokes emanating from
his jolly face are largely imaginative or hypothetical. It is true that
occasionally similar streaks, called sunbeams, are observed where
water-drops are suspended in the sky, each streak consisting of
visible straight rows of water-drops. But when the sun's disc is visible,
sunbeams are rather rarely seen. The young artists of such pictures,
however, put the streaks in the drawing whether they can see them

at the moment or not. What the streaks represent and what the picture assumes as radiating from the sun in all directions is called *light*; the sun is called a luminous body (Lat. *lumen*, light) and the

same term can logically be applied to any visible body. The same assumption is implied and represented with impressive artistic power in fig. 149. The science of optics is commonly referred to as the study of light. The term *luminous* is applied

FIG. 150. Dawn.

also to surfaces and points. Thus a luminous surface may be considered a system of luminous points. That the idea of light is ancient is shown by the fact that the word is found in some of the earliest writings, e.g. in the first chapter of Genesis, the word *o'er* (אוֹר). In the old Persian religion, Zoroastrianism, the supreme deity was the god of light, Ahura-mazda (fig. 152) after whom some modern lamps are named.

FIG. 151. Ra,
a sun-god

FIG. 152. Ahura-mazda.

Any line or path along which light passes is called a *ray*. There is a resemblance between this term and the name of Ra, the sun-god. The same term ray has two other related meanings: it signifies a line drawn in a diagram to represent the path of light, and also the light itself which goes along the path. One must determine from the context in each particular case which meaning is intended. A *beam* of light is a bundle of rays travelling in similar sense. Several early observations,

FIG. 153. Searchlights.

such as the straight edges of sunbeams and the difficulty of seeing around a corner, led to the idea that light rays in air or in any uniform medium are *straight*. The Egyptian sun-emblem of fig. 149 suggests this rectilinear propagation of light. Visible straight lines are characteristic of much human handiwork as seen in the forms of buildings, roads, canals, instruments, pyramids, etc., but, in nature, apart from man, the visible straight line is rare. Examples are the sunbeam, the edge of a crystal, the brow of a waterfall, a vertical spider-thread in a web where there is no breeze—with these few items the list is complete. So characteristic is the visible straight line of man's handiwork that the straight lines on the surface of the planet Mars, the so-called canals, have been proposed by some as evidence of human-like beings on that planet. All processes of aiming and sighting imply the straightness of light rays.

In general, we rely more on the evidence of our eyes than of our ears. There is a saying that "seeing is believing," but there is no such saying about hearing. In fact, we are advised not to believe all we hear. The ventriloquist exploits this preference in producing the illusion of the wooden dummy talking.

It was pointed out in 1672 by Otto von Guericke that *light rays are invisible*, or in other words, that light is invisible. When a man sees a house, the bystanders do not see any rays passing from it to him; nor does he. What he sees is the house. That is the fact: the hypothetical (and convenient) expression of the fact is in terms of light rays. If we could see light rays, we should hardly be assuming their existence, for seeing would be believing. When the moon shines at night, we see no rays or sunbeams reaching her from the sun, nor moonbeams from her to us. The fact is that we see the moon. According to the hypothesis, the space surrounding the moon is filled with a flood of sunbeams, but it looks quite dark to us. You may hear the claim that the rays of a search-light (fig. 153) are visible: as a matter of fact, we see, along the edge of the beam, a rectilinear series of particles which have been rendered luminous by a ray.

FIG. 154. Aten's voyage.

An Egyptian representation of the sun embarking on his daily voyage across the sky is given in fig. 154. This drawing was made about 1500 B.C. The sun's disc, called by the Egyptians Aten, rests in a self-propelled boat and floats through the sky under the all-seeing "Eye of Horus" or "Eye of Heaven." When Aten sinks in the west, a great host of bodies, such as fields, roads, trees, stones, birds, flowers, and so forth, become invisible bodies. They are then called dark bodies and are said to have darkened. They are also said to be eclipsed. But next morning when Ra returns from his hidden journey through the underworld, those dark bodies are transformed at his magic touch to visible or luminous bodies. We may also call them dependently luminous bodies with the sun as their source of light. What light they send to the eye, they have just received from the sun or from a cloud or some other body whose source of light is the sun. Thus a dependently luminous body can be a source of light to another dependently luminous body as, for example, when the moon illuminates a landscape. The beautiful old myth represents the earth as the Sleeping Beauty and the sun as the Prince.

The eclipse of the landscape after sunset is caused by the intervention of the "shoulder" of the earth. The latter is accordingly described as *opaque* (Lat. *opacus*, shadowy). When an opaque body intervenes between a luminous body and the eye, for example a

Fig. 155. Day and night.

book between a lamp and the eye, the lamp is no longer visible to that eye and is also said to be eclipsed; but it is not darkened nor is it eclipsed to all eyes. In terms of light, we express this fact by saying that the rays are stopped by the opaque body or that they cannot traverse it and do not reach the eye. Thus there are two types of eclipse and each is illustrated by a lunar or a solar eclipse. In the lunar eclipse (fig. 157), the earth comes between the moon and its source of light the sun, and the moon is darkened. In a solar eclipse (fig. 158), the moon comes between the sun, a luminous body, and the eye of an observer on earth. In both cases, a body, V, intervenes between two others. In the first type of eclipse, V comes between a dependently luminous body, D, and its source and thereby darkens D. In the second type, V comes between a luminous body, L, and the eye of an observer, E; in this case L becomes invisible to E but is not darkened. That both types of eclipse occur only when the three bodies are in a straight line lends further support to the idea that light rays in a uniform medium are straight.

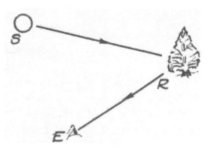

Fig. 156. Reflection of light.

The sun cannot be darkened by the interposition of any body between it and any other luminous body. If, after sunset, we telephone to friends in the west, we learn that where they are the sun is shining. If we describe a solar eclipse in a letter to friends who live two hundred miles north or south of us, they write back saying that on that day the sun was not eclipsed for them. Accordingly, the sun is called self-luminous. It emits light independently of other luminous bodies. Other examples of self-luminous bodies are a camp-fire, a grate fire, a fire-fly, a flame, a flash-light (when it "works"), a search-light, a glowing neon-tube, and so forth; but altogether, these are much less numerous than dependently luminous bodies.

FIG. 157. Lunar eclipse.

All this grand pageantry of changing appearances which depends on the coming and going of the sun, is expressed in the ancient monosyllables, Day and Night. In fig. 155 *T* is the twilight line, the boundary between night and day, where "rosy-fingered dawn" and the gloaming are seen.

The light from the tree in fig. 156 must pass through the air in order to reach us; and because the air is between us and the tree, it is called a *medium* (Lat. *medius*, between). The air is said to be transparent to light or, briefly, transparent. Similarly, the medium between the sun and the earth is transparent whether it be empty space or not.

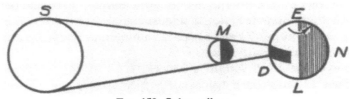

FIG. 158. Solar eclipse.

Since the tree sends us light which it has just received from the sun, the light must rebound from the tree *back into the same medium*, the air. Such a change of direction is called *reflection*; and the tree is called a reflector of light. It is dependently luminous by reflection.

When, at full moon (fig. 157), the earth *E* comes between the moon *M* and the sun *S*, observers situated where it is night see the moon darkened and eclipsed. At the same time, people on the opposite side of the earth experience daylight. This is an eclipse of the first type. The earth is opaque, the moon dependently luminous by reflection, and the sun its source of light.

Fig. 159. Palaeolithic artists at work.

When the new moon comes between the sun and the earth (fig. 158), the sun is eclipsed for those observers who are within a certain strip of the daylight surface of the earth, i.e. the hemisphere on the sunny side of the twilight line; but to most people on the "day-side" of the earth the sun is visible throughout the whole event, (except, of course, where clouds interfere). This is an eclipse of the second type. Chaldean astrologers observed as early as 3000 B.C. that solar eclipses occur at new moon and, of course, during daytime, whereas lunar eclipses occur at full moon during the night.

After sunset, when a white tent or any white, dependently luminous body is changing into a dark body, it goes through a series of changes indicated briefly by the words white, gray, and black. These are described as *colour* terms. The tent becomes less and less visible. The light from it becomes fainter and fainter or less and less brilliant. This change is called a decrease in *intensity*. The *intensity of illumination* of the tent decreases and so does the intensity of the light coming from the tent to the eye. The intensity of an utterly or ideally dark surface is zero, and as a surface darkens its intensity of illumination approaches zero. The intensity of a source of light is often referred to as its *power of illumination* or, briefly, its power. A common unit of power of illumination is one candle-power. A common unit of intensity of illumination is the brightness of a white screen placed one foot away from a source of one candle-power; this unit is, therefore, called one foot-candle.

Having found in other studies that man, the "weapon-animal," is the maker of machines, one naturally asks whether he has succeeded in devising any optical instruments. The increase which man has effected in the number and power of artificial sources of light is among the greatest

Fɪɢ. 160. Palaeolithic lamps, made of shell and stone. Mesopotamian, early 3rd millennium ʙ.ᴄ.

advances he has made in his control of nature. Instead of having to desist at sunset from activities requiring sunlight, he can virtually extend the length of day locally to any degree he wishes. The palaeolithic artists (fig. 159) who painted such pictures as the bison and mammoth of figs. 2 and 3, page 4, on the walls of caves which received no sunlight, used fires and shallow *lamps* (fig. 160) to obtain artificial lighting for their work. The illuminants used were fats and greases from meat and the wicks were wisps of hay or slivers of wood. Incidentally, this was an early use of capillarity (p. 72). Greek and Roman lamps (fig. 162), from which the design of the "lamp of learning" is obtained, were not very different from those of the Stone Age, although the intervening lapse of time was thousands of years. Nowadays, if we wish to see at night, we usually have only to press a switch, and night turns into day in our vicinity. For great boons, Nature is sufficiently niggard to exact exorbitant prices, and, in this instance, the oculists and optometrists warn us that we must beware lest the luxury of artificial lights be not at the cruel cost of our eyesight.

Fig. 161. Egyptian metallic mirror.

Images produced by reflection in water were well known in pre-historic times, as indicated by the myth of Echo and Narcissus. Mirror images were also known to gold- and silver-smiths even before the copper age, for when these metals are pure and molten their surfaces act as mirrors. In fact, one of the refiner's tests for freedom from dross was the appearance of his image in the liquid. It was also a prehistoric observation on the part of close friends that a person could see his or her image in the eyes of another person, under favourable circumstances. A metallic mirror like that of fig. 161 has been found in an Egyptian mummy-case dating about 2000 B.C.

Another old idea that is important in optics is represented by the word *shadow* (Anglo Saxon *sceadu*, dimness). The fact that the edges of shadows of posts or straight tree-trunks are straight was one source of the idea that light rays are straight lines. Several practical examples of this idea were known to primitive man, for instance that, where he entered a shadow, some object, a rock or a tree, intervened between him and the source of light, the sun, the moon, or a camp-fire. The event of the man entering the shadow of a tree, cast by a camp-fire, illustrates both types of eclipse at once. The tree intervening between him and his source of light, the camp-fire, darkened him (eclipse of type I). When he entered the shadow, he could no longer see the camp-fire; it was eclipsed to him (type II). Knowing that concealment is obtainable by hiding behind a bush or hill, he also knew by the same token that an unseen foe might lurk behind any opaque object of sufficient size. Such knowledge seems self-evident to us, for a baby learns it when playing "peek-a-boo" and a young child in the game of hide-and-seek; hence we cannot remember the time when we were

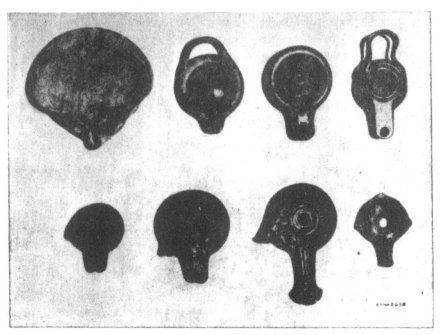

FIG. 162. Greek lamps, showing development of form. Sixth to First Centuries B.C.

not aware of such ideas. From experiences like these are abstracted such concepts as eclipse and opaqueness. In contrast with opaqueness is the quality of transparency, known to the savage by his discovery that stones and fish can be seen in water, and ferns when imbedded in ice, and also in the rarer observation that a camp-fire can be seen through a piece of quartz or rock-crystal. That objects can be seen through air was probably seldom if ever considered, for we still say that a tumbler is empty when it is full of air. There was of course the observation that objects are difficult or impossible to see through foggy air. A sheet of paper is translucent; it eclipses the source but becomes luminous by transmission. One factor of perspective is eclipse.

Primitive man was very much interested in his shadow,[1] in addition to his use of it as a clock. One Egyptian belief was that the human personality is a trinity, composed of the body, its shadow, and the spirit. They believed that in the resurrection these would be reunited if the body were still intact. Hence the importance of mummification. Kings of one dynasty, therefore, in order to insure their supremacy in the hereafter, regularly hired tomb-robbers to destroy the bodies of rulers of previous dynasties. The pyramid was a boast to the world, saying, "I, the great Pharaoh, am here but you cannot reach my body to destroy it. In the attempt, you would surely die a hideous death without my moving even a finger." Yet no pyramid escaped the ghouls; hence the acute interest in the tomb of Tutankhamen with its royal mummy and mortuary furniture fortuitously preserved. In optics and in other sciences we pay much attention to shadows but our interest is quite different from that of the Egyptians.

In early times, eclipses of the sun and moon were viewed as portents. In Mesopotamia, they were thought to presage the destinies of kings. Even now they are apt to produce eerie sensations and ominous forebodings in the beholder. It was the duty of royal Babylonian and Chaldean astrologers, even as early as 3000 B.C., to know of eclipses beforehand. The following quotation is from a cuneiform inscription found in the ruins of a palace in Mesopotamia. It is a report sent by a royal astrologer to the Assyrian king Sennacherib, about 700 B.C.

To the King, my Lord, thy servant Abil-istar; may there be peace to the King, my Lord. Concerning the eclipse of the moon of which the King, my Lord, sent an inquiry to me. . . . In the city of Akkad, . . . the observation was made and the Eclipse took place. This eclipse, to the King, my Lord, it sends peace. The eclipse of the sun . . . did not occur. That which I saw with my own eyes, to the King, my Lord, I send.[2]

From records extending over several centuries, the Babylonian astronomers discovered a cycle called the Saros cycle, of about eighteen

[1]See J. G. Frazer, *The Golden Bough*, I, p. 285.
[2]Quoted in G. Smith, *Assyrian Discoveries* (1875), p. 409.

years, or its treble, fifty-four years, in which solar and lunar eclipses occur. Thus they learned to predict the times of eclipses. Prediction, of course, has been from earliest times a chief goal of science. The most celebrated Babylonian astronomer was Kidinnu or Cidenas (343 B.C.) of Sippra on the Euphrates, who was a contemporary of Aristotle.

Our astronomers have calculated back to the time of Abil-istar, and they know that the solar eclipse which Sennacherib expected did occur, but Abil-istar was right, for it was not observable at Akkad.

Among the instruments invented and used by Chaldean astronomers were the water-clock or clepsydra, the sun-dial and a modification of it, the *polos*,[3] which served as a clock at night. The influence of those old astronomers is still present in modern affairs. The time required by the moon to make one "round-trip" about the earth, was called a "moonth," i.e. a month. It is about twenty-eight or twenty-nine days, depending on the method of measurement which is employed. Thus the moon was a "sky-clock," a measurer of time, as its name implies, for it is derived from the ancient root *me*, meaning "to measure," which is plainly seen in the word measure, and in the Latin word for month, *mensis*. It happens that the Latin name for the moon was *luna*, which is related to the words *lux*, light, and *lucere*, to shine.

At an early time the calendar of the Chaldeans or Babylonians had twelve months of thirty days each. In the convenience of having all months alike, they were actually better off than we are, after all these centuries of calendar reform. Such a calendar, of course, makes a year of 360 days. The Chaldeans discovered the ecliptic, the sun's annual path among the stars, along which eclipses occur. They divided the ecliptic into 360 equal parts called steps or degrees (Lat. *gradus*, step), one for each day, and thus inaugurated the division of the circle into 360 degrees. The first day when the dog-star, Sirius, whom the Egyptians called Sothis, showed on the eastern horizon before sunrise marked the end of one year and the beginning of the next. The interval between successive heliacal risings of Sirius was 365 days and a fraction (about one-quarter). There were different ways of taking up the slack of five days, which were called "the extra days." Some blotted the extra days from their memories by an orgy of wassail and feasting—chiefly wassail. In some regions, the slack was taken up by giving every sixth year an extra month.

That language is very conservative may be seen in the preservation of ideas from Chaldean astrology is some of our words and expressions,

[3]B. Farrington, *Science in Antiquity*, p. 29.

e.g. jovial, lunatic, mercurial, venereal, saturnine, mundane, disaster, influenza, zodiac, "under a lucky star," "my lucky stars," catastrophe, and so forth. And horoscopes can still be had (for a consideration).

Mesopotamian astronomers divided the ecliptic into twelve parts, one for each month, and displayed extremely nimble imaginations in giving a name to the cluster of stars (constellation) in each division. These constellations you have doubtless learned from your researches in *Dr. Chase's Almanac.* Here they are for reference:

Aries	the ram	♈	Libra	the balance	♎
Taurus	bull	♉	Scorpio	scorpion	♏
Gemini	twins	♊	Sagittarius	archer	♐
Cancer	crab	♋	Capricornus	goat	♑
Leo	lion	♌	Aquarius	water-carrier	♒
Virgo	virgin	♍	Pisces	fishes	♓

The zone including these is about 18° wide. The sun, moon, and planets are always in this belt. Three of these names are related to mechanics: the balance, the archer, and the water-carrier. The rest are zoological and so the zone is called the zodiac (Gk. *zōion*, animal).

Fig. 163.
Assyrian lens
of rock-crystal.

Some of the cuneiform inscriptions are in such minute characters that early archaeologists found it difficult to believe they could have been made without the aid of a *lens* or magnifying-glass of some sort. Finally, in 1885, in the ruins of a palace which King Sennacherib built at Nineveh about 700 B.C., the archaeologist Layard found a quartz lens. His report runs thus (fig. 163):

With the glass bowls was discovered a rock-crystal lens, ... the earliest specimen of a magnifying or burning-glass. I am indebted to Sir David Brewster ... for the following note: "This lens is plano-convex ... its length being 1 6/10 inch. It is about 9/10 ths. of an inch thick.... Its convex surface ... has been fashioned on a lapidary's wheel; the convex side is tolerably well polished and, though uneven, it gives a tolerably distinct focus at a distance of 4 1/2 inches.... It is obvious from the shape and rude cutting that it could not have been intended as an ornament: we are entitled, therefore, to consider it as intended to be used as a lens."[4]

The first lens was probably a natural one, a dew-drop. That a dew-drop can act as a magnifying-glass was probably first observed in prehistoric times. The word lens is derived from the name of the lentil plant, whose seed has the same shape as a double convex lens, convex on both sides and of circular outline. The derivation is seen more plainly in the adjective lenticular, meaning lens-shaped.

[4]Sir Austen Layard, *Nineveh and Babylon.*

There is good reason to believe that the Chaldeans, from their observations of the phases of the moon, knew that its bright part is always toward the sun and that it is, accordingly, a dependently luminous body with the sun as its source of light.

When a visible body first comes before our notice, we perceive not only its shape but also its *colour*—sometimes, the colour first. That is to say, in terms of the hypothesis, we note the colour of the light which the body emits. From the study of colour-blindness we obtain strong presumptive evidence that the first *hues* perceived by primitive eyes were blue and yellow, then, later, red and green (page 298). Objects appear to us in gray when perceived with the edges of the retina, especially the objects of a moonlit landscape. The fact that a considerable number of common colour-words are monosyllables: white, gray, black, blue, red, green, pink, bright, dark, or French *jaune, rouge*, German *gelb*, etc., suggests that they belong among the oldest words and signify old ideas. The myth of Iris, the beautiful messenger between Heaven and Earth, records the early observation of the hues of the rainbow, which was her ladder or speedway. Iris was a daughter of the legendary Electra, who was called "the shining one." Electra was brilliant but her daughter excelled her, for Iris was iridescent.

The colouring materials used in prehistoric paintings, on pottery and in fabrics, show that early man had an excellent practical knowledge of colour. Many evidences prove that there were in ancient Egypt, Babylon, and Assyria, artists and artisans whose practical knowledge and skill in colour were quite advanced, judged even by modern standards. No valuable theory of colour, however, was reached before the time of Newton.

The chief optical instruments invented in prehistoric times were the lamp, the mirror, and the lens. They underwent comparatively little improvement, and no other optical instruments were devised in the Babylonian and Egyptian cycles of culture. In the classical and Arabian cycles, we find the development of curved mirrors and some study of the eye and of lenses. It is in the Western cycle of culture, however, that we find an adequate theory of optics developed, the three prehistoric devices greatly improved, and many new optical instruments invented.

There are many passages in the Hebrew scriptures which refer incidentally to optical topics. In one the sacred writer describes the creation of light as one of the first of the Creator's fiats:

And darkness was upon the face of the deep. . . . And God said, Let there be light; and there was light. And God saw the light, that it was good: and God divided the light from the darkness. And God called the light Day and the

darkness he called Night. And the evening and the morning were the first day.
—Genesis i. 2-5.

This story of creation was probably carried down for centuries by oral tradition and first committed to writing about 1300 B.C., or a few centuries later.

Our second excerpt mentions plane mirrors, among the tabernacle furnishings (1500 B.C.):

And Moses said ... See, the Lord hath called by name Bezaleel ... and hath filled him with the spirit of God, in wisdom, in understanding, in knowledge and in all manner of workmanship.... Then wrought Bezaleel.... And he made the laver of brass, and the foot of it of brass, and the mirrors for the women assembling, which assembled at the door of the tabernacle.—Exodus xxxvi and xxxviii.

Even this old passage seems to single out a certain section of the human race as being more inclined to the use of mirrors.

In a third passage, Elihu, one of Job's "comforters," in attempting to reconcile the ideas of justice and omnipotence in the Deity, referred to a type of eclipse and to the mirror surface of molten metal:

I have yet to speak on God's behalf.... With clouds he covereth the light: and commandeth it not to shine by the cloud that cometh betwixt.... He directeth it under the whole heaven, and his lightning unto the ends of the earth. Hearken unto this, O Job, stand still and consider the wondrous works of God. Hast thou with him spread out the sky ... as a molten looking-glass?—Job xxxvi and xxxvii.

A fourth quotation refers to the rainbow. The date of this selection is about 2300 B.C.

And God said, This is the token of the covenant which I make ... for perpetual generations.... I do set my bow in the cloud, and it shall be for a token of a covenant.... And it shall come to pass, when I bring a cloud over the earth, that the bow shall be seen in the cloud.—Genesis ix. 13.

In the civilizations of the Near East, a practical knowledge of optics was acquired, but no theory of value was developed. This condition is fairly representative of many persons who enter modern elementary optics courses. They have a considerable store of optical facts, but their words disclose a vagueness and confusion not unlike that of ancient Egyptians. The quotations given in the remainder of this story will trace the path by which certain leaders, with many helpers, systematized and clarified the mass of data which had been gathered. We shall see that as an adequate theory was developed new vitality was evidenced by the production of many new optical instruments and by further advances in theory. We shall see that throughout this advancement a knowledge of mathematics was of crucial importance.

CHAPTER TEN

Optics in the Classical Cycle

THE springtime of a cycle of culture is characterized by a religion of the people, frequently a religion related to the soil, such as the worship of Demeter, the goddess of harvests, in the classical cycle. To this period belongs the Greek poet Hesiod of Helicon (700 B.C.). His *Cosmogony* gives an account of creation in which occurs the term *aether*, applied to the earth's uppermost and most tenuous atmosphere. In the story of mechanics Leonardo da Vinci referred to it as "the empyrean" or "region of elemental fire."

Verily, at first, Chaos came to be, but next. wide-bosomed Earth, the ever-sure foundation of all, the deathless ones who hold the peaks of snowy Olympus, and dim Tartarus in the depths of the wide-pathed earth, and Eros [Love], fairest among the deathless gods who unnerves the limbs and overcomes the mind and wise counsels of all gods and all men within them. From Chaos came forth Erebus [the Fury of discord] and black Night; but of Night, were born Aether and Day, whom she conceived and bare from union in love with Erebus.[1]

About twenty-four centuries after Hesiod, the term aether took on a new importance in the science of optics, as we shall find in our visits with Descartes, Huygens, Young, and others.

THE HEIGHT OF A PYRAMID

Diogenes Laërtius, in his biography of Thales, records an episode which signifies a useful contribution to optics:

Hieronymus informs us that Thales measured [*H*] the height of a pyramid [fig. 164], by the shadow it cast [*CR*], taking the observation at the hour when our shadow [*SW*] is of the same length [*h*] as ourselves.

As indicated in the diagram, Thales' assumptions were that the sun's rays are (i) straight lines, and (ii) parallel to each other. The geometrical theorem that he used was the principle of *similar triangles*, namely, that if two triangles are of the *same shape* (similar), all their

[1]From the translation by H. Evelyn White in the Loeb Classical Library.

corresponding linear dimensions are proportional to each other. They are mutually the miniature and enlargement of each other and if there is no distortion, the change in size must be in the same ratio throughout. In fig. 164, since $\triangle\ PCR$ is similar to $\triangle MSW$ or, in symbols, $\triangle\ PCR\ |||\ \triangle\ MSW$,

$$\therefore\quad \frac{H}{CR} = \frac{h}{SW} = 1,$$

where H and CR are respectively the height of the pyramid and the length of its shadow, and h and SW, the height of a man and the length of his shadow. $\therefore\quad H = CR.$

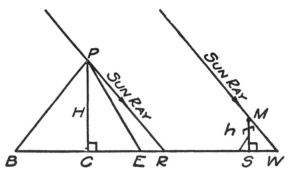

Fig. 164. Thales' determination of height of pyramid.

Thales' hosts, the Egyptian priests, were so delighted with this new trick which their pupil Thales had taught them in their own "dark subject," that they arranged to have Thales give a request performance before Pharaoh, Amasis II. As a matter of fact, it was not necessary to wait for any special time of day, for h can be measured at any time, and CR and SW whenever there is direct sunlight (or moonlight). Hence H can be calculated from the ratio $h:SW$, whether it is unity or not. It may have been, however, that, remembering the capabilities of his audience and the fact that his lecture was to be "popular science," Thales thought it better to wait. On the other hand, there are those who deem that he knew the principle of similar triangles only in particular simple cases and that for that reason he was obliged to wait until $h\ =\ SW$. Possibly we have ascribed too exact a knowledge to Thales. He may have merely argued intuitively that if the height of the man equals the length of his shadow, then the height of the pyramid must equal the length of its shadow.

Of all Thales' achievements none gained him more renown or gave more impetus to the advancement of science among the Greeks than his prediction of the total solar eclipse of May 28, 585 B.C. When the day arrived, it happened that the Medes and Lydians were indulging in a battle; but when the sun became darkened and day changed into night, they postponed their operations until they could have better light for their work. Then, having "cooled off," they reconsidered and contracted a lasting peace. Herodotus gives an account:

> There had arisen between the Lydians and Medes, a war in which the Medes often discomfited the Lydians and the Lydians often worsted the Medes.... In the sixth year, a battle took place in which it happened, when the fight had begun, that suddenly the day became night. And this change of the day, Thales, the Milesian, had foretold to the Ionians.

To predict an eclipse without knowledge of an eclipse cycle would be practically impossible; and to obtain such a cycle would be impossible for one man, since it would require a record of eclipses extending over a period longer than a single lifetime. Babylonian astronomers had discovered the Saros or eclipse cycle and, about 640 B.C., one of their sages, Berossus, founded a school where in all probability Thales studied. With a knowledge of the Saros, he was certainly in a position to predict an eclipse. His prediction, therefore, records not a Greek discovery but the westward spread of science from Mesopotamia to Asia Minor.

Once Thales was asked the question, "Which is older, Day or Night?" He replied rather neatly, "Night is the older, by one day."

A CORPUSCULAR THEORY OF VISION

Knowing from the story of mechanics that Democritus of Abdera was an atomist, we are hardly suprised to find that his theory of vision was also atomic or corpuscular. He assumed that a visible object emits streams of particles, each a replica or *image* (Gk. *eidolon*, idol) of the object, or as it were a chip off the old block, the very image of its dad. These images on entering the eye were supposed somehow to produce the stimulus by which the object is seen. Here is a brief outline of his theory written by Alexander of Aphrodisias, who was a successor of Aristotle as Dean of the Lyceum at Athens in A.D. 190:

> Democritus thinks that certain simulacra are being continually emitted by visible bodies. These particles have the same form as their parent bodies and upon penetrating the eye, produce vision. He cites as proof the fact that one observes always in the pupil of the eye of anyone who sees an object, the image or copy of the object seen.

This explanation of vision has not proved satisfactory, yet the term image has persisted in optics, though with a changed meaning. It may be added that Newton's corpuscular theory of light, Planck's quantum theory, and Einstein's term *photon* (or light-atom), are at least reminiscent of the corpuscular theory of Democritus.

THE BURNING-GLASS

Aristophanes' play, *The Clouds*, contains the following reference to a quartz lens used as a burning-glass to melt wax. Poor old Strepsiades has told Socrates about his debts but has not yet disclosed the horrible truth that his estate is being squandered at the horse-races by his renegade son:

SOCRATES. How did you get into debt with your eyes open?

STREPSIADES. A galloping consumption seized my money. Come now, do let me learn the unjust logic that can shirk debts: now do just let me learn that. Name your own price; by all the gods, I'll pay it. . . . I have found a very clever method of evading my lawsuit, so that you yourself would admire it.

SOCRATES. Of what description?

STREP. Have you ever seen that stone in the druggists' shops, the beautiful and transparent one by which they kindle fire?

SOCRATES. Do you mean the burning-glass?

STREP. I do. Come, what would you say, pray, if I were to take it when the clerk was entering my case and were to stand at a distance in the direction of the sun, and melt out the entry of my case [from the wax slate]?

SOCRATES. Cleverly done, by the Graces.

STREP. Oh! How delighted I am, that a debt of five talents has been paid! [Hugging himself.][2]

Fig. 165 shows an interesting example of burning-lens, for it is made of ice. The sun-rays which pass through it do not melt the ice but when brought together to a point or *focus* on the kindling, they set it on fire (Mariotte's experiment).

PLATO AND ARISTOTLE

About the year 387 B.C. the Greek philosopher Plato founded at Athens the college or small university called "The Academy." He is considered by many the greatest of philosophical writers and his teacher, Socrates, the greatest of philosophers. Plato's book, *Timaeus*, is an account of creation and it discusses some optical questions, two of which will be quoted here although they do not signify any great advance. It pays, however, to know some of the less fruitful attempts as well as the brilliant successes in the story of science; for one of the most vital and hopeful features of science is that scientists often wring profit even from their failures, mistakes, and negative results.[3]

It is well to realize that the ancient assumption of light rays is not the only hypothesis that can be made or that has been made.

[2]Translated by B. Rogers (Bell, 1916).
[3]See J. Jastrow (ed.), *Story of Human Error* (Appleton, 1936).

Plato considered the eye an active agent in vision, not a passive recipient, and language has no lack of expressions which imply the same view, for instance, words like *examine*, *inspect*, *peer*, and such expressions as, "The sailor spies a sail," "He holds him with his glittering eye," "He looked daggers at him," "If looks could kill, he would have died." Plato assumed that the eye has invisible sensory tentacles by which it perceives luminous bodies somewhat as a blind man examines the road with a cane. Here is part of his explanation of vision:

Of the sense-organs of the human body the gods first made light-bearing eyes. They caused the pure fire within us which is akin to that of day to flow

Fig. 165. Ice-lens.

through the eyes in a smooth stream. Whenever the stream of vision is surrounded by daylight, it forms a kindred substance along the path of the eye's vision. Wherever the vision stream collides with an obstructing object it thus brings about the sensation that we call "seeing."

Although the supposition of vision tentacles has not proven satisfactory, yet Plato had the right to make it and test it by use. It will appear now and then throughout the classical cycle.

The next quotation from Plato records the lateral inversion of any image in a vertical plane mirror:

In mirrors, the left appears as right, because contact takes place between opposite parts of the visual stream and of the object.

This fact of lateral inversion was probably known in Egypt at least as early as 2000 B.C., as indicated by fig. 161. We may safely trust the lady owning that mirror to have known about lateral inversion, and also "the women assembling, which assembled at the door of the tabernacle" (1500 B.C.). Plato's attempt at explanation does not seem very successful in showing a relation between this phenomenon and any other previously known.

The vertical inversion of an image in a horizontal plane mirror was known to prehistoric man, for it is seen at the shore of any pond. There the images of trees are upside down but not laterally inverted.

> But when the mirror is turned [to a horizontal position] it makes the whole face appear upside down, since it repels the lowest vision stream to the top.

Plato's explanation still has the same limp. He is not at his best in physics; yet he has had a strong influence on many physicists in their work.

Even as physicists we might find greater profit by turning to some of Plato's philosophic writings: for example, to his famous simile of the cave from Book VII of the *Republic*. The words of an idealist may be of special help to a realist in his struggle to distinguish between fact and inference. Here is a paraphrase of the simile or allegory of the cave. Socrates is addressing his friend Glaucon:

> Let us compare our condition to the following case. Think of men living in an underground cave which has a long corridor sloping up towards daylight. In this they lie chained from childhood, so that they cannot turn their heads. Higher up at a distance behind them, a fire burns and between it and them is a road with a wall built along it like the screen which showmen have for their puppets. Let us picture various articles carried along so as to project above the wall, effigies of men and of other creatures in various materials.
>
> "These are strange prisoners," said Glaucon.
>
> "They are like ourselves," Socrates replied. "Do you think they would have seen anything of themselves or of each other except the shadows thrown by the fire on the wall of the cave opposite them? Do you not think that they would suppose what they saw to be the real things?
>
> "Naturally, they would," rejoined Glaucon.
>
> "The only truth that such prisoners could conceive would be the shadows of those objects. Suppose a prisoner were released and obliged to stand up, turn his head and look and walk toward the light. If someone told him that what he saw on the wall were foolish phantoms whereas now he was nearer to Being and was turned to what actually IS, would he not be dazzled and perplexed and would he not think that the shadows were truer than the strange things he now perceived vaguely?"
>
> "Of course, he would," said Glaucon.
>
> "At first he would perceive shadows more easily, then, later, the reflections of objects in water and, at last, the things themselves. Then he would realize that the fire, namely, the sun, is the ultimate cause of what he used to see in his cave. Let his ascent along the corridor represent the upward journey of the soul. In the world of knowledge, the Form or Essence of the Good is perceived

last and with difficulty, but finally it is recognized as the cause of all that is right and beautiful and the source of all truth and reason. This Form of the Good must be apprehended by whosoever would act wisely in public or in private.

"We are led to realize," said Socrates, "that education is not what some professors claim. They profess to put knowledge in the soul where no knowledge was. But our argument shows that there resides in every man's soul the instrument wherewith he grasps *Truth* and that this faculty must be turned away from what is Becoming until it can behold Being and the brightest blaze of Being which we declare to be the Beautiful and the Good."

From a physicist's viewpoint, it is worth noting that Plato's word "Becoming" describes a variable, whereas "Being" stands for a constant. The physicist, by searching persistently for a constant, has, in a general way, followed Plato's advice. By the study of Becoming, Aristotle became an evolutionist and it was over this point that he broke away from the Academy and founded the Lyceum. It should be added that this development made no alteration in the warm bond of friendship and admiration between Aristotle and his old master Plato.

Aristotle was born at Stagira and is sometimes called the Stagirite. His mother was of Ionian extraction and there is much in him to remind one of the enquiring minds of Thales, Pythagoras, Anaximander, and other Ionians who had insatiable curiosity. Aristotle was, accordingly, a pioneer in several departments and among these was geography. He has been called the "founder of geography." One of his proofs that the earth is spherical in shape is given in his treatise *About the Sky*. (This argument is believed to have been first used by Pythagoras and Anaxagoras.) The proof is optical, for it deals with eclipses—it illustrates the fact that one natural science often aids in the development of another.

From these considerations, it is clear [?] that the earth does not move and is at the centre of the universe. Its shape must necessarily be spherical. In eclipses the outline is always round, and since it is the interposition of the earth that causes the eclipse, the form of this line will be determined by the form of the earth's surface which is, therefore, spherical. Hence one should not be too quick to doubt the view of those who hold that there is continuity between the regions of the Pillars of Hercules and that of India and that, in this way, the ocean is one.

One important effect of the last sentence on human affairs was the discovery of America, for, quoted by an Italian writer in 1450, it was a potent stimulus to Columbus to make his first westward voyage. Another result, of course, was the erroneous name West Indies and the name Indians for the aborigines of this continent.

EUCLID'S CATOPTRICA

About 280 B.C., Euclid whom we met as the professor of geometry
at the University of Alexandria, published two treatises on optics
entitled *Optica* and *Catoptrica*. From the latter we shall read several
passages. Its title indicates that it deals with reflection of light, that
is to say, with mirrors. It is the first treatise of real consequence in
the story of optics. About A.D. 400, the Alexandrine scholar, Theon,
the father of Hypatia, edited Euclid's *Catoptrica*, and we shall find one
item in it which must have been added after Euclid's day. It is quite
possible that Euclid was an Armenian, but he took some of his early
training at the Academy of Athens after Plato's time. He used Plato's
hypothesis of vision tentacles and called them eye-rays (Gk. *opseis*).
The term was later translated to Latin as radii or rays. The first postulate in his treatise is that in a uniform medium, for example air, a
vision tentacle or eye-ray is straight:

Vision is always along a straight line, whose parts lie evenly between
its end-points. All visible bodies are perceived along straight lines.

Euclid's opinion of reflection was that the vision rays emanate from
the eye, and then strike an opaque body by which they are reflected.
He reached the conclusion that this change of direction is similar to
the bounce of a round ball (which is not spinning). Fig. 166 is based
on Euclid's original diagram, which was labelled in Pythagorean style.

PROPOSITION 1. By mirrors whether plane, concave, or convex, rays are
reflected at equal angles. Let E represent the eye and MS a plane mirror.
Let the ray ER from the eye be reflected to the body B. I say that $\angle ERS = \angle BRM$.

Since Euclid's time, in accordance with the assumption that B
emits light rays, it has become customary to reverse the darts in
Euclid's diagram. BR is then called the *incident ray*, because it strikes
the mirror (Lat. *cido*, strike), and R is the point of incidence and
reflection. RE is the *reflected ray*. Draw RN, the perpendicular or
normal at R. This normal is not a ray. Then $\angle BRN$ is called the
angle of incidence (symbol, i). It is bounded by the incident ray and
the normal. The angle NRE is the *angle of reflection* (symbol, r); its
boundaries are the reflected ray and the normal. The angles r and i are
on opposite sides of the normal. The angle law of reflection, sometimes
called the first law of reflection of light, may be stated thus: the angle
of reflection equals the angle of incidence, or, briefly,

Eq. 1. $r = i.$

This equation, of course, is readily obtained from Euclid's, for if
$\angle ERS = \angle BRM$, then their complements, r and i, are also equal.

There is an alternative statement of this law which is so different that, at first glance, it does not seem equivalent. *Catoptrica* was first to give this second statement.

PROP. 19. In plane mirrors, the image and the visible body are equidistant from the mirror.

Or, in other words, the image is as far behind the mirror as the object is in front of it. If we represent the image *I* in Euclid's diagram (fig. 167), it is seen from the geometry of the case, that if either statement of the law is true, the other must also hold.

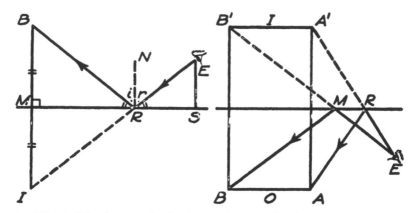

FIG. 166. Euclid's diagram of reflection.   FIG. 167. Euclid's diagram of lateral inversion.

It follows from the angle law of reflection that if a ray strikes a mirror at right angles, it is reflected *back along the same path*, for in that case $i = 0$ and $\therefore r = 0$. Euclid's statement was

PROP. 2. If a vision ray strikes any mirror at right angles, it is reflected back along the same path.

Catoptrica's treatment of lateral inversion is better than that of Plato's *Timaeus*. Here is the theorem (included as the second part of Prop. 19):

In plane mirrors, the right side [of the object] appears as the left side [of the image] and vice versa.

Fig. 167 is based on Euclid's diagram for this proposition. Except for its darts, it resembles closely a modern diagram used for tracing the "vision rays" by which the eye sees an image in a plane mirror.

The centre of the hollow sphere of which a spherical mirror is a part, is called the *centre of curvature* of the mirror *C* (fig. 168). Euclid called it simply "the centre."

PROP. 5. In concave [spherical] mirrors, if you place your eye at the centre, ... the reflected eye-rays are concurrent. Let *MVR* be a concave mirror, *C* its centre, and let the eye *E* be placed at *C*. The rays *CM*, *CV*, and *CR*, radiating from the eye at *C*, are incident upon the circumference. Then the angles formed at the joints, *M*, *V* and *R* are equal, for they are on a semicircle [Euclid. iii. 18] and therefore are all at right angles. Consequently, the rays are reflected back along the same paths, *MC*, *VC*, and *RC*, as was shown in PROP. 2. Therefore, they are concurrent.

Any point at which rays are concurrent may be called a *focus*, whether the rays converge or diverge. The ∠ *MCR* is called the angular aperture of the mirror. In this diagram, Euclid makes the

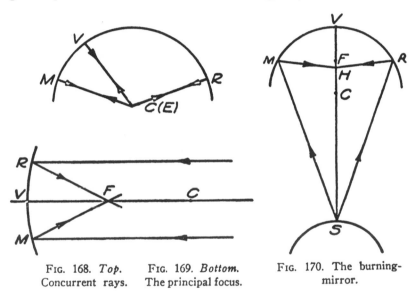

FIG. 168. *Top.*   FIG. 169. *Bottom.*   FIG. 170. The burning-
Concurrent rays.   The principal focus.   mirror.

error, so common with beginners in optics, of drawing this angle too large, i.e. greater than about 10°, in which case the mirror is hardly practical.

The last theorem in *Catoptrica* deals with the use of a concave mirror as a "burning-mirror." It speaks of rays emitted by the sun, not by the eye, and may indicate that the author, whether Euclid or Theon, was beginning to find the hypothesis of eye-rays unsatisfactory.

PROP. 30. By concave mirrors which face the sun, fire can be produced. Let *MVR* [fig. 170] be a concave mirror, *S* the sun, and *C* the centre [of curvature] of the mirror. Draw *SC* and produce it to *V*. The incident ray *SR* is reflected to *H*. Also the ray *SM*. Similarly, ... all rays from *S* incident upon the mirror. By these warm rays, therefore, fire is collected at a point *H* [a focus]. Hence, tinder placed at *H* is set on fire.

In this old diagram, Euclid drew *VCS*, the principal *axis* of the mirror. The point *V* is called the *vertex* of the mirror. It is chosen arbitrarily, but usually at a central position on the mirror. In this diagram, direct sunlight is represented as strongly divergent, but if two rays separate only a few centimetres in going 93 million miles, they are practically parallel. To the Greek, of course, the sun was nearer than it is to us. The pagan lived here and now; he avoided the infinitely far, whether in space or time. Heaven was only at the top of Olympus. Aristarchus was put in jail for saying that the sun was thousands of miles away.

Modifying fig. 170, we see that parallel light, parallel to the axis *VC*, is reflected as a beam, converging at the focus *F*, about midway between *V* and *C* (fig. 169). This focus is important enough to receive the special name, "principal focus of the mirror," or, briefly, "the focus." *VF* is called the *focal length* of the mirror. All the facts about the mirror centre about the point *F*. This explains the derivation of the term "focus" (the Latin word for hearth or fire-place—the lineal descendant of the old camp-fire). It is the point about which the whole life of a household centres. Mathematicians found a similarly important point in the parabola and other conic sections, and it was also given the name focus. Kepler considered the term especially appropriate for the point *F*, and later the term was broadened to include other points. As we have seen, Euclid used the term "point" to represent the idea of focus.

When Plutarch, in his *Marcellus*, tells that Archimedes set on fire a distant Roman trireme by means of a huge curved mirror, the story has at least this support, that Archimedes took his training at Alexandria where Euclid and others had studied concave mirrors and knew how to use them as burning-mirrors.

One observation recorded in the *Catoptrica* refers to an instance of *refraction* of light, that is, the bending or change of direction which occurs when light passes obliquely from one transparent medium to another, for example, from water to air. The experiment which the book outlines very briefly is sometimes called the "coin and cup" experiment. Its first performance has been credited to Ctesibius of Alexandria, the inventor of the force-pump. Here is a free translation:

> Let a body [e.g. a coin] be placed at the bottom of an [opaque] vessel so that it just cannot be seen [by a certain observer]. If the observer keeps the same position while water is poured into the vessel, the body comes into sight.

This item appears as a postulate and is not used in the treatise. The date of Ctesibius (50 B.C.), and the loose articulation of this item with the rest of the book, support the idea that its author was not Euclid.

Included in the works of Euclid is a treatise entitled *Theon's Recension of Euclid's Optics*, from which comes this excerpt:

Let us suppose that the eye-rays form a cone which has its vertex at the eye and its base on the visible body.

Theon's statement is represented by fig. 171. In terms of light rays, this diagram becomes fig. 172. The darts are reversed and each point

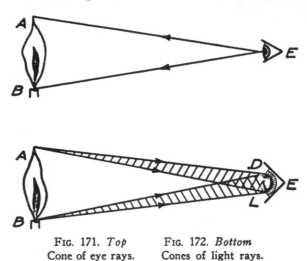

FIG. 171. *Top*   FIG. 172. *Bottom*
Cone of eye rays.   Cones of light rays.

on the luminous body *AB* becomes the vertex of a cone of light rays, whose base is the transparent pupil of the eye *DL*. It is interesting to note that it took about fourteen centuries to reverse those darts, i.e. from Plato to Alhazen the Arabian (A.D. 1026).

From Euclid's discussion it is plain that geometry is of great utility in optics, in fact indispensable, for he used a number of theorems from his own famous text, e.g. III. 18, "The radius of a circle touches its circumference at right angles." Later, we shall see two more of Euclid's theorems applied, viz.,

I. 32. If one side of a triangle be produced, the exterior angle equals the sum of the two interior opposite angles.

VI. 33. Angles at the centre of a circle are proportional to the arcs on which they stand.

MEASURING THE EARTH

By a clever application of optics and geometry, Eratosthenes of Alexandria succeeded in measuring the circumference and diameter of the earth with remarkably simple apparatus and astonishing accuracy. He was chief librarian at the Museum of Alexandria in the days of

Archimedes, and was a very learned and versatile scholar, quite capable of doing original research. Some cynical jokesters gave him the nickname "Beta," the second letter in the alphabet, to insinuate that he knew many subjects but was preeminent in none. Archimedes admired him sufficiently as friend and mathematician to maintain a correspondence with him and to dedicate to him his treatise *On Method*.

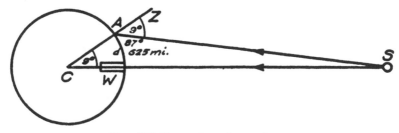

FIG. 173. Eratosthenes' experiment.

Eratosthenes knew that on the day of summer solstice at noon, each year the sun S (fig. 173) shone down directly to the bottom of a deep well at Assuan or Aswan (where there is now a huge and famous dam). In other words, the sun was directly overhead at that place and time. He knew also that Aswan was 5000 *stadia*, i.e. about 600 miles, south of Alexandria, A, where he made his observations. On that day at noon he used a *gnomon*, which is essentially a vertical rod (GN, fig. 174). Thus he measured $\angle NAS$, the elevation (e) of the sun, and found its complement $c = \angle VAS = 9°$ approx. or 1/40 of a circle (360°). By a theorem of his old master Euclid, I. 32, he knew that in fig. 173 $\angle C + \angle S = \angle ZAS$. He considered the sun to be so far away that $\angle S$ was small enough to be negligible in comparison with $\angle C$ and $\angle ZAS$. Hence $\angle C = \angle WCA = \angle ZAS = 9°$. Another theorem of Euclid (VI. 33) shows that angles at the centre of a circle are proportional to the arcs which subtend them. Since 9° at the centre of the earth stands on an arc of 600 mi., ∴ 360° must stand on an arc of $600 \times 40 = 24,000$

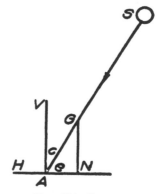

FIG. 174. Gnomon.

mi. But this arc will be the circumference of the earth. For convenience, these numbers have been rounded off. Eratosthenes' actual result was 24,662 mi. and the value now used for rough calculations is 25,000 mi. Using *Eq.* 2, p. 18, $c = \pi d$, Eratosthenes calculated the diameter of the earth thus:

$$d = c \div \pi = 24,662 \div 3.141 \ldots = 7,854 \text{ mi.}$$

The value now accepted by geodetists is 7,918 mi. or roughly 8,000 mi. Hence it is seen that Eratosthenes' error was only about one per cent. It must be granted that his achievement was exceedingly creditable and altogether a clever and valuable contribution. It would hardly seem possible to obtain so brilliant a result from a stick of wood and a protractor.

It was remarked previously that, in general, we rely especially on the evidence of our eyes. What value Eratosthenes set upon his eye-

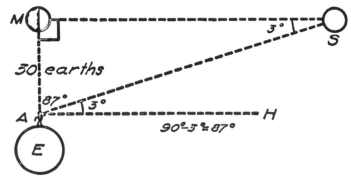

FIG. 175. Aristarchus' experiment on distance of sun.

sight may be judged from the fact that at the age of eighty, on realizing that he was turning blind, he died of voluntary starvation.

ARISTARCHUS OF SAMOS

In the story of mechanics, Archimedes mentioned a lost book on heliocentric theory by Aristarchus of Samos. Another treatise, written by the same author about 270 B.C., fortunately has survived. It shows from the moon's phases that the latter is dependently luminous by reflection, revolves about the earth, and has the sun as source of light. The treatise then attempts the difficult problem of finding the sun's distance from the earth. Here are three snippets dealing with this topic:

1. The moon receives its light from the sun.
2. The earth is at the central point of the moon's orbit.
4. When the moon appears to us halved [first quarter], its angular distance from the sun [fig. 175 $\angle MES$] is less than a quadrant [90°] by one-thirtieth of a quadrant [3°]. Hence it may be concluded that the distance of the sun from the earth is greater than eighteen times the distance of the moon from the earth.[4]

[4]Sir T. Heath, *Aristarchus of Samos.*

One method of reaching a conclusion from Aristarchus' observation is to draw a triangle *EMS* to scale and, by measurement, to find the ratio *ES* : *EM*. One chief difficulty in this problem is the smallness of the ∠ *ESM*. With the apparatus at his disposal, Aristarchus could not possibly measure this small angle with the necessary accuracy; for the error in his readings was greater than the angle he was trying to measure. Subsequent work during twenty centuries, with tremendous improvements in instruments, has shown that ∠ *ESM* = about 0° 8' instead of Aristarchus' 3° or 180'. When ∠ *ESM* is drawn to scale with ∠ *S* = 8', it is found that the ratio *ES* : *EM* becomes about 388 instead of 19.

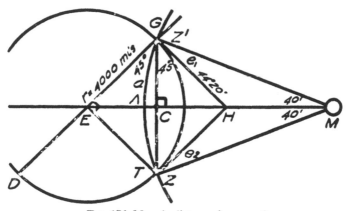

FIG. 176. Moon's distance from earth.

To find the sun's distance, it will still be necessary to find *EM*, the moon's distance from the earth. This is somewhat easier than measuring *ES*. One method is to aim at a point on the moon with two telescopes at the same time from, say Greenwich and Cape Town (*G* and *T*, fig. 176), which are about six thousand miles apart. The ∠ *GMC* is found to be about half a degree (42'). Drawing Δ *GMC* or Δ *GEM* to scale, as before, we find that *EM* / *EG* = 60, approximately. In other words the moon's distance from the earth is about 60 times the earth's radius, or 30 times its diameter *GD* (8,000 miles). Therefore, the sun's distance from the earth is about 8,000 × 30 × 388 = 93,000,000 miles approximately.

Have you seen anything in Aristarchus' statements to warrant his being jailed? Some of the ignorant office-holders in Samos did. They clapped him in jail on a charge of sacrilege and impiety, because he said that the sun was thousands of miles away and was as big as the whole Peloponnesus. Some clamoured for his death. It is horrible to think what might have happened if he had said 93,000,000 miles. He would have had only until they regained consciousness to make good his escape.

THE SHADOW

Lucretius frequently speaks of optical phenomena. Concerning a shadow, he says in *De rerum natura*:

That which we are accustomed to call a shadow can be nothing else than air devoid of light.

The term shadow has two related meanings; one refers to a space behind an opaque body within which dependently luminous bodies become eclipsed or darkened; the other refers to a surface that is darkened because it is in a space-shadow, for example, a man's shadow on the ground. Lucretius' definition refers to the former. On closer inspection however, one sees that it is really impractical. It does not enable one to determine whether a certain space is a shadow or not. It may be true, but as a definition it is useless. Our definition avoids the hypothetical term "light" and speaks in terms of facts. It can be applied in practice. It says, "Place a dependently luminous body in the given space, and if the body is eclipsed or darkened, then that space is a shadow."

Knowing that Lucretius is an atomist, we are not surprised to find him employing Democritus' doctrine of images. Here is his attempt to explain transparency in a corpuscular way:

We often observe a conversation carried on through closed doors [e.g. a bath-room door], doubtless because sound can safely traverse crooked pores whereas images refuse to pass through; for they are disrupted unless they traverse straight pores such as those in glass through which they can fly.

This passage should warn us as to what a hypothesis can do to the logic of even as good a man as Lucretius, if he gives free rein to speculation and imagination with comparatively little weighing of facts.[5] To explain transparency, he finds it necessary to bring in several assumptions, each lacking factual foundation and more difficult to explain or credit than transparency itself: crooked pores in opaque bodies, straight pores in glass, sound-atoms limber and slippery for navigating crooked pores. Light images he assumes to be straight, like light rays, or spear-shaped, and incapable of turning sharp corners, and, wisely enough, refusing the attempt for fear of being "disrupted." Thus he tacks on the defenceless atoms any property that his imagination conjures up and he lands in a ludicrous position. As one imagines

[5]G. M. Stratton in *Story of Human Error*, p. 328.

the slithery sounds wriggling through crooked pores like a cat going under a fence, the "explanation" becomes even comical.

But it comes to pass that we receive the thunder in our ears after our eyes discern the lightning, because [?] impulses always travel more slowly to our ears than those which stimulate our sense of sight.

Lucretius has the cart before the horse this time. How does he know that impulses travel more slowly to the ears than to the eyes? It is because the thunder is heard after the lightning is seen. His only defence would be that the general rule had been previously induced from a number of other examples and then the lightning and thunder case deduced as in the quotation. Or perhaps poor Lucretius was forced into this false position by the exigencies of the metre.

The source of the term *refraction* occurs in a vivid passage in which Lucretius describes the oars of a trireme at Ostia, Rome's seaport.

For all the part of the oars which is above the salt sea-spray is straight and the rudders are straight above; but every part that is immersed beneath the surface of the liquid, seems to be broken back [Lat. *refracta*] and bent round, yes, even to turn upwards again and to twist back so that it almost floats at the surface of the liquid.

As the story unfolds, the phenomenon of refraction will receive much attention.

One might well expect that this poet and scientist would devote some verses to the beauties of the rainbow.

When at such times the sun amid the dark tempest has shone out with its rays full against the spray of the storm clouds, then among the black clouds shine forth the hues of the rainbow.

THE SHORTEST AND QUICKEST PATH

We have seen the temporal power of Greece decline and the seat of empire moving westward to centre at Rome. But Rome contributed little to the practice of optics and less to its theory. Meanwhile Alexandria remained a great centre of learning and culture. We shall therefore leave Rome and fly back to Egypt for our last visits there before the final curtain descends to mark the close of the classical cycle of civilization. The date of our first visit there is A.D. 50, about a century after Lucretius. Our host is the erudite geometer and physicist, Hero (or Heron) of Alexandria.

Hero wrote a book, *Catoptrica*, on reflecting surfaces, and in it gave a remarkable statement of Euclid's angle law of reflection. He said that reflection takes place along the best short-cut, i.e. along the *minimum path*. By an examination of fig. 166, p. 235 one can see that this is equivalent to the law $r = i$. Thus light seems to be a lazy agent in the best sense of the term; and since the speed is uniform, light seems to be also economical of its time, for it appears to "choose" not only the shortest path but also the quickest route to its goal. By this observation Hero planted a seed which eighteen centuries later bore such a crop of fruit as Hero himself could not have imagined. It leads to the law called the principle of least action.

PTOLEMY'S *OPTICA*

In our last visit in Alexandria our host is an old friend, the great Greek astronomer Claudius Ptolemy, the author of *Almagest* and the founder of the Ptolemaic system. He lived in the temple of Serapis in a suburb of Alexandria called Canopus, and there he conducted some excellent researches on refraction, the bending of light that occurs when it passes obliquely from one transparent medium to another or, in other words, through a transparent surface. Like Euclid, he spoke in terms of eye-rays. His view was that refraction is the bending of the eye-rays that occurs when the eye (usually in air) sees a body in another transparent medium and the eye-rays strike the separating surface obliquely. Of the treatise, *Optica*, which Ptolemy wrote, there is no copy extant. During the heyday of the Arabian empire, it was translated into Arabic and in the twelfth century a certain admiral, Eugenius Siculus, translated the Arabic into Latin. That Latin version was published at Turin in 1885 with the title, *L'Ottica di Claudio Tolomeo, ridotta in Latino da Eugenio, pubblicata da Gilberto Govi, Torino*. The Library of Congress, Washington, D.C., has a copy of this book, in whose preface one reads this explanation:

In section V, part of which, unfortunately, is lost, Ptolemy discussed the bending of eye-rays, which always occurs at unequal angles.

Ptolemy devised and used the apparatus of fig. 177 which is employed in optics to this day and is called a Ptolemy refraction tank. His diagram is given in fig. 178; it resembles those used nowadays in standard texts. His argument contains the following passage:

When the arc *az* was ten ninetieths of the quadrant *ab*, the arc *gh* was eight ninetieths, nearly. When *az* was 20°, *gh* was 15½° etc. as shown in the table [i is angle of incidence and R, angle of refraction]:

i	R	i	R	ratio, i/R	
x	viii	xx	xv & ½	1.125	1.29
xxx	xxii & ½	xL	xxviiii	1.33	1.42
L	xxxv	Lx	xL & ½	1.42	1.48
Lxx	xLv & ½	Lxxx	L	1.54	1.6

From an excellent series of such experiments, Ptolemy induced the following laws of refraction of light:

1. When an incident ray strikes the separating surface at right angles, it enters the second medium without change of direction. In this case $i = 0$ and $R = 0$. It is the only case in which $R = i$. In general $R \neq i$, whereas in reflection $r = i$ always.

2. The angle in water is smaller than that in air, no matter which way the light is going. Hence if a ray is entering water from air obliquely, the bending is toward the normal in the water, the second medium. But if a ray goes obliquely from water to air, the refraction occurs away from the normal in the second medium (air).

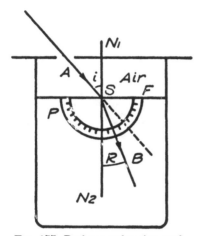

FIG. 177. Ptolemy refraction tank.

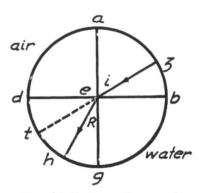

FIG. 178. Ptolemy's diagram of refraction in water.

3. If a refracted ray strikes a mirror at right angles, it *retraces* its *whole path* in both media so that i and R interchange places; what was i becomes R and vice versa.

From this table of data it is seen that Ptolemy was reading angles only to the nearest half-degree. Evidently the ratio i/R is not constant; it varies about 30 per cent. Thus R is not directly proportional to i as some students erroneously assume. It is true, however, that for values of i up to about 40°, the ratio i/R varies only about 3 per cent.

To express a general relation between R and i, Ptolemy used an equation of the form

Eq. 2. $$R = \mu(i) - k(i)^2$$

where μ and k are constants. From the fourth and fifth pairs of readings (where the error was least) we obtain by simultaneous equations the values $\mu = 0.825$ and $k = 1/400$. These give values of R that agree fairly well with modern readings for small angles. For $i = 80°$, however, they predict that $R = 0.825 \times 80 - 6400 \div 400 = 66 - 16 = 50°$, whereas the modern reading is 47° 36′, about 5 per cent below Ptolemy's result.

Here are some of Ptolemy's results on the refraction of glass:

Let *tkl* be the circumference of a semicylinder of pure glass (fig. 179),

i	*R*	*i*	*R*	*ratio, i/R*	
x	vii	xx	xiii	1.43	1.48
xxx	xviiii & ½	xL	xxv	1.66	1.60
L	xxx	Lx	xxxiiii & ½	1.66	1.74
Lxx	xxxviii & ½	Lxxx	xLii	1.84	1.9

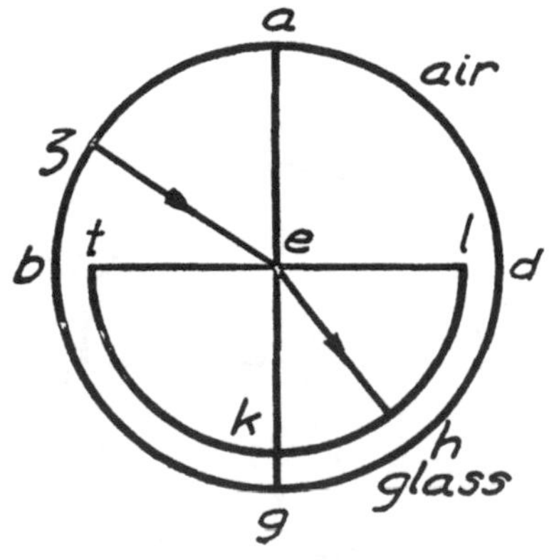

FIG. 179. Refraction by glass. 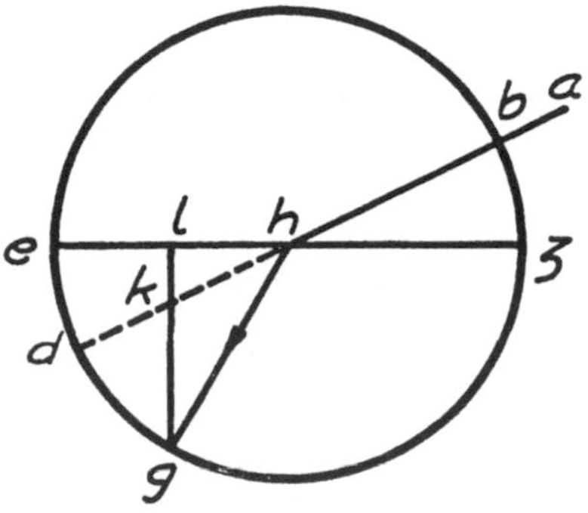   FIG. 180. Coin and cup experiment.

The diagram, discussion, and readings, in this case, all seem very similar to a modern demonstration with an optical disc.

Ptolemy discusses the coin and cup experiment of Ctesibius which we found in Euclid's *Catoptrica*. If it was inserted by Theon, then the editor would have Ptolemy's treatise available. Ptolemy's discussion makes some progress (fig. 180):

> If we place a coin at *g*, which is in the lower part of the vessel, as long as the vessel is empty [i.e. full of air], the coin is invisible [at *a*], since the vision-ray [Lat. *visibilem radium*] which can reach it along a straight line [*qui posset ad illum recte procedere*] touches *b*, the lip of the vessel. When water is poured in, however, until the surface comes up to the line *zhe*, the ray *abh* is flexed to *gh*, which is steeper than *ah*, and so the location of the coin is seen at *g*. The perpendicular *lkg* intersects the line at the point *k*, and will be the situation of the coin which ultimately is seen nearer the surface of the water at *k*.

The apparent position of the coin is actually somewhat to the right of *k*.

Ptolemy also made a study of refraction of sunlight by the earth's atmosphere at sunrise and sunset, that is, at the twilight line, and drew therefrom an estimate of the height of the atmosphere. We shall find his work in this topic further pursued by the Arabian Alhazen. Ptolemy failed to discover the exact relation between i and R, and for fifteen centuries it remained one of the unsolved riddles of optics. The branch of mathematics which is the key to the mystery is trigonometry, and since Hipparchus and Ptolemy were founders of that science, the latter had a particularly good chance to make the discovery. One may wonder why he did not succeed when the solution was right up his alley. As we have noted before, it seems easier to us who look back over the field than it did to the pioneers who were looking forward to the solution.

Thus ended the development of optics in the classical cycle of culture.   Ptolemy's work was the expiring flicker of that lamp of learning.

It would be rash to challenge those who observe in our civilization many symptoms which were present in other cycles when their final doom was imminent. It seems highly probable that if our civilization is to escape the common descent into the grave, its life will be prolonged only by a Herculean and concerted effort. Not easily can the hand of fate be stayed. But while the symptoms of softening and decadence are seen in many places, in science it appears that there is greater vitality than ever. The scientists in general have remained true to their principles and have not shunned work and sacrifice. We have found many of them strongly religious and ethical. They have not succumbed to the allurements of idleness and amusement. They now have opportunity, in collaboration with others, of reading and interpreting the warning written in the history of the world, of seeing what measures should be adopted to rescue their civilization from extinction, and of helping to lead humanity in adopting those measures.

For about eight or ten centuries after Ptolemy, that is, during mediaeval times, the science of optics had few votaries in Europe and made little progress; it was in virtual eclipse. It must be remembered, however, that during that mediaeval period, the brilliant cycle of Arabian culture flourished and became the torch-bearer of learning, spreading as far westward as Morocco and Spain. Our next visit, therefore, will take us to Cairo and through an interval of nine centuries, to make the acquaintance of the greatest student of optics in the Middle Ages.

Our stay in Cairo, however, will be comparatively brief, for the Arabian Empire was of relatively short duration and did not have many outstanding contributors to optics.   The tide of culture will then continue the general drift westward which we have already noted, and as veritable camp-followers we shall betake us to Paris and Florence to witness what progress is afoot.

CHAPTER ELEVEN

"Westward Ho"

THE greatest physicist of the Arabian Empire was called, at home, Abu Ali Al-Hasan Ibn Alhasan, but in our brusqueness we usually cut it down to Alhazen. He was a contemporary of Al Biruni who introduced the term specific gravity in mechanics. About the year A.D. 1025, Alhazen wrote a famous book on optics with the title *Opticae Thesaurus* ("Treasury of Optics")[1] which records a number of advances in optics.

ALHAZEN'S DISCOVERIES

Alhazen was the man who reversed the darts in Euclid's diagrams. He discarded the theory of eye rays, which had been used in the classical cycle from Plato to Ptolemy, and adopted openly the ancient and tacit assumption that luminous bodies emit light rays which enter the eye and produce stimulation and vision. He repeated Ptolemy's experiments on refraction with greater accuracy and showed that Ptolemy's formula failed to express exactly the relation between the angles i and R, especially when $i > 30°$. He failed however, to discover the true relation between i and R, and the problem had to wait another six centuries for solution.

Alhazen discovered a second law of reflection of light which is named after him: the incident ray, the reflected ray, and the normal at the point of incidence are all in the same plane (coplanar). The bounce of a round ball (provided it is not spinning) follows the same rule, and players apply this rule intuitively in such games as tennis and ping-pong to send the ball into a part of the opponent's court where he isn't, and thus afford him maximum exercise. If the opponent is in court A, the tennis player drives the ball down toward the outer edge of court B, knowing by experience that the ball (if not spinning) will bounce in the same vertical plane as the line of drive.

[1] The Library of Congress, Washington, D.C., has a copy of this rare volume.

If it is spinning, that introduces other complexities which do not come under Alhazen's law. Thus both laws of reflection are illustrated by the bounce of a round ball. Alhazen discovered that his law holds also for refraction of light: the incident ray, the refracted ray, and the normal are coplanar.

The names of some parts of the eye, e.g. retina, cornea, aqueous humour, and vitreous humour, were introduced by Alhazen as a result of his dissection of the eye.

To the old problem, "Why are the sun and moon larger when near the horizon?", Alhazen gave the answer that they subtend the same angle at the eye whether close to the horizon or high in the sky, but close to the horizon, they are seen near such objects as trees, houses, and hills and by comparison they seem larger.[2] He called the appearance an illusion. In recent times, some have claimed that this explanation is not completely satisfactory.

Alhazen fused colours in the eye by spinning them on a top, as may be seen from the following quotation from his book:

> The colour of the surface of a top, when it is spun, is perceived as one colour, compounded from all the colours which are on its surface.

The *colour-top*, as we shall see, later assumed great importance in the study of colour. Alhazen also discovered that the colour of any surface depends partly on the colour of surround-ing bodies.

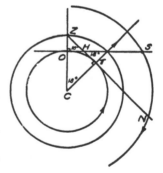

From his study of twilight, Alhazen reached the conclusion that the earth's at-mosphere must be at least forty or fifty miles high. Twilight is caused by reflection of sunlight from particles in the air. Its duration varies according to latitude, sea-son, and the distribution of particles. This calculation, therefore, is really to find how high the particles reach. The higher they reach, the longer will twilight endure, as indicated in fig. 181. Ptolemy and Alhazen measured the duration of twilight and an

Fig. 181. Height of the atmosphere.

average value of their results is about half an hour (36 min.). During 36 min., the earth turns through

$$\frac{360}{24} \times \frac{36}{60} = 9°.$$

[2] Sir William Bragg, *The Universe of Light* (Macmillan, 1933), p. 63, plate ix.

∴ in fig. 181, in which O is the observer, S the sun, and $OZ = h$, the height of the atmosphere,

$$\angle SHN = \angle OHZ = \angle OCT = 9°.$$

∴ height of atmosphere $h = CZ - CO = CT \cdot \sec 9 - CT$
$$= CT (\sec 9 - 1) = 4,000 (1.0125 -- 1.0)$$
$$= 4,000 \times 0.0125 = 50 \text{ miles.}$$

A better average value for the duration of twilight is about 72 min. During this interval the sun sinks through 18° and the value of h obtained from this reading is 200 miles. When Torricelli wrote to Ricci at Rome about the new instrument which he and Viviani had invented and which Boyle later named the barometer (Gk. *baros*, weight), he referred to the height of the atmosphere:

Those who have made a study of twilight have observed that the atmosphere is visible to a height of about 50 or 54 miles.

Torricelli was using a value obtained by Alhazen.

BACON'S PREDICTIONS

From his studies of refraction, based on the work of Ptolemy and Alhazen, Roger Bacon prophesied the invention of the telescope and the microscope.

Greater things than these can be performed by refracted vision. For we can give such figures [shapes] to transparent bodies [lenses] and dispose them in such a manner with respect to the eye and the objects that the rays will be refracted ... toward any place we please, so that we shall see the object near at hand or at any distance ... we please. Thus from an incredible distance [telescope], we may read the smallest letters; and may number the smallest particles of dust and sand [microscope], by reason of the greatness of the angle under which we see them ... and many things of the like sort, which persons unacquainted with these principles would refuse to believe.

Some have credited Roger Bacon with the invention of spectacles, but the claim is disputed. His book, *Opus Majus*, goes no further than the invention of a reading-glass, as may be seen from these words:

If a man looks at letters through the medium of a crystal or a glass or some other transparent body placed above the letters and if it is the smaller part of a sphere whose convexity is toward the eye [plano-convex], the letters will appear larger to him. This instrument is useful to the aged and to those with weak eyes.

In 1629, Charles I of England granted a charter to the "Spectacle Makers' Guild" and in 1760, Benjamin Franklin of Philadelphia invented "bifocals."

THE PIN-HOLE CAMERA

No better illustration of Leonardo da Vinci's happy combination of theory and practice could be met than his invention of the camera. His priority regarding this instrument is disputed by some and well-founded credit is given to others as far back as Vitruvius, Euclid and Aristotle. As a painter, da Vinci considered that he should make a study of light and in so doing a knowledge of the nature of the eye would be fundamental. Accordingly, following Alhazen's example, he dissected the eye, although such activities were frowned upon by both the Church and public opinion. Applying the knowledge thus gained, Leonardo invented the pin-hole camera which is a direct copy of the eye (fig. 182). One of his manuscripts says:

A small aperture in a window-shutter projects on the inner wall of the room an image of the bodies which are beyond the aperture. This will explain how the eye acts. For I have learned the actual interior structure of the eye [by dissection].[3]

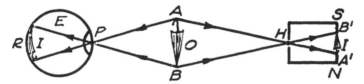

FIG. 182. The pin-hole camera.

The first camera, then, was a room with the scientist inside; and *camora* is an Italian word for room. It is related to the old English word *chamber*, meaning a room. This instrument is also called a *camera obscura*, which is simply the Latin rendering of "darkened chamber." One modification of the camera obscura is the modern camera and another is the periscope.

The action of the pin-hole camera is quickly outlined by classifying its parts as to their penetrability by light: the walls are opaque, the pin-hole, being of air, is transparent, and the screen, SN, is translucent. The luminous point A emits light rays in all directions; of these AHA' traverses the pin-hole and makes A' a luminous point of the same colour as A. Thus a system of luminous points $A'B'$, which is the replica of AB, is projected on SN. It is called a *real image* of AB, since the rays by which it is formed reach its location on SN. The diagram predicts that this image is inverted vertically and laterally. Its magnification, $M = A'B'/AB$ by definition, and depends on the relative distances of O and I from H. Experiment bears out these

[3]Quoted in H. Grote, *Leonardo da Vinci als Ingenieur*, p. 54.

predictions. Plato's eye-ray hypothesis gives little help in explaining this phenomenon.

The action of Alhazen's colour-top depends on the feature of vision called persistence, whereby impressions made on the eye persist for some time after the stimulus is removed. Da Vinci refers to this topic:

> The motion of a single fire-brand whirled rapidly in a circle causes this circle to appear as one continuous and uniform flame. The drops of rainfall seem continuous threads descending from their clouds; and so, herein, one may see that the eye preserves [temporarily] the impressions of the moving things that it sees.

Other applications of persistence of vision are the movies, a twenty-fourth of May "sparkler" or pin-wheel, and two that will come later in the story of optics, the stroboscope and the colour-discs of Newton, Maxwell, and Munsell.

Since Leonardo da Vinci was, above all else, a painter, it is hardly surprising to find him discussing the subject of colour, and his remarks about the blue of the sky disclose a remarkably keen insight, far ahead of his times. His marvellous skill in the use of colour is still the envy of artists. The expression, "region of fire," that he uses in the following passage refers to the uppermost part of the atmosphere. The term came from the mediaeval belief that the highest heaven was composed of the element of fire. Hence it was called "the empyrean."

> I say that the blue which is seen in the atmosphere . . . is caused by the heated moisture having evaporated into the most minute imperceptible particles which the beams of the solar rays . . . cause to seem luminous against the deep intense darkness of the region of fire that forms a covering above them. We may take the case of the smoke produced by old dry wood, for as it comes out of the chimneys, it seems to be a pronounced blue. But if this smoke comes from new green wood, then it does not assume a blue colour, because . . . it is heavily charged with moisture.[4]

Similarly smoke from the lighted end of a cigarette looks bluish and that from the mouth-end brownish or reddish. Smoke particles in passing through the cigarette become coated with moisture. Light from larger, heavier particles has lower frequency and from smaller ones, higher frequency. Blue light has higher frequency than red or brown. How remarkable that da Vinci in 1508 should come so close to the explanation of this effect.

Da Vinci was probably the first to suggest that light is transmitted in waves. He reached the idea by considering the analogy between sound and light.

[4] J. P. Richter, *The Works of Leonardo da Vinci*, x, 161.

THE TELESCOPE

Most of Galileo Galilei's renown in non-scientific circles was gained from his discoveries made with the telescope. In 1610 he published a Latin treatise[5] from which the following selections are taken:

About ten months ago, a report reached my ears that a certain Dutchman [*Belga*] had constructed a telescope [Lat. *perspicillum*], by the aid of which visible objects though very distant . . . may be seen distinctly as if near. . . . I plunged into an intensive study of refraction and in a short time, accomplished

Fig. 183. Galileo Galilei.

my purpose; I prepared myself a tube . . . of lead, in the ends of which I fitted two glass lenses, each plane on one side but one spherically convex on the other side and the other, concave. . . . Finally, sparing neither labour nor expense, I succeeded in preparing for myself an instrument so excellent that objects through it, appear magnified almost a thousandfold in area. . . .

The "Dutchman" was probably Hans Lippershey, a spectacle-maker of Middleburg. With the instrument referred to, Galileo made several

[5]Galilei, *Opere*, III, *Astronomicus Nuncius*, p. 59.

exciting discoveries, sun-spots, volcano craters in the moon, and phases of Venus like those of our moon, but, above all, Jupiter's moons:

There still remains the disclosure of what seems to me . . . the most important contribution in this memoir, namely, . . . the discovery . . . of four planets which from the first foundation of the world up to our own times, have never before been seen by anyone.

On the seventh day of January, . . . in the present year, one thousand six hundred and ten, in the first hour of the following night, when scanning the celestial constellations through my telescope, Jupiter presented itself to my view; and . . . I discovered what I had never before been able to see, owing to the inferior power of my other telescopes, namely, that three little stars stood near the planet which though small, were nevertheless very bright. At first I took them to belong to the category of fixed stars and yet they made me wonder somewhat; for they appeared arranged exactly in a straight line parallel to the ecliptic and brighter than other stars of the same magnitude.

After three centuries this report of a discovery still glows with fervid rapture:

Their arrangement . . . was as follows: Toward the east were two stars and

East   ✳   *   (♃)   ✳   *West*

a single one on the west side. . . . On Jan. 8, when, led by I know not what lucky chance, I had returned to view this sight again, I found a very different configuration; for there were now three little stars, all to the west of Jupiter,

East   (♃) ✳ ✳ ✳   *West*

nearer to each other than on the previous night, and at equal distances from each other. . . . I began to feel perplexed as to how Jupiter, contrary to all the received predictions, could be found to the east, when, on the previous day, it had been west of two of them, and forthwith, I had a misgiving that possibly Jupiter had moved in a direction contrary to astronomical computations and thus, by its own motion, had moved past these stars. Accordingly, it was with the most impatient longing that I waited for the next night; but alas, my hopes were dashed, for the sky was shrouded with clouds in every quarter [astronomers' bugaboo!]. . . . But on the tenth, those same stars appeared in the following orientation with respect to Jupiter. For there were only two of them now and

East   ✳ ✳ (♃)   *West*

both east of Jupiter, the third one, as I surmised, being occulted by the planet. They were situated, as before, in the same straight line with Jupiter along the zodiac. When I had beheld these phenomena, realizing that such vagaries could not by any stretch of the imagination be attributed to Jupiter, and perceiving, moreover, that throughout it had been the same group of stars that I had been observing, . . . at last, changing from doubt to astonishment, it began to dawn on me that the motions I had witnessed resided not in Jupiter but in the stars I have been describing. Consequently I considered that they should be observed thenceforward with greater attention and accuracy.... Then on the thirteenth day of January

East   ⋆ ♃ ⋆ ✦ ⋆   *West*

and Jan. 15th:

East   ♃ ⋆ ✷ ✸ ✵   *West*

> I therefore, concluded and declared beyond the peradventure of a doubt that there are in the heavens three stars revolving around Jupiter, as Venus and Mercury revolve around the Sun. This relation was established more clearly than noon daylight by many subsequent observations . . . also the fact that there are not only three but indeed four erratic stars performing their revolutions around Jupiter. For now . . . our sense of sight presents to us four satellites circling around Jupiter as the Moon does about the Earth, while the whole system travels along a mighty orbit around the Sun in the interval of twelve years.

Galileo's reverence is seen in his words, "I am quite beside myself with wonder, and am infinitely grateful to God that it has pleased Him to permit me to discover such great marvels."

It is not difficult to guess whether or not Galileo used this discovery to lambaste the peripatetics or "paper philosophers" as he called them, and to champion the Copernican theory. The Aristotelian doctrine considered the heavens as celestial, perfect, and unchangeable, but this very disturbing rebel, Galileo, was discovering mundane features in the Moon, Venus displaying herself in a new set of costumes, the sun breaking out in a batch of black spots and pirouetting on his axis, and Jupiter disclosing a whole bevy of consorts. These discoveries had a telling effect in jolting Aristotelian beliefs from the entrenched positions they had held for centuries. Some of Galileo's attitude is seen in the following snippet from one of his letters to Kepler[6]:

> Oh, my dear Kepler, how I wish that we could have one hearty laugh together! Here, at Padua, is the principal professor of philosophy, whom I have repeatedly and urgently requested to look at the moon and planets through my glass, which he pertinaciously refuses to do. Why are you not here? What shouts of laughter we should have at this glorious folly! And to hear the professor of philosophy at Pisa labouring before the Grand Duke with logical arguments, as if with magical incantations to charm the new planets out of the sky.

THE MICROSCOPE

Few instruments have contributed more to the advance of science than the microscope. It was invented in its primitive form about the

[6]Quoted in Lodge, *Pioneers of Science.*

same time as the telescope. Although priority in this case is disputed, yet the balance of evidence seems to favour the Dutch spectacle- and telescope-makers, Zacharias Joannides and his father. As its name implies, the microscope enables us to examine bodies that are too small to be seen by the naked eye (Greek, *mikros*, small; *skopein*, examine) and we found its invention predicted in the writings of Roger Bacon. The theory of its action is rather too complicated to be discussed at this juncture. It consists essentially of two convex lenses and hence is called a compound microscope, whereas a single lens used as a "magnifying glass" is called a simple microscope.

CRITICAL ANGLE

Kepler published in 1611 a famous Latin treatise on optics, entitled *Dioptrici*. It contains an excellent study of refraction as indicated by the following excerpts,

II. Axiom. Rays passing into a denser medium obliquely are refracted toward the normal drawn in that medium at the point of incidence. Also those which emerge from the denser medium are refracted away from the normal in the lighter medium.

III. Axiom. The refraction of the rays is the same whether they are entering the medium or emerging from it.

As we found in Ptolemy's work, if a refracted ray is reversed, it retraces its path in both media and the angles i and R become interchanged.

VII. Axiom. The angle of refraction in glass is approximately proportional to the angle of incidence if the latter is not greater than about 30°.

VIII. Axiom. Within the said limits, the angle of refraction in glass is very nearly two thirds of the angle of incidence in air.

XIII. Proposition. In a glass body, no incident ray which is inclined to the normal to one of its surfaces, at an angle greater than 42°, can penetrate that surface.[7]

Consequently, this angle, $i = 42°$, is called the *critical angle* of glass. Its existence and its magnitude were discovered by Kepler. The critical angle for water is about 48°. It is plain to see, however, that even Kepler failed to find the exact relation between i and R; but the time for that discovery was near at hand.

Kepler's treatise was written in Euclidean style and is replete with interesting passages, for its author had remarkable clarity and vividness of expression. He improved the camera by placing a lens at its aperture. He also devised the astronomical telescope by substituting a convex lens for the concave lens of Galileo's instrument.

[7] J. Kepler, *Opera Omnia* (ed. Frisch, 1859), *Dioptrici*, II, 528.

Because the latter instrument produces an upright image and is shorter for equal magnification, it still persists in the form of the opera glass. It is no wonder that Kepler's book was one of the texts that kindled in young Isaac Newton a lifelong enthusiasm for optics, so that when as a professor he had a choice of topics, he chose to lecture on this subject. These lectures were given in Latin and are preserved in the book *Lectiones Opticae.*

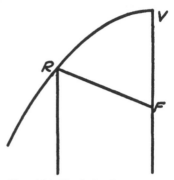

Fig. 184 is Kepler's diagram of a concave mirror with the lettering modified. As we saw in Euclid's work, there is in connection with any curved mirror a certain point F about which most of the facts and action of the mirror centre. Because a fire-place is similarly the central point of a household and the

FIG. 184. Kepler's diagram of concave mirror.

Latin word for hearth is *focus*, Kepler applied this name to the point F. He also introduced the terms vertex and axis.

> There are, however, among these lines certain points of special importance for which there is a clear definition but no name. . . . In the study of light and with the subject of mechanics in mind, we shall call these points foci. . . . Let F be the focus, V the vertex, and VF the axis.[8]

SNELL'S LAW

About the year 1620, Willebrord Snell (von Guericke's professor of physics at Leyden), discovered the law of refraction,[9] which had been sought in vain since the time of Ptolemy. In fig. 185, i and R are the angles of incidence and refraction respectively and $SA = SB$. Snell stated the law thus:

> For any given pair of transparent media, e.g. air and water, the ratio $\dfrac{\text{cosec } i}{\text{cosec } R} = \mu$, a constant.

A brief word here about cosec. Let $\angle\ ASN'$, or briefly, i (fig. 185) be any acute angle, contained by the hypotenuse AS $(=h)$ and base SN' $(=b)$ of a right-angled $\triangle\ ASN'$, whose altitude is AN', $(=a)$. The cosecant of i (written cosec i), is the ratio $h:a$. Its reciprocal, $a:h$, is called the sine of i or briefly, sin i. Snell's statement, therefore, was

$$\frac{SA}{AN'} \div \frac{SB}{BN''} = \frac{BN''}{AN'} = \mu \qquad (\because\ \ SA = SB)$$

[8]*Opera Omnia*, II, *Astronomiae pars optica*, iv, 186.
[9]See W. F. Magie, *Source Book of Physics* (McGraw-Hill, 1935).

The use of μ (the Greek letter for m) for this constant may well be a reflection of Ptolemy's equation (p. 245).

Descartes expressed the same law somewhat differently in his treatise on optics, *La Dioptrique* (1637). He discussed the subject in terms of the mechanical analogy of driving a ball with a racquet so that it strikes *GSE* (fig. 185), the surface of a pond:

> Let us now think of the ball which moves from *A* toward *U*, as encountering at the point *S* . . . water whose surface *GSE* deprives it of half [or part] of its velocity. Since the action of light follows in this respect, the same laws as the ball, . . . we must say that when its rays pass obliquely from one transparent body into another, . . . they bend in such a way that they are less inclined to

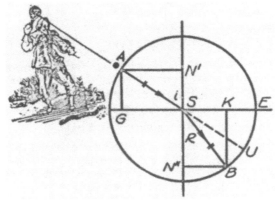

FIG. 185. Index of refraction.

the separating surface *GSE*, . . . in the medium which receives them more easily [i.e. $\angle GSA < \angle ESB$ or $i > R$], and in proportion to the ease with which they are received by the two media. But we must note that this inclination should be measured by the lengths of the straight lines *SG* or *AN'* and *SK* or *BN''*. . . not by the magnitudes of the angles i and R. For the ratio between the two angles changes with different values of i [Ptolemy], while the ratio $AN':BN''$ is constant for all refractions caused by the same media [Snell's law].

The constant just mentioned is called Snell's constant, or the *index of refraction* of the given pair of media. If air is one of the pair, it is often unmentioned. Snell, therefore, is one of that large group of scientists who have gained fame by discovering a constant in a world full of variables and Kepler was another by reason of his discovery of the critical angle. The index of refraction might well have been represented by Snell's initial, but the symbol usually used for it is μ. Dividing the terms of Descartes' ratio respectively by *AS* and *SB* which are equal, we obtain

$$\frac{AN'}{AS} \div \frac{BN''}{BS} = \frac{\sin i}{\sin R} = s = \mu$$

which is seen to be equivalent to Snell's statement. A number of scientists have felt that it would have been more gracious of Descartes to have mentioned his indebtedness to Snell in connection with this law. Descartes was little inclined to accord credit to others. To view his omission in this case as an oversight is charitable and to consider it as evidence of an independent rediscovery, probably indulgent. Technically Descartes is acquitted by the fact that although Snell showed his manuscript to Huygens, he did not publish.

Descartes concluded that, by a corpuscular theory, light must travel more swiftly in water and in glass than in air. This item was later of crucial importance in choosing between a corpuscular or emission and a wave or undular theory of light. Descartes' derivation of the law of refraction was almost wholly theoretical and by deduction. He performed few if any experiments in the subject. In this instance he committed the very error for which he severely criticized the Schoolmen. He expressed a poor opinion of the work of Galileo, who was an experimentalist. Very little of Descartes' own work has been of lasting value in physics whereas the bulk of Galileo's work has remained of permanent value. By him the foundations of dynamics and acoustics were firmly and truly laid. The lesson which he taught was that although speculation and theory can be powerful and indispensable tools, nevertheless theoretical inferences must always be verified by experiment. As Horace said, it is good for the scientist to let his head bump around amongst the stars, yet he should keep his feet on earth. The optimum is neither theory nor practice but a happy union of theory and practice.

The following passage from *La Dioptrique* has several points of interest. It marked the advent of a new theory and the passing of two old ones. In it Descartes assumed the existence of *luminiferous aether* and, for the last time in the literature, a scientist felt it necessary to combat Plato's idea of eye rays. It was also the last instance of a scientist employing the Aristotelian principle of "Horror of the Void" in support of an argument:

This will prevent you at first from finding it strange that light can transmit its rays instantaneously from the sun to us. Light rays are merely the lines along which this action takes place. In general, people see only the action which comes from luminous bodies, for experience shows us that these bodies must be self-luminous or illuminated to be seen and it is not at all that our eyes actively perceive them. Realize then that, since there is no void in Nature, as nearly all philosophers avow, and since there are nevertheless plenty of pores in all bodies that we see around us, as experiment can demonstrate very clearly, it is, therefore, necessary that these pores must be filled with some sort of very subtle and very fluid matter which extends without interruption right from the Stars to us.

Descartes' claim that the transmission of light through the luminiferous aether is instantaneous was disproved in 1673 by Olaus Roemer, a Danish astronomer. Galileo tried in vain to make the determination. At night he blanketed a lantern with an inverted pail. At a prearranged time he lifted the pail. A friend similarly equipped and stationed three miles away was to flash his lantern likewise at the instant when he saw Galileo's light signal. Then Galileo was to measure the time interval between lifting his pail and seeing the return flash. The results were inconclusive. The idea of the method was sound but the order of magnitude of the distance far too small.

DIFFRACTION AND INTERFERENCE

The professor of mathematics at the University of Bologna in 1660, Francesco Grimaldi, wrote a Latin treatise on light, entitled *Physical and Mathematical Study of Light*[10] from which the following quotation is translated. This book was first to recognize that light can bend around a corner; Grimaldi named the phenomenon *diffraction*.

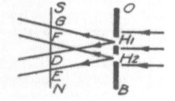

FIG. 186. Grimaldi's diagram.   FIG. 187. Diffraction.

The same book was also first to record an *interference* effect produced by superposing two beams of light. Grimaldi followed in the steps of his great countryman, da Vinci, for the apparatus he used in his experiment was essentially a pin-hole camera with two minute pin-holes. He did not live to see the publication of his book in 1666 nor to witness the far-reaching consequences of his discovery. Fig. 186 is a copy of one of Grimaldi's original diagrams.

PROP. 22. A body which is receiving light may actually become darker upon receiving still more light. .... Let two small apertures H_1 and H_2 [fig. 187] be made in the shutter of a dark room, far enough apart from each other that the cones of light which they admit begin to interpenetrate each other only at a certain distance from the apertures. Upon projecting the light on a white card, placed perpendicular to the axes of the two cones, somewhat beyond the place where the two surfaces intersect, two circular "images" are seen which

[10]F. Grimaldi, *Physico-Mathesis de Lumine*, of which there is a copy in the Library of Congress, Washington, D.C.

are partly superposed upon each other. . . . When the two apertures are opened, . . . the segment $ADCF$, which is common to the two cones, has greater intensity of illumination than the surface of the two bases in general. But what deserves the closest attention is that the two arcs, ADC and AFC, have a remarkable darkness although they are receiving more light than the rest of the circumferences of which they are parts. . . . If the card is brought nearer to the two apertures so as to diminish the segment $ADCF$, a position is reached at which the arcs ADC and AFC change to a red colour. This is because the light is formed of rays which are less able to transmit images because of their diffraction.

Grimaldi also observed rainbow colours reflected from a metal plate on which he had scratched a series of fine parallel lines.

Knowing from acoustics the significance of interference produced by superposition of waves, one can see that Grimaldi's experiment could readily lead to a wave theory of light. In fact, Grimaldi's article suggests that "the circumstances have imparted to the rays an *undulation* or a fluctuation which they retain after reflection." So great was Newton's prestige that his preference for a corpuscular or emission theory of light eclipsed Grimaldi's work for over a century. It took the work of Young and Fresnel (1810) to break the spell and to convince scientists of the occurrence of diffraction, although it had been observed, demonstrated, and named in 1666 by Grimaldi.

CHAPTER TWELVE

The School of Newton and Huygens

DOUBLE REFRACTION

E RASMUS BARTHOLINUS (1625-92) was professor of mathematics and medicine at the University of Copenhagen. He was a contemporary of Newton, and one of his students and collaborators was Olans Roemer, who first determined the velocity of light in solar space. In 1669 Bartholinus discovered the double refraction of Iceland spar, a rhomboid crystal form of calcium carbonate which occurs in Iceland, the largest of Denmark's dependencies. Bartholinus' results on double refraction were published in a Latin memoir from which we shall read:

> Objects which are viewed through this crystal . . . appear double. . . . We place . . . a dot A . . . on clean paper . . . and place upon it the lower surface of the Rhomboid. . . . Through the upper surface of the crystal . . . we see a double image of A. . . . The distance between the two images . . . increases in proportion to the size of the Crystal.

In discussing his results Bartholinus introduced three terms which have remained in use, *double refraction*, *ordinary ray*, and *extraordinary ray*.

> Letting the eye and the object remain at rest, we turn the prism. . . . We perceive that while one of the images remains fixed in position . . . the other follows the motion of the prism [describing a circle]. By assuming that . . . refraction is the cause of the phenomenon involved, it is admissible to draw the inference that the double image is produced by double refraction. . . . We designate as ordinary refraction that which produces the fixed image and as extraordinary that which produces the other image. The crystal itself is described as doubly refracting.[1]

NEWTON'S *OPTICKS*

The Preface of Newton's *Opticks* allows one to estimate how fond Newton was of controversy:

[1]*Experimentis Crystalli Islandici*, 1669. Quoted in fuller detail in Magie, *Source Book in Physics.*

Part of the ensuing Discourse about Light was written at the Desire of some Gentlemen of the Royal Society, in the Year 1675 and then sent to their Secretary.... To avoid being engaged in Disputes about these Matters, I have hitherto delayed the printing and should still have delayed it, had not the Importunity of Friends prevailed upon me.... I have here publish'd what I think proper to come abroad, wishing that it may not be translated into another language without my Consent.

Let us look at its opening passage and a number of its axioms.

My Design in this Book is not to explain the Properties of Light by Hypotheses, but to propose and prove them by Reason and Experiments: In order to which I shall premise the following Definitions and Axioms.

Axiom I. Angles of Reflexion and Refraction lie in one and the same Plane with the Angle of Incidence.

Axiom II. The Angle of Reflexion is equal to the Angle of Incidence.

Axiom III. If the refracted Ray be returned directly back to the Point of Incidence, it shall be refracted into the Line before described by the incident Ray.

Axiom IV. Refraction out of the rarer Medium into the denser is made towards the Perpendicular: that is, so that the Angle of Refraction be less than the angle of Incidence.

Axiom V. The Sine of Incidence is either accurately or very nearly in a given Ratio to the sine of Refraction.

These axioms give a résumé of previous findings and single out the foundation stones which form the basis of the theory of optics. Axiom I is Alhazen's law and Axiom II is Euclid's angle law of reflection. Axioms III and IV we met as discoveries of Claudius Ptolemy; and V is Snell's law of refraction with a slight demur. These axioms are arranged in approximately chronological order and the sixth brings us to Kepler's work with some developments by Newton. It is Newton's definition of focus. Kepler defined principal focus; Newton broadened the term. The second part of the axiom is a beautiful example of Newton's ability to say much in few words. What takes a number of weeks of study in an elementary course, he brushes off with a few strokes of the pen. In reading his work, one should keep lead-pencil and paper at hand for filling in the steps, and the same is true to a considerable extent for all mathematical treatises. Fig. 188 is given to help in grasping the full significance of the words:

Axiom VI. The Point from which Rays diverge or to which they converge may be called their Focus. And the Focus of the incident Rays being given, that of the reflected or refracted ones may be found by finding the Refraction of any two Rays.

In fig. 188, two rays are drawn from A and their intersection after refraction locates the focus A'. A and A' are called *conjugate foci*. The two rays ARA' and ACA', respectively parallel to the axis and through

C the centre of the lens, are called *critical rays* because they analyse
the situation; but Newton says any two rays will do. Instead of giving
a separate discussion for curved mirrors, he simply remarks that for
these one merely uses the same ideas over again. He next derives the
fundamental mirror-and-lens formula. Have the lead-pencil and paper
ready for action.

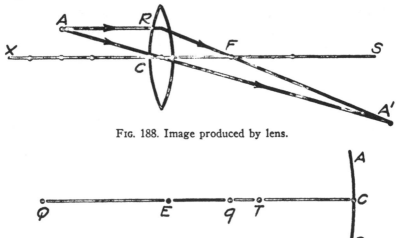

FIG. 188. Image produced by lens.

FIG. 189. Image produced by mirror.

CASE 2. Let *ACB* in fig. [189] be the reflecting Surface of any Sphere whose
Centre is *E*. Bisect any Radius thereof (suppose *EC*) in *T* and if in that Radius
on the same side of the Point *T*, you take the points *Q* and *q*, so that *TQ*, *TE*,
and *Tq* be continual Proportionals and the Point *Q* be the Focus of the incident
Rays, the Point *q* shall be the Focus of the reflected ones.

In examining this passage, we must, as usual, beware of Newton's
disarming simplicity of expression and must fill in any steps in the
argument which he takes for granted and omits. He does not show two
reflected rays but merely says that

Eq. 1. $TQ : TE = TE : Tq$, or $TQ.Tq = (TE)^2$
             Let $QC = u$, $qC = v$, and $TC = TE = f$ (the focal length)
             Then $TQ = (u - f)$, $Tq = (v - f)$ and
             $\therefore (u - f)(v - f) = f^2$
             $\therefore uv - uf - vf + f^2 = f^2$.

Eq. 2. $\therefore$ by dividing by uvf, $\dfrac{1}{f} = \dfrac{1}{u} + \dfrac{1}{v}$.

This is a mirror-and-lens formula. The symbols *u* and *v* represent the
distances of two conjugate foci from *C*, the vertex of the mirror or the
centre of a lens. Since *u* and *v* are, respectively, the distances of an

object and its *image* from the instrument, they are sometimes symbolized by o and i. From *Eq.* 2, p. 264, are obtained the following useful equations:

Eq. 3. $f = \dfrac{oi}{o+i} = \dfrac{uv}{u+v}$ and

Eq. 4. $i = \dfrac{of}{o-f}$ or $v = \dfrac{uf}{u-f}$.

By *Eq.* 3 we can find f if o and i are known, and *Eq.* 4 finds i if o and f are known. Fig. 190 fills in some details in Newton's diagram. It

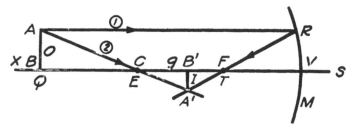

FIG. 190. Image produced by concave mirror.

shows C the centre of curvature of the concave spherical mirror, V the vertex, F the principal focus and two critical rays. AB is a luminous body standing on the principal axis XS, and $A'B'$ is its real image, projected on a screen between F and C. In axiom VI, Newton gives no hint as to how he reached the idea but in his *Lectures in Optics*[2] he gives a full demonstration using similar triangles.

By similar triangles, $\dfrac{QC}{CB'} = \dfrac{AB}{A'B'} = \dfrac{RV}{A'B'} = \dfrac{FV}{FB'}$

$\therefore$ $\dfrac{QC}{FV} = \dfrac{CB'}{FB'}$

$\therefore$ $\dfrac{QC+FV}{FV} = \dfrac{CB'+FB'}{FB'}$

$\therefore$ $\dfrac{QC+CF}{FV} = \dfrac{CF}{FB'}$

$\therefore$ $\dfrac{FQ}{FC} = \dfrac{FC}{FB'}$

$\therefore$ $FQ : FC = FC : FB'$.   (Q.E.D.)

Newton's former student, Dr. Edmund Halley, Astronomer Royal, was probably the first to derive this useful mirror-and-lens formula.

[2]I. Newton, *Lectiones Opticae* (1669), I, iv, Prop. 29.

A suggestion of the power with which Newton attacked a problem may be gathered from the fact that when he took up the following experiment it was already celebrated and "known to the antients," yet his work was such that ever since the experiment has been called "Newton's experiment with a prism." The terseness with which he introduces the term "spectrum" in the following letter is characteristic of him.

Cambridge, February 6, 1670

MR. HENRY OLDENBURG, S.R.S.
LONDON.

Dear Sir,

To perform my late promise to you, I shall without further ceremony acquaint you that in the year 1666, I procured me a triangular prism, to try therewith the celebrated phaenomena of colours. And in order thereto, having darkened my chamber and made a small hole in my window-shuts, to let in a convenient quantity of the Sun's light, I placed my Prisme at his entrance that it might be thereby refracted to the opposite wall. It was at first a very pleasant divertissement to view the vivid and intense colours produced thereby, but after a while, applying myself to consider them more circumspectly, I became surprised to see them in oblong form which according to the received laws of refraction, I expected should have been circular.... Comparing the length of this coloured spectrum, with its breadth, I found it about five times greater; a disproportion so extravagant that it excited me to a more than ordinary curiosity of examining whence it might proceed....

I took two boards and placed one of them behind the prism [P_1] at the window so that the light might pass through a small hole, made in it for the purpose, and fall on the other board which I placed at about 12 feet distance, having first made a small hole in it for some of that incident light to pass through. Then I placed another prism [P_2] behind this second board so that the light trajected through both boards might pass through that also and be again refracted before it arrived at the wall. This done, I took the first prism in my hand and turned it to and fro slowly about its axis so much as to make the several parts of the image cast on the second board successively pass through the hole in it that I might observe to what places on the wall the second prism would refract them. And I saw by the variation of those places that the light tending to that end of the image towards which the refraction of the first prism was made [i.e. violet] did in the second prism suffer a refraction considerably greater than the light tending to the other end [red]. And so the true cause of the length of that image was detected to be no other than that [sun]light is not similar or homogeneal but consists of difform rays some of which are more refrangible than others; and that according to their particular degrees of refrangibility they were transmitted through the prism to divers parts of the opposite wall.

The phenomenon which Newton has just described in a few sentences (!), namely the separation of rays of different hues by refraction, is called *dispersion*. One objection to lens telescopes was the colour fringes of their images due to dispersion. Newton thought, wrongly[3] but very excusably, that this defect could not be obviated and since all rays, of whatever hue, obey the same law of reflection ($r = i$), he invented a reflecting telescope. His first paper read before the Royal Society was a description of this invention. At the same time (1672), he was elected a Fellow of the Society at the age of thirty. One of the Royal Society's treasures is a reflecting telescope which Newton made and presented to the Society.

To support his demonstration of analysis of sunlight, Newton gave two proofs by synthesis; in the first, he utilized persistence of vision to fuse the hues of the spectrum in the eye by means of a rotating disc,

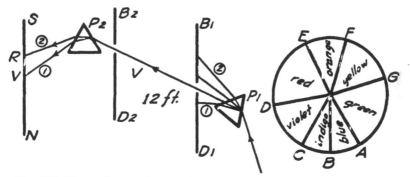

FIG. 191. Newton's experiment with a prism.   FIG. 192. Newton disc.

which was a modification of Alhazen's colour-top. Newton's *Opticks* refers to this disc which is now called a Newton disc:

> In a mixture of Primary Colours, the Quantity and Quality of each being given, to know the Colour of the Compound.
> With the centre O and Radius OD, describe a circle ADF, [fig. 192] and distinguish its Circumference into seven Parts.... Let the first Part, DE, represent a red Colour, the second, EF, orange ... FG, yellow, ... GA, green, AB, blue, ... BC, indigo, ... and the seventh, CD, violet.
> If a burning Coal be nimbly moved around in a Circle with Gyrations continually repeated, the whole Circle will appear like Fire [da Vinci]. By the Quickness of the Successions, the Impressions of the several Colours are confounded in the Sensorium and out of the Confusion, ariseth a mix'd Sensation.

When a Newton disc is rotated, the colour seen is generally an approach to gray or white, usually with some faint hue, e.g. pinkish gray, greenish gray, etc., but who could expect to "paint the colours of the rainbow" on the disc? If a Newton disc is made of sectors of coloured gelatine or

[3]In 1758, John Dollond, a London optician, prepared an achromatic lens composed of two parts, one of crown glass and the other of flint glass. This was an excellent contribution to the development of the telescope and the microscope.

glass, it appears, when spun swiftly, almost as colourless as window-pane. Objects seen through it have nearly their natural colours. His second proof was by inserting a duplicate prism in reversed position; the light emerging on a white card gave it the same appearance as if in sunlight. A lens serves the same purpose.

If a piece of ruby glass is inserted on either side of the prism in Newton's experiment, only the red part of the spectrum appears on the screen. The rays of other hues are said to be absorbed (probably converted to heat). Ruby glass, then, must be transparent to red light and to other hues as opaque as lead. It is therefore called a red *filter*, and the light which passes through it, being of only one hue, is called *pure* red or monochromatic (Gk. *mono*, one; *chroma*, colour). If a piece of green glass is used, we may see on the screen green prominently and, with less intensity, yellow and blue. This is a green filter but not pure green. It is said to be impure and no opprobrium attaches to the term; it might also be called complex. Most colours in practice are impure or complex. This green we can say has two *partials* (borrowing shamelessly but gratefully from acoustics) and green is its *fundamental* hue. When a colour has two partials—and most colours have more—they are apt to be the two *spectrum neighbours* of the fundamental but not necessarily so.

If a drop of red ink falls on a white blotter, the fibres at that spot are stained red and become red filters. At that place incident sun-rays penetrate a certain distance into the surface and are then reflected (in part) for they have traversed red filters which transmitted the red rays and absorbed all others. The process (or effect) is called a *subtraction* phenomenon, for the eye is affected by what is left after the filters exact their commission. The part subtracted is complementary to that reflected, for together they make up a full spectrum.

Robert Hooke combined two filters and Newton referred to his experiment:

From hence also is manifest the reason of an unexpected experiment which Mr. *Hook*, somewhere in his *Micrographia*, relates to have made with two ... transparent vessels, filled the one with red, the other with a blue liquor: namely, that though they were severally transparent enough, yet both together became opake; for if one transmitted only red, and the other only blue, no rays could pass through both.

That Newton favoured a *corpuscular theory* of light in preference to a wave theory is indicated by the following query in his *Opticks:*

Query 29. Are not the Rays of Light very small Bodies emitted from Shining Substances? For such Bodies will pass through uniform Mediums in right [straight] Lines without bending into the Shadow [diffraction] which is the Nature of the Rays of Light.

Supporters of the emission theory "explained" Grimaldi's observation of alternate light and dark bands on a screen (or explained them away) by assuming, without much experimental evidence, that the bands were produced by mutual attractions and repulsions between light corpuscles and the edges of the narrow aperture through which the light passed. This argument was refuted by the work of Fresnel (1810).

Of the many interesting passages of the *Opticks* we shall find space for only two more. The first contains the germ of the term *polarization* of light, which later became very important.

And lastly, the unusual Refraction of Island-Crystal looks very much as if it were perform'd by some kind of attractive virtue lodged in certain Sides both of the Rays and of the Particles of the Crystal.

The fundamental meaning of polarization is "two-sidedness." Thus the poles of the earth distinguish its northern and southern sides and similarly the poles of a magnet. The joists which support a floor are two-sided and must be placed with their sides vertical if they are to give the floor adequate support. Their mechanical properties are not the same at a side and at an edge. The credit for discovering polarization of light goes to Huygens.

NEWTON'S RINGS

If a convex lens is squeezed against a piece of plate glass so that sun-rays come through an air wedge between the two pieces of glass, concentric rings of rainbow hues are observed. These are named Newton's Rings (or Newton Rings) after their discoverer. The wave theory of light gives a satisfactory explanation of the phenomenon by employing the idea that opposite phases if superposed counteract each other. In elaborating a corpuscular explanation of these colour fringes Newton made use of the fact that when light strikes a transparent surface obliquely both reflection and refraction occur. He assumed that light particles have pulsating fits and starts of reflection and refraction.

Nothing more is requisite for putting the Rays of Light into Fits of easy Reflexion and easy Transmission than that they be small Bodies ... which stir up Vibrations ... which agitate them so as by turns to increase and decrease their Velocities, and thereby put them into those Fits.

This assumption comes near to borrowing some of the wave-theory's thunder for it approaches the idea of like and opposite phases. From this and other statements in the *Opticks* one might surmise that with five or ten years' more work, Newton might have come to prefer the wave theory. Between the lines he sometimes seems to say, "Huygens, Huygens, almost thou persuadest me to be a wave-theory man." Had that been the case, progress could have been accelerated by half a century or more. But since the great Newton said corpuscles, corpuscles it was until the work of Young and Fresnel showed the advantages of the undular theory.

WAVE THEORY OF LIGHT

Although Huygens (1629-1695) was a great admirer and friend of Newton, yet he differed from him in preferring a wave theory of light to a corpuscular theory. In dealing with assumptions and hypotheses it is legitimate to have a preference as did Newton and Huygens in this case; but in stating facts, no such choice is possible. Huygens was the chief founder of the undular theory of light. Here and there he had a few supporters, such as Robert Hooke, Benjamin Franklin, and Leonard Euler. In 1690 he published a treatise on light, entitled *Traité de la Lumière*, which contains the following introduction of his theory. Its opening sentences afford an interesting explanation of the word "explanation." You may have noticed already that the word "explain" is not the easiest to explain. Subsequent work has shown that we cannot follow Huygens' advice altogether in attempting to make mechanical explanations; in some cases we cannot follow it at all.

In true philosophy, . . . we conceive the causes of all natural effects in terms of mechanics. That is what we must do, according to my opinion, or renounce all hope of ever understanding anything in physics.

When we consider (1) the extreme speed with which light is transmitted in all directions, and (2) the fact that when it comes from different directions, even from opposite ones, the rays cross each other without hindering each other, we realize that when we see a luminous body, this could not be by the transport of matter which is projected to us from that body as a ball or an arrow traverses the air, for assuredly, that is too much at variance with these two qualities of light and especially the latter. It is then in another manner that it travels, and the knowledge which can lead us to an understanding of it, is the knowledge we have of the propagation of sound in air.[4]

That is to say, the knowledge of wave-motion. It is reasonable and legitimate to assume that light travels in the form of a train of waves; but assumptions, being products of the imagination, are like falsehoods in that one assumption often leads to another. In the case of the wave theory of light we shall see that we are forced to make other assumptions which seem impossibilities or at least contrary to the laws of mechanics for macroscopic bodies. Since light traverses solar space, if that is empty space or vacuum, then light waves are waves of nothing which on reaching the earth can do something. Thus we are forced to assume a medium, the luminiferous aether, and its properties must be

[4]C. Huygens, *Œuvres complètes*, XIX, 461.

radically different from those of any substance we know. If our reasoning is to be sound we must scrutinize all assumptions closely; and
above all we must strive to distinguish between fact and hypothesis.

Huygens has just mentioned the enormous speed of light and his
book tells of a famous series of observations made by the Danish
physicist Olans Roemer, professor of mathematics at Copenhagen, from
which he determined the speed of light in solar space. In determining
the speed of sound in air by a direct method, we found that the length
of an ordinary laboratory was too short. It was necessary to go out
where the distance involved miles. Similarly, in determining the speed
of light one difficulty is to obtain a laboratory that is long enough.
Galileo tried the experiment with a distance of three miles; but in vain.

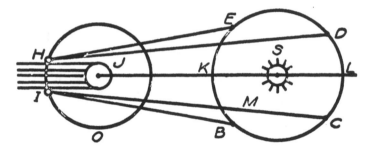

FIG. 193. Huygens' diagram of Roemer's experiment.

In fact it turns out that if a laboratory were as long as the whole earth,
it would still be far too short. This looks like an insurmountable difficulty. One may wonder how in the world a physicist could obtain a
laboratory longer than the earth. But an astronomer did accomplish
the trick:

M. Roemer makes use of the eclipses [or occultations] experienced by the
moons that revolve around Jupiter and frequently enter his shadow. This is
his argument. Let S in the figure [fig. 193] be the sun, $BCDE$, the Earth's
annual orbit, J, Jupiter, Io, the orbit of the inner satellite [named by Galileo
after poor Io]. Let H be the moon entering Jupiter's shadow, I, the same
satellite emerging from the shadow. If the Earth remained at B, after 42½
hours an observer would see a similar emergence, for that is the period in which
Io makes one journey around its orbit. And if the Earth remained always at
B, during 30 revolutions of Io, the observer would once more see Io emerging
from the shadow after an interval of 30 times 42½ hours. But since the Earth
moves during that time to C, away from Jupiter, it follows that if light requires
time for its transmission, the light of the small moon will be seen later at C
than at B, and it would be necessary to add to the interval 30 times 42½ hours,
the time light takes to traverse the distance MC, the difference between the
distances IC and IB.

Now by a series of observations of these occultations, made during ten consecutive years, these differences have been found to be considerable, as much as ten minutes and more; and it has been discovered that to travel the whole diameter of the earth's orbit *KL*, which is twice the distance from there to the Sun, light requires about 22 minutes.

Subsequent work has corrected this value to about 16 minutes and 40 seconds, or about 1000 seconds for light to travel the diameter of the earth's orbit, i.e. 2 × 93,000,000 miles, or 186,000,000 miles. Therefore, in one second light travels through solar space about 186,000 miles. In the metric system, this terrific speed is

$$186,000 \times 5280 \times 12 \times 2.54 = 3 \times 10^{10} \text{ cm. per sec.,}$$

or about 7 times around the earth in one second.

In employing the undular theory Huygens made use of a principle which is called *Huygens' wave-front principle*.

We have still to consider, in studying the spreading out of these waves, that each particle of the matter in which a wave proceeds not only communicates its motion to the next particle, ... which is on the straight line drawn from the luminous point, but that it also necessarily gives a motion to all the others which touch it and which oppose its motion. The result is that around each particle there arises a wave of which this particle is the centre.

By means of his wave-front principle and the assumption that light travels *more slowly* in water and in glass than in air, Huygens was able to deduce Snell's law of refraction. He also showed that the ratio between the two velocities equals Snell's constant, the index of refraction, or in algebra,

Eq. 5. $$\frac{v_a}{v_g} = \mu_g^a = \frac{3}{2}$$

where v_a and v_g are the velocities of light in air and in glass.

∵ $v_a = 3 \times 10^{10}$ cm.p.s.   ∴ $v_g = 2 \times 10^{10}$ cm.p.s.

if light is a wave phenomenon.

Here then was a crucial difference between the emission and the wave theories of light. The latter required that light travel more slowly in water and glass than in air, whereas the former required that it travel faster. About 1850, Arago and his two assistants, Foucault and Fizeau, settled this point experimentally and their findings were a victory for the wave theory.

By a further brilliant and powerful use of his wave-front principle, Huygens succeeded in obtaining an explanation of the *extraordinary ray* of double refraction. Ordinary refraction he ascribed to the transmission through the aether in the interstices of the crystal, and "more slowly within the crystal than outside it." He assumed that the speed of light transmitted by the particles of the crystal had different values according to whether it was travelling parallel to the axis of the crystal or at right angles to it. Consequently the wave-fronts would not be spherical but ellipsoidal (or spheroidal). In this way he reached an explanation entirely in accordance with the facts known about extraordinary rays. Such a feat he could not have achieved had he not been a master mathematician with an excellent knowledge of solid geometry.

At the close of his famous treatise, Huygens recorded his discovery of the *polarization* of light, admitting that as yet he had not succeeded in devising an explanation. He showed that the ordinary and extraordinary rays emerging from a crystal traversed a second crystal with greater or lesser intensity according to which way the latter was turned. Thus their properties were different from side to side. In other words they were polarized.

THE PHOTOMETER

When Count Rumford (1753-1814) entered Bavaria as efficiency expert to Prince Maximilian, he found the region overrun with beggars. His attitude toward this disorder, he expressed in these words:

> To make vicious and abandoned people happy, it has generally been supposed necessary first to make them virtuous. But why not reverse this order? Why not make them first happy and then virtuous?

By building a large factory and offering employment with good wages, and good working and living conditions, he solved the problem of mendicancy completely. In solving the problem of furnishing good lighting for his workers, he invented the *photometer*[5] and here is his report:

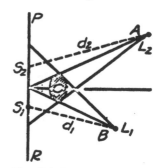

FIG. 194. Rumford's shadow photometer.

> Being employed in making a number of experiments to determine, if possible, the most economical method of lighting up a very large public manufactory . . . a method occurred to me for measuring the relative quantities of light emitted by lamps of different constructions, candles, etc. . . . . Let the two lamps or other lights to be compared, A and B in the figure [fig. 194], be placed at equal heights upon two movable stands in a darkened room. Let a sheet of white paper be fastened upon the wainscot and let the lights be 6 to 8 feet from it in such a manner that a line drawn from the centre of the paper, perpendicular to its surface, shall bisect the angle formed by the lines drawn from the lights to that centre. When this is done, a small cylinder of wood, C, about ¼ of an inch in diameter and 6 inches long, must be held in a vertical position about 2 or 3 inches before the centre of the paper, and in such

[5]See D. B. Hammond, *Stories of Scientific Discovery.*

a manner that the two shadows of the cylinder S_1 and S_2 may be distinctly seen. If these shadows be found of unequal densities [intensities of illumination] then that light whose corresponding shadow is the denser, must be removed further off or the other must be brought nearer to the paper, till the densities of the shadows appear to be exactly equal. The distances of the lights from the centre of the paper being measured, the squares of those distances will be to each other as the real intensities [i.e. powers of illumination] of the lights in question at their sources. I have made many improvements in the apparatus, and I have now brought the principal instrument to such a degree of perfection that, if I might, without being suspected of affectation, I should dignify it with a name, and call it a photometer.[6]

[6]Rumford, "Intensities of Light," read before the Royal Society of London in 1794. Or see *Life and Works of Count Rumford* (Macmillan, 1876), v. 5, p. 7.

CHAPTER THIRTEEN

The School of Young and Fresnel

INTERFERENCE

A MONG Rumford's many services to science, his instrumentality
in founding the Royal Institution of London was not the least.
For its professor of physics, he selected a very brilliant man, Dr.
Thomas Young. In his Bakerian Lecture on Light and Colours,
delivered before the Royal Society in 1801, Young showed the significance of Grimaldi's experiments, and with them and many others of
similar nature, brought strong support to the wave theory of light. In
this passage we meet the source of the term *interference*, also of *fringes*,
used to describe striae or alternate light and dark bands first observed
by Grimaldi:

The optical observations of Newton are yet unrivalled . . . and they only
rise in our estimation as we compare them with later attempts. . . .

A further consideration of the colours of thin plates as they are described in
the second book of Newton's *Optics*, has converted that prepossession which I
before entertained for the undulatory system of light into a very strong conviction of its sufficiency. . . .

Suppose a number of equal waves of water to move upon the surface of a
stagnant lake with a certain constant velocity and to enter a narrow channel
leading out of the lake. Suppose then, another similar cause to have excited
another series of equal waves which arrive at the channel at the same time. . . .
One series of waves will not destroy the other but their effects will be combined.
If the elevations . . . of one series coincide with the other, they must together
produce a series of greater joint elevations; but if the elevations of one series
are so situated as to correspond to the depressions of the other, they must exactly
fill up those depressions and the surface of the water must remain smooth. . . .

Now I maintain that similar effects take place whenever two portions of
light are thus mixed, and this I call the general law of the interference of light.

I made a small hole in a window-shutter and covered it with a piece of thick
paper which I perforated with a fine needle. I placed a small looking-glass
without the window-shutter, in such a position as to reflect the sun's light in a
direction nearly horizontal, upon the opposite wall. I brought into the sunbeam
a slip of card about one-thirtieth of an inch in breadth and observed its shadow.

Beside the fringes of colour on each side of the shadow, the shadow itself was divided by similar parallel fringes [striae, fig. 195] . . . leaving the middle of the shadow always white. Now these fringes were the joint effects of the portions of light passing on each side of the slip of card [CR, fig. 197] and inflected or rather diffracted into the shadow. It is well known that a similar cause produces in sound that effect which is called a beat: two series of undulations of nearly equal magnitude co-operating and destroying each other alternately.[1]

Young was a man of whom we might say that he "had everything." All the blessings were his. He had a splendid and handsome physique and an independent and scintillating intellect. At two he could read fairly well. At fourteen he knew twelve languages. The whole world of knowledge in all its departments was an open book to him, as it was to Imhotep in the days of Zoser. Young was a man of culture and

Fig. 195. Striae.

grace. His profession was medicine and his finances were comfortable. He had a beautiful accomplished wife and a happy home life. He had plenty of interesting problems and everything to make life happy. The deciphering of the Rosetta tablet was merely a fascinating avocation for him. If there was one fly in the ointment it was that his associates were less talented and were hardly on the same wave-length; as a leader he was sometimes too far ahead of the troops.

For a century Newton's prestige had been so great that his preference for a corpuscular theory of light had established it in the scientific world to the exclusion of Huygens' wave theory, which had many advantages over the corpuscular—and some disadvantages. Young's work, together with that of Augustin Fresnel of Paris, was so well done that it turned the tide of opinion, the wave theory rolled to shore and swept the corpuscular theory out of vogue—for a century.

For several years Young's work on wave theory brought him little but vilification. He was showered with such epithets as illogical, absurd, visionary, and worse. A writer whose intellect, in comparison with that of Young, was of pygmy calibre, referred to his work as "feeble lucubration," and an innovation "that can only serve to check the progress of science." Young replied to this attack in a very able pamphlet but effected little change in scientific opinion, since only one copy was sold. It was two Frenchmen, Arago and Fresnel, who brought Young the

[1]T. Young, *Miscellaneous Works*, I, 140.

support and recognition he deserved. His was not the first instance of a prophet being not without honour save in his own country.

The career of Augustin J. Fresnel (1788-1827) was somewhat different from that of Young. His health was so delicate that he did not learn to read until he was eight; but he was diligent and thorough and

FIG. 196. Thomas Young.

he lived in the exhilarating atmosphere of the brilliant French school of mathematical physics that flourished for half a century before and half a century after the days of Napoleon. Fresnel made some of Young's discoveries independently. When he took up the cudgels for the wave-theory in France, his first paper was judged by Arago, the Astronomer Royal, who became Fresnel's first convert and actually requested that his name appear as co-author of the paper. Now Arago never grappled with a problem by halves, and soon he showed Fresnel some of Young's work. Fresnel wrote to Young in 1819: "If anything could console me for losing priority, it is to have met a savant who has enriched physics with so many important discoveries and who has at the same time strengthened my confidence in the wave theory of light."

Young replied, "I return a thousand thanks, Monsieur, for the gift of your admirable memoir, which surely merits the very highest rank amongst papers which have contributed most to the progress of optics." Young and Fresnel became close friends and collaborators. One of Fresnel's advantages was his mathematics, which he used to such purpose that we can say of him that he brought the science close to that of modern college texts in optics.

To scrutinize the emission "explanation" of Grimaldi's results, Fresnel performed a classic experiment which bears his name. He allowed a beam of light to fall on two plane metallic mirrors inclined to each other at nearly 180°. Where the two reflected beams were super-

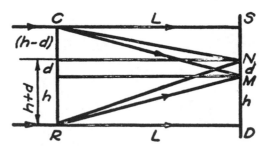

FIG. 197. Calculation of wave-length.

posed on a screen, he observed alternate light and dark bands. Here there could be no talk about attraction and repulsion between light corpuscles and edges of an aperture. These facts rendered the emission explanation valueless, cut away the ground from underneath his opponents and overwhelmed them. For the first time a considerable number of scientists saw some of the advantages of the undular theory, admitted that Grimaldi was justified in claiming the occurrence of diffraction, and realized that Newton was wrong in his opinion that light does not bend into the shadow.

From the phenomenon of striae or interference fringes in a shadow let us try to calculate the wave length of monochromatic light, e.g. yellow.

In fig. 197, L is the distance from the opaque object CR (a needle) to its shadow SD. M is the bright streak at the middle of SD, and N is the next stria. Then RN is one wave-length (λ) longer than CN. Let $MN = d$ and $MD = h$.

$\therefore \lambda = (RN - CN) = \sqrt{L^2+(h+d)^2} - \sqrt{L^2+(h-d)^2}$ by the Pythagorean theorem. Expanding this expression by binomial theorem gives the value of λ.

$$\therefore \lambda = L\left[\left\{1 + \left(\frac{h+d}{L}\right)^2\right\}^{\frac{1}{2}} - \left\{1 + \left(\frac{h-d}{L}\right)^2\right\}^{\frac{1}{2}}\right]$$

$$= L\left[1 + \tfrac{1}{2}\left(\frac{h+d}{L}\right)^2 + \ldots -1 -\tfrac{1}{2}\left(\frac{h-d}{L}\right)^2 -\ldots\right]$$

$$\therefore \lambda = L\left[\frac{h^2 + 2hd + d^2 - h^2 + 2hd - d^2}{2L^2}\right] = \frac{2hd}{L}$$

or more briefly,

$$\lambda = \frac{RN^2 - CN^2}{RN + CN} = \frac{L^2 + (h+d)^2 - L^2 - (h-d)^2}{2L} = \frac{4hd}{2L} = \frac{2hd}{L}.$$

For yellow light, if $L = 5$ m., and $h = 3$ mm., it is found that $d = 0.5$ mm.

$$\therefore \lambda = \frac{2 \times 3 \times \frac{1}{2}}{5 \times 1000}\text{ mm.} = \frac{3 \times 10^7}{5000}\text{ Ångström units } = 6000\text{ Å.}$$

1 mm. = 1000 microns = 1000 μ
1 μ = 10,000 Å $\therefore$ 1 mm. = 10^7 Å.

Similarly it is found that λ for violet light is 3800 Å,
and λ for red light is   7600 Å.

Hence it can be said in acoustical terms that the interval between red and violet light is about an octave.

Of Young's many optical researches we shall mention only three more here. He described and explained the defect called *astigmatism*. He also advanced the idea that colour perception is achieved by three kinds of nerve endings in the retina which respond respectively to red, yellow-green and blue-violet light.

It would be far from astonishing to find that the beautiful and fascinating phenomena of polarized light intrigued such a man as Young. Nor would it be surprising to learn that he hurdled some difficulties which were too much for his predecessors. The great Huygens had failed to reach an explanation for polarization. Young proposed the assumption that *light waves are transverse*.[2] That idea may have occurred to Huygens: but if so, he refused it countenance and bade the temptation get behind him, for pond-waves cannot travel through the interior of a pond; they are restricted to surfaces. But light waves traverse the interior of the medium. This then seems to be an assumption contrary to ordinary mechanics; but if adopted it organizes the subject of polarization beautifully.

About the same time the idea of transverse waves came independently to Fresnel, but when he broached the assumption to Arago, the

[2]Robert Hooke had also proposed this idea.

latter protested that he could never summon the temerity to publish such an incongruous idea. Accordingly when the next part of their memoir was published, Arago's name was omitted at his own request.

That light is *polarized by reflection* from a non-metallic surface was discovered in 1809 by Etienne L. Malus (1777-1812). He was a young Parisian military engineer and like most physicists of that day was a staunch subscriber to the emission theory of light. The French Institute had proposed a prize question on double refraction and Malus won the prize in 1810 at the age of 33, only two years before his untimely death. His friend Biot, another ardent corpuscularist, tells us that Malus at the open window of his room in the rue d'Enfer saw the setting sun reflected from the windows of the Luxembourg Palace and viewed the image through a doubly refracting crystal. As he rotated the crystal, at a certain position, he was astonished to see one of the two images disappear. Later in the evening he found the same effect with candle-light reflected by a water surface. Thus he discovered polarization by reflection. With remarkable insight and savoir faire he performed that same night a number of experiments which discovered a whole set of fundamental facts and principles in polarization. Seldom indeed has any man rivalled the achievement of Malus in founding a new department in physics in one night.

THE KALEIDOSCOPE

Sir David Brewster invented the kaleidoscope about 1817 and the excerpts quoted below are from a paper he published two years later—partly to vindicate his patent rights, which had been treacherously infringed.

The name Kaleidoscope which I have given to a new Optical Instrument for creating and exhibiting beautiful forms, is derived from the Greek words, καλός, beautiful, εἶδος, form, and σκοπεῖν, to see. . . . The principal parts of the kaleidoscope are two reflecting planes, made of glass or metal. When these two plates are put together at an agle of 60°. . . and placed in a tube . . . and the eye placed at the end E in the figure [fig. 198], it will observe the opening AOB multiplied six times and arranged around the centre O. The kaleidoscope is capable of creating beautiful forms from the most ugly and shapeless objects. It produces symmetrical and beautiful pictures by converting simple into compound forms . . . and arranging them into one perfect whole. . . .

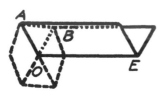

Fig. 198. Kaleidoscope.

Upon the principles of combination . . . 24 pieces of glass may be combined 139,172,428, 887,252,999,425,128,493,402,200 times, an operation the performance of which, even upon the assumption that 20 of them were performed every minute, would take hundreds of thousands of millions of years. This system of endless change is one of the most extraordinary properties of the kaleidoscope. With a number of loose objects, it is impossible to reproduce any figure which we have admired. When it is once lost, centuries may elapse before the same combination returns. We have heard of many cases where the tedium of severe and continued indisposition have been removed, and where many a dull and solitary hour has been rendered cheerful by the unceasing variety of entertainment which the Kaleidoscope afforded.[3]

Those who follow Sir David's calculations on combinations may rejoice that he rounded off the last period in his answer. Apparently Sir David, before obtaining a patent, gave his specifications to a jeweller to make a copy. Then, presto, the instruments were selling by thousands in London and Paris—a million in a few months! A hint to young inventors is that if Brewster had made the instrument himself, the jeweller could not have tricked him. For a decade or more, the kaleidoscope was all the rage. Fifty years ago, many a parlour had a centre table with a plush-covered photograph album on its shelf and on the top, on a lace doily, a kaleidoscope for helping some young man to endure tedium while her ladyship made ready for the concert.

The same treatise introduced the term *complementary colours*:

There are indeed few minds that are not alive to the soothing and exhilarating influence of musical sounds or that do not associate them with the dearest and most tender sympathies of our nature. But the ear is not the only avenue to the heart, and though sorrow and distress are represented by notes of a deep and solemn character and happiness and gaiety by more light and playful tones, the same kind of feelings may also be excited by the exhibition of dark and gloomy colours and by the display of bright and aërial tints. Even those who are ignorant of these principles [of harmony] will acknowledge the superior effect which is obtained by the exclusion of all other colours except those which harmonize with each other. In order to enable any person to find what colours harmonize with each other, I have drawn up the following table which contains the harmonic colours.

Deepest Red	—Blue and Green equally mixed.	Yellow	—Violet and Indigo in nearly equal proportions.
Red	Blue, unmixed.	Greenish Yellow	—Pale Violet
Orange Red	—Blue mixed with much Indigo.	Green	—Violet
Orange	—Blue and Indigo the Indigo predominating.	Greenish Blue	—Violet and Red in equal proportions.
Orange Yellow	—Indigo unmixed.	Blue	—Red.
		Indigo	—Orange Red.
		Violet	—Green.

These colours are also called *complementary colours* because the one is the complement of the other or what the other wants of white light; that is when the two colours are mixed, they will always form 'white light' by their combination.

[3]D. Brewster, *Treatise on the Kaleidoscope*, 1819.

WHEATSTONE'S ROTATING MIRROR

The rotating mirror, invented in 1830 by Sir Charles Wheatstone (1802-75) of King's College, London, is an example of an instrument that failed (at least at first) in the particular purpose for which it was devised but proved successful in other problems. Wheatstone wished to find whether an electric spark is oscillatory, but failed to find the answer by means of the mirror. Strangely enough, Lord Kelvin found the answer by algebra and Savary by an experiment. In the story of

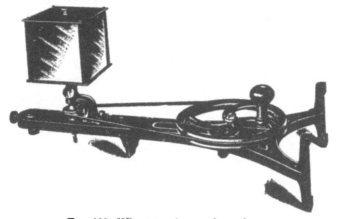

Fig. 199. Wheatstone's rotating mirror.

acoustics, however, we saw that König used Wheatstone's mirror to excellent purpose in solving the mystery of timbre. Here is Wheatstone's account of his failure—which proved a success:

The path of . . . an illuminated point in rapid motion, it is well known, appears as a continuous line in consequence of the after-duration of the visual impression. . . . It occurred to me that the motion of the reflected image of the electric spark in a plane mirror would answer all the purposes; the apparent motion of the reflected image in a small moving mirror would be equal to an extensive motion of the object itself: the same mirror might be presented to any object to be examined; thus forming with its moving machine an independent and universally applicable instrument. Following up this idea I made a series of experiments relating to the oscillatory motions of sonorous bodies. The satisfactory results thus obtained made me desirous to ascertain whether . . . some information might not be gained respecting the direction and velocity of the electric spark; the method by which I then proposed to effect this purpose was first announced in a lecture delivered by Dr. Faraday at the Royal Institution in 1830. . . . When the apparatus was made to revolve rapidly, . . . no deviation . . . of the . . . sparks was observed. The apparatus revolved fifty times a second and as a difference of the twentieth part of the circumference could easily have been observed, we may safely conclude that the spark passed through the air . . . in less . . . than the thousandth part of a second [50 × 20 =

1000]. I propose to employ a mirror with polygonal faces symmetrically placed with respect to the axis of rotation, a hexagon for instance [fig. 199].

The instantaneousness of the light of electricity of high tension . . . affords the means of observing rapidly changing phenomena during a single instant of their continued action. . . . A rapidly moving wheel seems perfectly stationary when illuminated by the explosion of a charged jar. Insects on the wing appear, by the same means, fixed in the air. Vibrating strings are seen at rest in their deflected positions. A rapid succession of drops of water, appearing to the eye a continuous stream, is seen to be what it really is, not what it ordinarily appears to be.[4]

FIG. 200. Stereoscope.

One might wonder why Wheatstone did not give his own lecture. As a matter of fact, he was so shy that he disliked lecturing because of stage-fright and brainstorm, whereas Faraday was a marvellous lecturer. Wheatstone shared Young's fondness for deciphering codes and was successful in deciphering a number of manuscripts in the British Museum which had been too difficult for others. That he was a prolific inventor and a successful patentee is hinted in his phrase, "universally applicable instrument."

Another of Wheatstone's optical inventions was the *stereoscope* (fig. 200), by which we see pictures in depth or "solid" (Gk. *stereos*, solid). It has several modern modifications and more are likely to follow. Among scientists who have been aided to success by Wheatstone's mirror are Foucault and Michelson, in the determination of the speed of light in water, glass, and air. On p. 200 we met Mayer's clever simplification of the rotating mirror invented by Wheatstone.

SPEED OF LIGHT

We have seen that the corpuscular theory of refraction required that the speed of light be greater in water and in glass than in air, whereas the wave theory assumed the reverse. Arago seized upon this

[4]C. Wheatstone, "Velocity of Electricity," *Phil. Trans. Roy. Soc. Lond.* (1834).

difference as a crucial test of the relative merits of the two hypotheses. He set two of his assistants, Foucault and Fizeau, to determine the velocity of light in air, water, and glass, or at any rate to compare them. These enthusiastic young researchers differed in their opinions as to the best method for conducting the investigation. Soon they agreed to go their separate ways and the two researches became a race to see who would solve the problem first. Both methods were good and were related to that of Wheatstone's rotating mirror, Foucault's more directly for it used a rotating mirror. Foucault was first to make his report (1850). Their experiments showed that light travels faster in air than in glass or in water. This brought strong support to the wave theory and militated against the emission theory. Their value for the speed of light in air was in the same order of magnitude as that obtained by Roemer—about 3×10^{10} cm.p.s. Their values for the speed in glass and water supported Huygens' position, namely, for glass, 2×10^{10}, and for water, 2.25×10^{10} cm.p.s. The velocity of light in vacuo (or in air) has proved to be of such fundamental importance that it has been determined again and again by over a dozen men. Supreme among all these is Albert Michelson of Chicago, whose results will be considered in their chronological place.

THE OPHTHALMOSCOPE

By the labours of such men as Galileo, Kepler, Newton, Huygens, Young, Fresnel, and Brewster, we have seen an adequate theory of optics developing for the first time in the story. Springing from this advancement in theory, now comes the invention of a bevy of new or improved instruments, the telescope, camera, kaleidoscope, rotating mirror, and stereoscope. Others will follow, such as the ophthalmoscope and a group of allied instruments, also the spectroscope, the stroboscope, and many others.

Helmholtz's invention of the ophthalmoscope (fig. 201), in 1851, was a great boon, for it has helped oculists to examine eyes and furnish proper spectacles. It has also been of great service in diagnosis; for the retina is the only part of the brain which can be examined without incision through bone. Helmholtz refers to his ophthalmoscope (Gk. *ophthalmos*, eye) in one of his papers:

The ensuing article contains the description of an optical instrument by means of which it is possible to see clearly and examine in the living eye, not only the retina but also the images of luminous bodies that are projected on it. An instrument of this purpose has essentially two different problems to solve. In the first place, all that we can see of the background of the normal eye, appears absolutely dark. . . . Accordingly, it is necessary . . . to find a method of illumination by which the particular part of the retina which we can see through the pupil will be sufficiently illuminated. . . . We need, then, besides a proper method of illumination, also some optical means of making possible for the observing eye a suitable accommodation for the situation which it is to observe.[5]

In fig. 201, S is a source of light, A an aperture in a concave mirror, CV. The diagram shows that an observer at A sees at I an inverted real image of R, the retina, of magnification greater than unity.

The laryngoscope and frontal mirror are derivatives of the opthalmoscope and were invented at about the same time. Helmholtz also invented an opthalmometer for examination of the cornea of the eye.

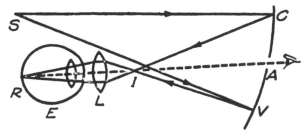

FIG. 201. Ophthalmoscope.

His monumental book on *Physiological Optics* is preeminent in its own field. It supports Young's hypothesis concerning the primary colours, red, yellow-green, and blue-violet. It also gives an analysis of colours on the basis of three features, hue, value, and saturation, which later became the basis of the best colour systems. A pure spectrum hue he described as saturated; as other hues are added the saturation decreases and approaches zero value at neutral gray. This feature has since been termed *purity* by Maxwell and *chroma* by Munsell; by its definition it is seen to be analogous to the quality or timbre of sound.

THE MAXWELL COLOUR DISC

Although the principal life-work of Professor James Clerk Maxwell of Edinburgh and Cambridge (1831-79) was in the field of electricity, yet his keen and active mind led him also into optics and other sciences. His first paper was read for him before the Royal Society of Edinburgh when he was only fifteen, and for more than half of his brief life he stood in the foremost rank of scientists. Even at twenty-three he was recognized as one of the ablest mathematicians in the world. At a party he would be found amusing a group with some cartooning, or reciting impromptu doggerel, or carrying on some other merry antics. He lec-

[5]H. Helmholtz, *Wissenschaftliche Abhandlungen* (1856), II, 286. Or, *Archive, für Ophthalmologie* (1856).

tured with a twinkle in his eye and a quizzical humour that several of his old students and admirers have tried in vain to capture in print.

From 1855 to 1872, he investigated the subject of colour and was the first to give that department of optics a firm and lasting basis of theory. The Royal Society expressed its appreciation of his labours by the presentation of the Rumford Medal. The instruments he devised and used in these researches were simple but effective and bore the marks of a genius that combined profound knowledge with shrewd ingenuity. By a development of Alhazen's colour-top he brought the study of colour to a quantitative stage:

The object of the following communication is to describe a method by which every variety of visible colour may be exhibited to the eye in such a manner as to admit of accurate comparison; to show how experiments so made may be registered numerically and to deduce from these numerical results certain laws of vision.

The different "tints" are produced by means of a combination of discs of paper, painted with pigments commonly used in the arts, and arranged around an axis so that a sector of any required angular magnitude of each colour may

be exposed. When this system of discs is set in rapid rotation, the sectors of the different colours become indistinguishable and the whole appears of one uniform tint. . . . The form in which the experiment is most manageable is that of the common top (fig. 202). An axis . . . carries a circular plate which serves as a support for the discs of coloured paper. The circumference of this plate is divided into 100 equal parts. The resultant tints of two different combinations of colours may be compared by using a second set of discs of a smaller size and placing these over the centre of the first set so as to leave the outer portion of the

FIG. 202. Maxwell colour disc.

larger discs exposed. The resultant tint of the first combination will then appear in a ring round that of the second. . . . As an example of the method of experimenting, let us endeavour to form a neutral gray by the combination of Vermilion, Ultramarine and Emerald Green. . . . The numbers as given by an experiment on the 6th March 1855 are:

$$0.37\ V + 0.27\ U + 0.36\ EG\ =\ 0.28\ SW + 0.72\ B$$
$$(SW = \text{snow-white};\ B = \text{black}).$$

We may also make experiments in which the resulting tint is not a neutral gray. . . . Experiments of this sort are more difficult both from the inability of the observer to express the difference which he detects in two tints which have perhaps the same hue and intensity, but differ in purity: and also from the complementary colours which are produced in the eye after gazing too long at the colours to be compared [psychological or physiological colours].

Hence it appears that the nature of a colour may be considered as dependent on three things. . . . Consider how two tints, say two lilacs, may differ. . . . In the first place, one may be lighter or darker than the other. . . . Secondly, one may be more blue or red than the other, that is they may differ in hue. Thirdly, one may be more or less decided in its colour, it may vary from purity on the one hand to neutrality on the other. . . . This method of combining colours has been used since the time of Newton [first used by Alhazen] to exhibit the results of theory. . . . Professor Forbes endeavoured to form a neutral tint by the combination of three colours only. For this purpose, he combined the three so-called primary colours, red, blue, and yellow, but the resulting tint could not be rendered neutral gray by any combination of these colours; and the reason was found to be that blue and yellow do not make green but a pinkish tint when neither prevails in the combination. It was plain that no addition of red to this could produce a neutral tint. This result of mixing blue and yellow was, I believe, not previously known. It directly contradicts the received theory of colours and seemed to be at variance with the fact that the same blue and yellow paint when ground together, do make green.[6]

The disc described in this article is called a Maxwell disc. Its simplest form is the kindergarten colour-top, a very instructive and profitable toy, by a proper use of which a student can, if he has a few necessary books and accessories, give himself a thorough training in theory of colour. Maxwell can say much in a few words. One of these sentences gives a complete analysis of colours under three features: (1) hue, (2) intensity (lightness or darkness), and (3) purity. If some of his successors had heeded his description of the third feature, their definitions and explanations would have been clearer.

The answer to Maxwell's yellow and blue problem is that the mixing of paints is a subtraction phenomenon, whereas the fusing of colours on a colour disc is an addition phenomenon. Paints are filters; yellow paint absorbs blue and purple and reflects yellow, and possibly its spectrum neighbours red and green to a less degree. Blue paint absorbs yellow and transmits blue, and, less intensely, its neighbours purple and green. What gets through after both subtractions is green. This can be shown neatly by Hooke's experiment with two square bottles, one containing a purple-blue solution of ammonio-copper sulphate and the other a yellow solution of picric acid (alcoholic). Place a candle flame behind them singly and combined.

If the same pigments are spread on separate cards and spun in proper proportion so that the sum of all the colours entering the eye is a full, balanced spectrum, the result is neutral gray, for the colours are complementary.

Maxwell's greatest contribution to optics was of a higher order than any we have discussed. Nor shall we say much about it here because it belongs partly to the science of electricity and magnetism and because of the mathematics involved. Employing his great ability in mathe-

[6] J. Clerk Maxwell, *Collected Papers*, I, 126. Also, *Trans. Roy. Soc. Edin.* (1885), XXI, ii.

matics, Maxwell reached the conclusion that light waves are electro-magnetic phenomena. By means of his electro-magnetic theory of light the subject of optics became organized into a consistent science with unity and continuity. Thus he performed for optics the same kind of service as Newton achieved for mechanics.

THE SPECTROSCOPE

The spectroscope is essentially Newton's prism experiment modified to convenient laboratory form. Dr. Robert Bunsen of Heidelberg (1811-99), with Gustav Kirchhoff (1824-87) as collaborator, invented it in order to analyse light obtained in flame-tests. How powerful an aid it has become in the advance of science could hardly have been imagined even by its own inventor in his most sanguine moments. Bunsen describes the instrument:

It is well known that many substances have the property when they are brought into a flame, of producing in its spectrum certain bright lines. We can found "on these lines" a method of qualitative analysis that greatly broadens the field of chemical research and leads to the solution of problems heretofore beyond our reach. . . . The gas lamp, described by one of us, gives a flame of very high temperature and very small luminosity. . . . The potassium compound [chloride] used for the investigation was obtained by heating chlorate of potassium which had been . . . six or eight times recrystallized. In fig. [203] is represented the apparatus that we have used for the examination of the spectra. A is a box blackened on the inside. . . . Its two inclined sides . . . carry two small telescopes, B and C. The ocular of B is replaced by a plate in which is a slit formed by two brass blades. . . . The lamp D is placed before the slit. . . . The end of a fine platinum wire, bent into a small loop and supported by the holder E, passes into the flame; on this hook is melted a globule of the chloride previously dried. . . . Between the objectives of the telescopes B and C is placed a hollow prism, F, with a refracting angle of 60° and filled with carbon disulphide. The prism rests on a brass plate that can be rotated about a vertical axis. . . . This axis carries on its lower end the mirror G, and above it the arm H, which serves as a handle for turning the prism and the mirror. Facing the mirror, a small telescope is arranged to give an image of the horizontal scale, placed at a short distance. . . . By rotating, one can make the entire spectrum of the flame to pass before the vertical thread of the telescope C. . . . To every point in the spectrum, there corresponds a certain reading of the scale.[7]

In this experiment carbon disulphide and a Bunsen flame were near each other, a combination calling for knowledge and cautious technique. In one of Bunsen's early experiments, an explosion blinded him in one eye; but with one eye he seemed to see more than many with two. Bunsen was a great chemist, a splendid citizen, and an inspiring teacher. Other scientists liked to work with him. Frankland, Tyndall, and other

[7]*Poggendorf Annalen* (1860), cx, 161. Also, *Science News Letter*, Feb. 25 (1928), xiii, No. 359, p. 121.

eminent scientists were proud and grateful for the privilege of taking
their doctorates under him.

The spectroscope has become so important that there is now a
special science called *spectroscopy* and a group of scientists (spectro-
scopists) who devote all their labours to the study of data obtained by
this instrument. With it, Bunsen and Kirchhoff discovered two elements, caesium and rubidium. One specially interesting discovery to
which it led was that of the gas helium, so named because it was found
in the corona of the sun (Gk. *helios*, sun) at eclipse by means of the

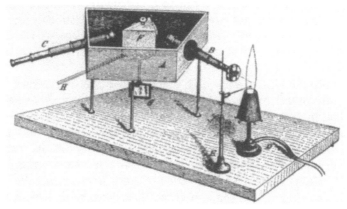

FIG. 203. Bunsen's spectroscope.

spectroscope, before it was known to exist on the earth. Three men
share the credit for this discovery (1868), J. Jansen, J. N. Lockyer, and
E. Frankland.

The fundamental purpose of the spectroscope is the analysis of
light. Let us use it to analyse the achromatic colours, white, gray, and
black. The original meaning of white is illustrated by the daylight
appearance of chalk, milk, or clean snow. It is defined as the colour of
pure magnesia in full sunlight. When sunlight (or arc-light) strikes a
mirror or a white card and thence reflects into a spectroscope, the
observer sees in either case a full spectrum, that from the card being
slightly less intense. Hence by an extension of the original meaning of
white, sunlight is often called "white light," although sunlight must
be as colourless as air, water, or ordinary glass. The mirror just mentioned, or a pond in sunlight, reflects "white light," but neither is
white. Sunlight reflecting from a neutral gray card or from a black one
into a spectroscope also gives in each case a full spectrum; that from
the gray card is less intense than from the white one and that from
black is still weaker. Hence black and white might well be called re-

spectively very dark and very light neutral gray. It was because of their lack of hue that Ostwald called the neutral grays "achromatic colours" (hueless colours). They form the basis of all classifications of colours.

FRAUNHOFER LINES

In 1821 the German physicist Joseph von Fraunhofer (1787-1826) of Munich prepared a new type of instrument by winding a helix of fine wire on a flat bobbin. He thereby became the first (after Grimaldi) to see a spectrum produced by such a corrugated surface. Because of its form the instrument was called a *grating*. Although its modern descendants, composed of parallel scratches on glass or metal, differ greatly from their prototype, they still retain the ancestral name of grating.

Fraunhofer was a very able scientist—quite competent in both theory and practice. Much of his work had to do with the manufacture of optical instruments and especially with achromatic lenses.[8] In 1814 he discovered in the light from a lamp the well known double orange line of the sodium flame and as he said in his report, he "wished to ascertain whether a similar line is to be seen in the spectrum of sunlight."[9] What he found was a series of dark lines, now called Fraunhofer lines. He labelled some of these dark lines with letters, e.g. the sodium line *D*, but he failed to find an explanation for his important observation. Indeed, forty years elapsed before anyone deciphered this riddle. Such inattention is better understood if we recall that the physicists were then locked in combat over the corpuscular and wave theories.

It was the German physicist, Gustav R. Kirchhoff of Heidelberg, who found the explanation of the Fraunhofer lines, and a remarkably important contribution that was, for it opened the spectroscopic study of celestial bodies. Here are the significant words:

While engaged in a research carried out by Bunsen and myself in common on the spectra of coloured flames . . . I made some observations which unexpectedly give an explanation of the cause of Fraunhofer lines and allow us to draw conclusions . . . about the composition of the sun's atmosphere and perhaps also of those of the fixed stars.

Fraunhofer observed that in the spectrum of a candle flame two bright lines occur which coincide with the two dark lines *D* of the solar spectrum. We obtain the same bright lines in greater intensity from a flame into which common salt is introduced. I arranged a solar spectrum and allowed the sun's rays, before they fell on the slit, to pass through a flame heavily charged with salt. When the sunlight was sufficiently weakened there appeared in place of the two dark

[8]Invented in 1758 by John Dollond, a London optician. See page 267.
[9]J. von Fraunhofer, *Gesammelte Schriften* (Munich, 1888).

D lines two bright lines; if, however, its intensity exceeded a certain limit, the two dark *D* lines showed much more plainly than when the flame charged with salt was not present.

From these observations I conclude that a coloured flame in whose spectrum bright lines occur, so weakens rays of the colour of these lines if they pass through it, that dark lines appear in place of the bright ones, whenever a source of light of sufficient intensity, in whose spectrum these lines are somewhat absent, is brought behind the flame. I conclude further that the dark lines of the solar spectrum [Fraunhofer lines] which are not produced by the earth's atmosphere, occur because of the presence of those elements in the incandescent atmosphere of the sun which would produce in the spectrum of a flame bright lines in the same positions. We may infer that the bright lines corresponding to the *D* lines in the spectrum of a flame always arise from the presence of sodium. The dark *D* lines of the solar spectrum, therefore, permit us to conclude that sodium vapour is present in the sun's atmosphere.[10]

The eye of the imagination can readily see that research of this kind could lead to a knowledge of the composition of the distant heavenly bodies, a knowlege which previously we might well have thought to be hopelessly and forever beyond our ken.

THE STROBOSCOPE

Automobile wheels in the movies sometimes seem to turn backwards or to be seized when the car is moving forward. The following excerpt holds the key to the puzzle. It is taken from a paper by two scientists, McLeod and Clarke, which appeared in the *Philosophic Transactions* of the Royal Society of London in 1877 under the title, "Rate of Vibration of Tuning-Forks."

If the image of a point of light . . . be observed in a vibrating mirror, the motion of which may be produced by a tuning-fork, . . . the point, in virtue of the retention of the image on the retina, will appear as a straight line. . . . If the luminous point be moving . . . at right angles to the plane in which the fork vibrates and parallel to the plane of the mirror, the combination of the two motions will produce a sinuous line or wave-form. . . .

When a series of equidistant points attached to a rotating disk . . . is employed, the properties of the wave differ very much according to the velocity of the moving disk. If the points are in a circle . . . rotating with such a velocity that the time occupied . . . in passing over a distance equal to that between two consecutive points is exactly equal to the period of the fork, . . . a continuous stationary figure is perceived: but if the point passes over a distance slightly greater than the intervals, the figure will show a slow progression in the direction of the moving circle: and when slightly less, . . . the motion . . . will be in the opposite direction.

From the foregoing it will be seen that the formation of these figures may be employed for determining the speed of revolution of the disk . . . if the period

[10]G. R. Kirchhoff, *Monatsbericht der Akademie der Wissenschaften zu Berlin* (Oct. 1859).

of the fork is known, and . . . the number of points on the rotating body. . . .
Since this is an optical method for investigating rotation, we suggest cyclo-scope as a name for the instrument.

The name of the modern instrument based on the principle outlined in this paper is *stroboscope* (Gk. *strobos*, twisting). The first scientist to use the stroboscopic principle was the famous blind Belgian physicist, Joseph Plateau (1801-83).

Suppose the shutter of a movie camera opens 24 times a second, and it is photographing a vehicle each of whose wheels has, say, 24 spokes and no distinguishing marks. If a wheel turns exactly once a second (or twice), then every time the shutter opens, the camera "sees" exactly the same picture of the wheel. Therefore, the wheels will not appear to turn although the vehicle is seen in motion. Now let the wheels turn a trifle more slowly than before. Every time the shutter opens, it sees each wheel slightly behind its previous position. These pictures combine to show a "slow-motion" picture of the wheel turning backward while the vehicle moves forward.

The performance of the modern stroboscope is often astonishing, and its applicability wide and growing. For example, by it an engineer can examine a wheel that is making hundreds of rotations (or revo-lutions) per minute. He adjusts the speed of the stroboscope until it synchronizes with the wheel, i.e. until the wheel appears stationary. Now he can examine it as if it were not rotating.

CONCAVE GRATINGS

The invention of ruled gratings on spherical concave surfaces has been a fundamental factor in the technique of modern spectroscopy. This invention was made by Henry A. Rowland, professor of physics at the Rensselaer Polytechnic Institute of Troy, N.Y., and from 1876, at Johns Hopkins University of Baltimore. Rowland was very able in both theory and practice and had a special flair for performing experi-ments requiring great mechanical skill and a technique of superlative deftness and precision. His paper on concave gratings was published in the *Philosophic Magazine* in 1882. The machine used for ruling gratings is called a dividing engine.

Professor Young was the first to discover that some of the gratings of Mr. Rutherford showed more than any prism spectroscope which had been con-structed. . . . One of the problems to be solved in making a machine [dividing engine] is to make a perfect screw, and this, mechanics of all countries have sought to do for over a hundred years and have failed. . . . I devised a plan whose details I shall soon publish. . . . I set Mr. Schneider, the instrument maker of the university, at work at one. . . . I have now had the pleasure since

Christmas of trying it. . . . By this machine I have been able to make gratings with 43,000 lines to the inch.

All gratings hitherto made have been ruled on flat surfaces. Such gratings require a pair of telescopes for viewing the spectrum. . . . I found that if the lines were ruled on a spherical surface the spectrum would be brought to a focus without any telescope. This discovery of concave gratings... reduces the spectroscope to its simplest proportions. . . . With one of my new concave gratings I have been able to detect double lines in the spectrum which were never before seen. . . . Its principal use will be to get the wave lengths of the lines of the spectrum . . .; to divide lines of the spectrum which are very near together . . .; to photograph the spectrum . . .; to investigate the portions of the spectrum beyond the range of vision; and lastly, to put in the hands of any physicist at a moderate cost such a powerful instrument as could hitherto only be purchased by wealthy individuals or institutions.[11]

While on the staff of Rensselaer Rowland made an excellent study of magnetic permeability. While studying abroad under Helmholtz in 1876 he performed a difficult and brilliant experiment showing that a charge revolving around a compass has the same kind of effect as an electric current.

Rowland had an impatient dislike for sham and pretense in any form. While investigating a quack's motor which was claimed to be moved by mystic force, Rowland suddenly stepped close to sever a wire which he suspected of being a compressed air tube. The quack intervened abruptly and presto, the mystic and the realist were closely engaged in settling an argument about a physical problem by physical violence, applying their knowledge of impact, the law of the lever, and the properties of matter.

VELOCITY OF LIGHT IN AIR AND IN VACUUM

The Nobel laureate in physics for 1907 was Dr. Albert A. Michelson (1852-1931), director of the department of physics at the University of Chicago. When he studied abroad one of his professors was Helmholtz. Throughout his career he carried on experiments to determine the speed of light in air and in vacuo with ever increasing accuracy. One of his ideals seemed to be to follow accuracy like a sinking star. In striving toward his receding goal of greater precision, he improved older instruments and techniques such as those of Fizeau and Foucault, and invented new precision instruments such as the Michelson interferometer. This instrument in the hands of an expert can distinguish between two lengths which differ by only 0.000002 cm. A scientist once said facetiously that it is a wonderful instrument if operated by Michelson.

[11]*Phil. Mag.* (1882), Ser. 4, XIII, 469.

The plan of his experiment on speed of light was somewhat as follows. Light from a powerful arc lamp situated on Mt. Wilson, California, reflected from a rotating octagonal mirror R to a mirror on Mt. San Antonio, 22 miles distant. Thence it was reflected back to strike another facet of R and reflect into an eye-piece if the facet were in a certain position. Suppose R were rotating at the terrific speed of 529 rotations per second and turned through 45° while the light ray went there and back, 44 miles. Then the velocity of light would be

$$c = 44 \times 529 \times 8 = 186,208 \text{ mi. per sec.}$$

The actual experiments and apparatus were so beset with multitudinous complications and difficulties that only the persistence and skill of a Michelson could win through to success. By averaging the results of hundreds of experiments Michelson ultimately (1926) reached the value

$$c = 299,796 \pm 1 \text{ km. per sec. for velocity of light in vacuo}$$

or $c = 186,284$ mi. per sec. and, in air, about 40 mi.p.s. less, which is a difference of only 4 in 18,000 and, in general, negligible. As we have seen these values are generally rounded off at 3×10^{10} cm.p.s. and 186,000 mi. per sec.

In 1887 Michelson and his colleague, Professor E. W. Morley, performed a celebrated experiment which bears their names. It was performed to measure if possible the relative velocity of the earth and the aether ocean through which it is supposed to move. The idea was to let two light-rays race each other, one across the aether stream and back, and the other parallel to it. By means of the interferometer the lead gained by the winner could be measured. But the race was a dead heat and the interpretation of this negative result was one starting point from which came Einstein's Theory of Relativity.

If a cadmium salt is introduced into a colourless flame and the light passed through a spectroscope a characteristic red line is observed. In 1892, Michelson, using his interferometer, determined at Paris the wave-length of this monochromatic light and found it to be 6438.4722Å. (15°C. and 760 mm. pressure). This number claims an accuracy of 1 in 60 million. Wilkinson then determined the length of the standard metre at Sèvres in terms of the cadmium wave-length. If now by some vicissitude the standard metre were damaged or destroyed it could be replaced, for we can think of no event which could alter the wave-length of the red cadmium light. This determination by Michelson seems to contain a response to the words of Lucretius, "for there must be at least something in the world which is changeless."

COLOUR NOTATION

An important book on the study and application of colour was published in 1905 by Albert H. Munsell (1858-1918), art instructor at the Normal Art School of Boston. Its title is *A Color Notation*. It follows Maxwell's lead and puts the findings of both investigators into practice. Here are a few passages from this valuable little treatise:

The incongruous and bizarre nature of our present color names must appear to any thoughtful person. Baby blue, peacock blue, Nile green, apple green, lemon yellow, straw yellow, rose pink, heliotrope, royal purple, Magenta, Solferino, plum, and automobile are popular terms, conveying different ideas to different persons and utterly failing to define colors. The terms used for a single hue, such as pea green, sea green, olive green, grass green, sage green, evergreen, invisible green, are not to be trusted in ordering a piece of cloth. They invite mistakes and disappointment. Not only are they inaccurate: they are inappropriate. Can we imagine musical tones called lark, canary, cockatoo, crow, cat, dog, or mouse, because they bear some distant resemblance to the cries of those animals?

Music is equipped with a system by which it defines each sound in terms of its pitch, intensity, and duration without dragging in loose allusions to the endlessly varying sounds of nature. So should color be supplied with an appropriate system, based on the hue, value, and chroma of our sensations, and not attempting to describe them by the indefinite and varying colors of natural objects. The system now to be considered portrays the three dimensions of color, and measures each by an appropriate scale. It does not rest on the whim of an individual, but upon physical measurements made possible by special color apparatus. The results may be tested by anyone who comes to the problem with "a clear mind, a good eye, and a fair supply of patience."

Having used the familiar structure of the orange as a help in classifying colors, let us substitute a geometric solid, like a sphere, and make use of geographic terms. The north pole is white. The south pole is black. The equator is a circle of middle reds, yellows, greens, blues, and purples. Parallels above the equator describe this circuit in lighter values [intensities], and parallels below trace it in darker values. The vertical axis joining black and white is a neutral scale of gray values, while perpendiculars to it (like a pin thrust into the orange) are scales of chroma. Thus our color notions may be brought into an orderly relation by the color sphere. Any color describes its light and strength [?] by its location in the solid or on the surface, and is named by its place in the combined scales of hue, value, and chroma.

The color tree is made by taking the vertical axis of the sphere, which carries a scale of value, for the trunk. The branches are at right angles to the trunk; and, as in the sphere, they carry the scale of chroma. Colored balls on the branches tell their Hue.

The notation used in this system places Hue (expressed by an initial) at the left; Value (expressed by a number) at the right and above the line; and Chroma (also expressed by a number) at the right, below the line. Thus R $\frac{5}{10}$, means HUE (red), $\dfrac{\text{VALUE} \quad (5)}{\text{CHROMA} \ (10)}$, and will be found to represent the qualities of the pigment vermilion.[12]

Even from these few excerpts, it is plain that Munsell was a clear and vivid writer. He was a prize winner at painting and had a distinguished career; his work as a teacher was also of the highest order. That he was a scientist is indicated by his invention of a photometer capable of comparing the intensities of two sources of different hues and also by his invention of such models as the colour sphere and tree. The former when spun shows beautifully three values of neutral gray. Recalling the failures of all investigators from Newton to Forbes to obtain this result, Maxwell being first to succeed, one is impressed by the performance of the colour sphere, with the conviction that whoever achieved this brilliant success must have had a masterly knowledge of colour and a superlative ability in its application. Another useful device described in Munsell's book is the Munsell colour circle which is exceedingly convenient both in theory and practice.

Fig. 204. Albert H. Munsell.

Munsell was invited to demonstrate his system before the Royal Society of London. His lectures were so popular that he was overworked and his strength overtaxed. On the return voyage he contracted a cold which led to his untimely death. His widow and their son are carrying on the useful service of dissemination of the Munsell Colour System.

SEEING THINGS RIGHT SIDE UP

From the study of lenses, it is seen that the lens of the eye must project on the retina an inverted image of surrounding objects, just as in the pin-hole camera. One may wonder how we see objects right side up when the retinal image is upside down. A youngster on a top veranda step, lowering his head and peering out through his legs at an

[12]A. H. Munsell, *A Color Notation*, 7th edn. (Munsell Color Co., Baltimore, 1926), p. 10. By permission of the publishers.

inverted world including the grinning passers-by, is likely investigating the same problem in his own peculiar way. The book entitled *The Universe of Light*, written by Sir William Bragg, director of the Royal Institution of London, cites a neat little experiment for asking nature this question. It was first devised by the British scientist Silvanus Thompson.

Let the observer prick a hole H in a card and bringing the card close to the eye as in fig. [205], look through the hole ... at a bright lampshade. ... Now let the pin P be held so that the head comes between the hole and the eye E. The shadow on the retina must be right way up, but the interpretation of the brain will be that it is upside down.

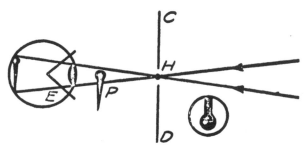

FIG. 205. Inverted image.

Suppose H is at the principal focus of the eye-lens (F). Then the light from the lens L to the retina R is parallel and a simple diagram will show that the shadow must be right side up. If H is a few millimetres from F, and the light from L to R slightly convergent or divergent, the argument still holds. Among scores of optical topics that the same book discusses clearly and entertainingly is Alhazen's problem about the size of the sun and moon when near the horizon.

X-RAYS THROUGH CRYSTALS

Since Sir William was a world authority on X-ray examination of crystals, a special department of optics, the following brilliant example of popular scientific exposition is of special interest:

The X-rays are a form of light from which they differ in wave-length only. The light waves which are sent out by the sun or an electric light or a candle and are perceived by our eyes have a narrow range of magnitude. The length of the longest is about a thirty-thousandth of an inch and of the shortest, about half as much. Let us remember that when we see an object, we do so by observing the alterations which the object makes in the light coming from the source and reaching our eyes by way of the object. We may be unsuccessful, however, if that object is too small; and this is not only because a small object necessarily makes a small change in the light. There is a second and more subtle reason;

the nature of the effect is changed when the dimensions of the object are about the same as the length of the wave or are still less. Let us imagine ourselves to be walking on the seashore, watching the incoming waves. We come, in the course of our walk, to a place where the strength of the waves is less, and when we look for the reason, we observe a reef out to sea which is sheltering the beach. We have a parallel to an optical shadow; the distant storm which has raised the waves may be compared to the sun, the shore on which the waves beat, is like the illuminated earth and the reef is like a cloud which casts a shadow. The optical shadow enables us to detect the presence of the cloud, and the silence on the shore makes us suspect the presence of the reef. Now the dimensions of the reef are probably much greater than the length of the wave. If, for the reef, were substituted a pole planted in the bottom of the sea and standing out of the surface, the effect would be too small to observe. . . . Even, however, if a very large number of poles were so planted in the sea so that the effect mounted up and was as great as that of the reef, the resulting shadow would tell us nothing about each individual pole. The diameter of the pole is too small compared with the length of the wave to impress any permanent characteristic on it; the wave sweeps by and closes up again and there is an end of it. If, however, the sea were smooth except for a tiny ripple caused by a breath of wind, each pole would cast a shadow which would persist for at least a short distance to the lee of the pole. The width of the ripples is less than the diameter of the pole, and there is, therefore, a shadow to each pole. . . . Just so, light waves sweeping over molecules much smaller than themselves, receive no impressions which can be carried to the eye and brain so as to be perceived as the separate effects of the molecules. . . . But the X-rays are some ten-thousand times finer than ordinary light.

Although the single molecules can now affect the X-rays, just as in our analogy, the single pole can cast a shadow of the fine ripples, yet the single effect is too minute. In the crystal, however, there is an enormous number of molecules in regular array, and it may happen that when a train of X-rays falls upon a crystal, the effects on the various molecules are combined and so become sensible. Again we may make use of an analogy. If a single soldier made some movement with his rifle and bayonet, it might happen that a flash in the sunlight caused by the motion was unobserved a mile away on account of its small magnitude. But if the soldier was one of a body of men marching in the same direction in close order, who all did the same thing at the same time, the combined effect might be easily seen. The fineness of the X-rays makes it possible for each atom or molecule to have some effect and the regular arrangement of the crystals adds all the effects together.[13]

By the method here intimated, Sir William Bragg and his son, Professor Lawrence Bragg, and other scientists have been able to determine the arrangement of the particles in crystals. They have also obtained a relation between the wave-length of X-rays and the distances between particles or atoms.

COLOUR BLINDNESS

Wilhelm Ostwald (1853-1932), professor of chemistry at the University of Leipzig and Nobel Laureate in 1909 for his service to

[13]Sir William Bragg, *Concerning the Nature of Things* (Bell, 1937), p. 130.

chemistry, was also an amateur artist of sufficient ability to have some of his paintings displayed in art gallery exhibitions. Thus he and Munsell were complementaries, for each was scientist and artist, Munsell chiefly an artist and Ostwald primarily a scientist. In 1917, Ostwald issued an elaborate colour atlas of some three thousand colours, and in 1930 he published a book, *Science of Colour*,[14] in which he describes what is now called the Ostwald Colour system. It discusses colour blindness:

Anatomists long ago observed two sorts of terminal organs in the retina of the human eye . . . called rods and cones, . . . cones in the *fovea centralis* and rods in the lateral area. . . . M. Schultze's investigations [1866] . . . led him to suppose that the rods merely distinguish between light and dark while the cones are instrumental in producing the sensation of hue. . . . The rods are looked upon as the original organ. . . . From this the cones were later developed . . . and in these a gradual adaptation to colour sensation has probably occurred: so that, at first, the stronger distinction between Yellow and Blue . . . came into existence and later, the weaker one between Red and Green. . . . The totally colour blind with whom the rods only are active, are extremely rare: those with an imperfect yellow-blue sense, occur . . . more frequently: while the red-green blind are commonest of all, in accordance with the general law that acquisitions that are later in point of historical development are the ones more readily lost.

Ostwald's term, achromatic colours, was mentioned on a previous page:

The most ancient department of our colour world . . . still occupies an exceptional position . . . because it forms the spine or axis of the whole body of colour. . . . The members of this group comprise the Achromatic Colours, or those which pass from White through Gray into Black. . . . The puzzling fact that we perceive a mixture of rays of all sorts of frequency and wave-length as a simple sensation of white or gray has, since the time of Newton, repeatedly engaged attention. . . . We have seen that the primitive eye could distinguish nothing but stronger and weaker radiation. . . . There was thus a time when no living being could perceive a difference of frequency. . . . This condition is still retained in our present-day eyes . . . and one of the most essential functions of these portions is the . . . perception of White, Gray, and Black. . . . The achromatic colours can be definitely measured, . . . if an ideal white surface is available or a surface of known whiteness. Measurements of this kind may be made with any photometer. Every [neutral] Gray is accordingly distinguished numerically by its white content, i.e. by the fraction of "white light" that it rejects. And further, all these numerical values are proper fractions lying between zero and unity. If w is the content of White and b that of Black, then the equation $w + b = 1$, holds good for every Gray. If $w = 0$, an ideal Black surface is present and $b = 1$. If $b = 0$, then the surface is an ideal White. In the former case the value of the neutral gray $g = w = 0$, and in the latter, $g = w = 1$. Ideal White and Black are extreme values of neutral gray.

[14]W. Ostwald, *Science of Colour*, trans. J. Scott Taylor (Winsor & Newton, 1931). By permission of the publishers.

Another passage in Ostwald's book discusses colour measurement:

With the colour-top . . . a method of mixing measured quantities of given colours additively was available. . . . But all such measurements contained unknown quantities. . . . It is here that the new colour science . . . comes to the fore. Already, . . . very considerable results have been attained. Results incomparably greater, however, are within the reach of those who will take this subject up, and the future historian of colour science should be able to date from the second decade of the twentieth century, . . . the same sudden advance as that assigned to the end of the eighteenth century by the historians of chemistry. And it will readily occur to him that, in both cases, this rapid development . . . is accounted for by the passing over of the qualitative into the quantitative era.

THE STORY OF
Thermics

CHAPTER FOURTEEN

Fingers Before Thermometers

DISCOVERING how to make fire was probably the greatest single advance ever made by the human race in its struggle to gain mastery over nature. That discovery was made in prehistoric times, even before the Stone Age, and in Greek mythology was symbolized by the story of Prometheus, whose name means "foresight." The first fire seen by primitive man was likely the burning of a dead tree that had been kindled by lightning. Early man also saw, occasionally, fire caused by friction between two swaying tree-limbs. The first fire which he contrived was an imitation of this, produced by rubbing a stick swiftly back and forth along a groove on a block of dry wood. Fig. 206 shows some methods of making fire by friction, and incidentally discloses a new use of the archer's bow. So once again, we see the "weapon animal" proceeding true to form in the contrivance of new implements. In the poem *Prometheus Bound*, composed by the Greek poet Aeschylus (about 475 B.C.), Prometheus refers to the fire-stick when he declares:

> Yea, in fennel stalk,
> I snatched the hidden spring of stolen fire
> Which is to men a teacher of all arts,
> Their chief resource.[1]

The ashes and charcoal found in palaeolithic caves and in other deposits prove the ability of early man to make fire. By its use he gained protection from the weather and from foes, "both the unseen and the seen"—from bacteria to cave-tigers. As man's knowledge advanced, fire enabled him to extract copper and iron from their ores and to fashion metal weapons and tools. Had it not been for man's practical knowledge of fire, he could not have risen from barbarism. Without fire, he would relapse into degradation and probably perish. Hence the study of a science that deals with this topic is of basic and crucial importance.

[1]From the translation by E. P. Plumptre, *Harvard Classics*, VIII, 156.

During the Stone Age or even earlier, man began to bury his dead
and to place with the bodies of the departed containers with supplies
of food and drink, also utensils, ornaments and weapons. Sometimes
the archaeologist finds in an ancient grave a piece of *flint* and a piece
of the mineral *pyrite,* placed together. This reveals a second prehistoric
method of making fire—by *percussion.* Pyrite, when skilfully struck a
glancing blow with a sharp flint, emits one or more sparks that can
kindle a fire if caught on dry moss or tinder. This ancient and rather
tedious technique is recorded in the name pyrite, which means literally

Fig. 206. Making fire by friction.

"fire-stone," being derived from the Greek words, *pur,* fire, and *lithos,*
stone. The root "pyr" is seen in such words as funeral-pyre, pyrometer,
pyrine, empyrean, pyrotechnics, etc. The flint method of making fire
persisted as late as 1830 in the flint-lock musket and in the flint, steel,
and tinder-box of our great-grandfathers' days. It persists even to
modern times in the cigarette-lighters which, like their ancestors, sometimes require laudable skill and patience.

One of a young boy's prehistoric symptoms is a passing urge to
play with fire: he is merely "recapitulating" the stage at which his
species made its greatest invention and took the upward path toward
mastery of nature. He should be treated kindly but firmly. The phase
will pass. The patient will recover; especially if his craving can be
sublimated or turned into useful channels such as burning some rubbish
(under supervision), or doing combustion experiments in a laboratory.
Tending a furnace will sometimes quell his passion for fire-making in
a few days.

The difficulty of making fires by the fire-stick or by fire-stones,
especially in moist weather, and the necessity of having a fire always
available, led to the practice in many tribes of assigning to some of the
girls the duty of keeping the home-fires burning. One survival of that

ancient tribal custom was the Roman institution of the Vestal Virgins—
of whom one modern counterpart is a box of matches. Some brands of
matches are called Vestas. Another memento of those early times is the
fondness we all share for seeing a cheery fire blazing in the parlour
grate, or for sitting about a camp-fire. Fig. 207 gives an artist's conception of the interior of the Temple of Vesta, the hearth-goddess, at
Rome. There, among other items of interest, we see the sacred fire that
burned ceaselessly on the altar night and day, with one exception. At

FIG. 207. Interior of Temple of Vesta.

the New Year (March first) the fire was extinguished and rekindled
by the ancient method of friction of wood. Entry into this sacred pre-
cinct by anyone other than a vestal virgin, the king, or the *pontifex
maximus*, was punishable by death.

"COLD" AND "HOT"

The preliminary stage of the science whose history we are tracing
dates farther back, however, than the making of fire, for it is as old as
the human race. This branch of physics has to do with all facts that are
perceived through the *thermal nerves*. These are situated chiefly in the
skin and by means of them we ascertain qualitatively the *temperatures*
of bodies, and describe them as hot, warm, cool, cold, or by similar
adjectives. In special cases we sometimes qualify such adjectives by
adverbs, similes, or expletives. The richness of human vocabulary in
this respect is further testimony to the practical importance of the
subject. Since acoustics and optics are named with reference to the

ear and the eye, it would be consistent to name this department of physics after the thermal nerves (Gk. *thermos*, temperature). We might use the name "thermics," although that is not customary. As acoustics is called sound, and optics has the common name light, so thermics is generally called heat. We are not yet in a position to define heat, and it is extremely important to distinguish between temperature and heat. That the distinction was not clear in early times is indicated by the fact that the Greek word *thermos* is sometimes best translated as heat and sometimes as temperature. In these days, one often hears such loose expressions as "having the oven at a baking heat" whereas the proper word would be temperature. At a later stage, we shall find the temperatures of bodies ascertained by *indicators* called thermoscopes or thermometers.

The science of thermics is the only one whose facts are commonly matters of life and death. A man can be deaf or blind and yet live; but no one bereft of thermal nerves could live. Any derangement of one's body temperature is a danger signal not to be ignored. Hence the practical knowledge which has been compulsively gained from earliest days is very great. A student regularly brings to an elementary course on heat a rich experience and a copious vocabulary but, as in the study of optics, he discloses a radical need of organization of his facts, expressions, and ideas. The individual cannot remember when he began the practical study of heat but his parents can and so can some of the neighbours. During his first days on this planet his wants were few, but one of his chief demands was thermal. If he was not kept comfortably warm (and dry), he indicated dissatisfaction, and if this was not heeded, he proceeded to turn on the acoustics with abandon. As the first remarks of two citizens when they meet are often about the temperature of the air, so the earliest thermal observations of primitive man were on the same topic; for the weather we have always with us and it commands attention. Thermics seems to be, however, the homely sister of the sciences. Acoustics and optics have their beautiful arts, music, painting, and sculpture, whereas poor thermics has none and ministers only to our comfort. But few of us could enjoy a symphony concert or the pictures in an art gallery if the place were bitterly cold. Heat, it seems, is a strongly utilitarian subject.

One of the most ancient observations was that the temperatures of most bodies commonly change. The temperature of the weather or of the ground changes from night to day and from season to season. When a body warms, it is said to gain heat and when it cools, to lose heat. This does not say what heat is, but merely assumes its existence. Thus a fire kindled in a cave warmed the cave or gave it heat. We have noted that a number of religions, Egyptian, Zoroastrian, Aztec, and others, worshipped the sun, and this was not only because of its light but also because of its warming effect. The practice of sacrificing burnt offerings originated in the idea that the sun-god is a god of fire and that such gifts might placate him. Fig. 208 is an old Egyptian picture, dating about 1500 B.C., which represents Ra, the sun, commencing his daily journey across the sky to bring warmth to the world. It may look as though Thoth and Isis are toasting their hands at the fire; but from

FIG. 208. Voyage of Ra, a sun-god.

the ideograph on page 9, it is seen that their attitude signifies worship. In our own times, old Sol receives considerable recognition: the first day of our week, like that of the Chaldeans, is named in his honour, and it is kept in a manner distinct from all others.

Among modern conveniences whose origins have thermal aspects is the group called containers. Of these we have myriads: baskets, pots, crocks, tumblers, flasks, test-tubes, and thus ad infinitum. There was a time when man's only containers were his body and his cupped hands. The latter are still used in washing the face, in catching a ball, or in an emergency to obtain a drink from a brook. Let us consider the thermal history of the basket and the skillet. Woman, the home-maker, likely took an active part in the invention of containers that were cooking utensils. The first method of cooking meat was by broiling it on embers or roasting it on a spit as at a barbecue. Later the method of stewing was discovered. The meat was placed in hot water in a hole in the rocks. The water was heated by rolling into it hot stones from a fire. This procedure foreshadowed the so-called "method of mixture," which we shall find of fundamental importance in the study of heat. When the hot stone and the cooler water came in contact with each other, primitive man observed two changes of temperature: the water warmed and the stone cooled. This is a particular case of a fundamental principle in thermics. In general, when two bodies of different temperatures are placed in contact with each other, the warmer one cools and the cooler one warms until they have the same temperature (provided they are left together long enough). This fact, or this set of facts, may be expressed hypothetically and conveniently by saying that heat flows from the warmer body into the cooler one (fig. 209). Such a transfer of heat is called *conduction*. When a finger touches a warmer body B, the latter cools at the point of contact and the finger warms or, in terms of the hypothesis, heat flows into the finger. The thermal nerves are thus stimulated and we have learned to interpret the sensation

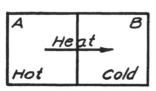

FIG. 209. Flow of heat.

by inferring that the body is warm. If now we touch a hot body H, the same kind of process occurs but more rapidly and by this variation in *rate of change* we have learned to infer that H is warmer than B. If the finger touches a cold body, C, the finger cools or loses heat. We infer that C is cold and by the rate of cooling, gain a qualitative estimate of how cold C is.

The blood stream of a normal healthy person is one of the rare bodies the temperatures of which remain constant (isothermal), or nearly so. A healthy person can travel from the Equator to the Arctics without harm: the change is relatively slow and his system adjusts itself marvellously to the varying conditions so that the temperature of his bloodstream remains unchanged. What the system cannot undergo safely is too sudden a change of temperature. The thermal nerves are sentinels to warn us that a certain change of temperature is too sudden for safety, or that it is going beyond a safe range.

THE BEGINNINGS OF POTTERY

Hollow gourds were natural containers for liquids but not suitable for heating liquids at a fire. In the marshy lands on the banks of the Euphrates, early man learned to plait rushes. He wove them into mats and sun-shades, and he daubed them with mud to make the walls of his hut. Baskets of plaited rush-leaves or matted reeds and grass were also devised in both the Nile and Euphrates valleys for holding cereals (fig. 210), but to make them hold liquids, they had to be plastered over with clay mud. When someone tried to use one of these mud-plastered

FIG. 210 (*Left*) Coptic basket, Egyptian, 4th to 6th century A.C.
(*Right*) Geometric pot, Greek, 9th century B.C.

baskets for heating water over a fire, it was discovered that the clay protected the reeds from scorching or burning. Far more important, however, was the discovery that the clay baked hard. Hence arose the process of "burning" or kilning clay, which led to the manufacture of pottery. The same process was applied to gourds and led to the invention of various small pots and ladles. Much ancient pottery declares its ancestry by its imitation of the basket and of the gourd in form and pattern. The basket pattern in pottery, of course, is still quite common in the present day. Even yet the tale of the basket's contribution to thermics is not completely told.

THERMICS OF CLOTHING AND DWELLINGS

Archaeologists have brought to light plenty of bone or ivory needles, some of which antedate even the Stone Age. Their thermal significance becomes apparent when one realizes that they record the custom of sewing together pelts and hides to make fur garments, tent-walls, and wind-screens for protection from the cold, from the wind, and from the sun. Such garments were prototypes of a vessel or instrument (fig. 245, p. 356) called a *calorimeter* (Lat. *calor*, heat and *metiri*, measure) which figured prominently in the story of thermics when the quantitative stage was reached. Essentially, the calorimeter is a double-walled vessel with an air-space between the two walls. The purpose of such a wall is to slow the rate at which heat flows through it, either into or out of the contents of the inner vessel. At a glance the resemblance between these walls and those of a house is readily seen—two solid layers with air-space between them. At the windows there is normally only one layer—but, when winter approaches, the householder puts on an extra window with air-space enclosed between. A suit of clothes has

Fig. 211. Gourd with incised decoration, modern Cyprus, and two gourd-shaped
vessels, Cyprus, 3rd millennium B.C.

the same essential structure, and so has a bed. It is a fact, then, that
we live a considerable fraction of our time in calorimeters or containers
whose walls are poor conductors of heat. In ancient times the use of a
cave as a dwelling was partly a thermal consideration since in it the
residents were not subjected to such uncomfortable thermal conditions
as when out in the open. The cave-walls were slow conductors of heat.
The list of devices constructed like the calorimeter is lengthy: thermos
bottles, refrigerators, incubators, and so forth. The thermal protection
of fur robes, woollen blankets, or of the snow-blanket on the earth, is
largely owing to the slow conduction of the air enclosed among the
fibres or the snow crystals.

The study of the basket seems to lead into several thermal topics.
From basket-weaving came the handicrafts of cloth-weaving and the
spinning of thread, and the consequent improvement in clothing by
the substitution of cloth for fur. If cloth is examined with a lens, the
basket weave is plain to see. As early as 3000 B.C., the Egyptians could
spin linen thread and weave linen cloth. They became so proficient at
the weaving of fine linen that they have never been excelled and seldom
equalled. Figs. 212 and 213 show spinning, dyeing of threads, and
weaving, as done in Egypt about 2400 B.C. The "waisted stones"
mentioned on page 6 indicate the existence of cloth-looms in Mesopotamia as early as 3500 B.C.

The task of spinning was often assigned, not only in Egypt but also until long afterwards in Europe, to unmarried women, who were, therefore, called "spinsters." Such is still their legal designation, although few spinsters now do much spinning of textile yarns. Thus the story of the basket leads from its lowly invention on the mud-flats of the Euphrates through many stages until, finally, the word "spinning" may summon to mind the beautiful musical description of that handi-

FIG. 212. Egyptian spinning and dyeing.

craft given by Mendelssohn in his *Spinning Song*. Throughout this development a considerable proportion of the factors involved were of thermal significance. There are plenty of other topics which the reader can pursue for himself in similar manner, for example, the history of pottery or of the extraction of metals from their ores.

FIG. 213. Egyptian weaving.

THERMICS IN THE SCRIPTURES

There is no need to wonder whether the Hebrew Scriptures make any reference to thermal topics: there is no literature that does not; and it is difficult to imagine how there could be one. In one passage scripture refers to a sacrifice of burnt offerings:

And Noah builded an altar unto the Lord: and took of every clean beast, and of every clean fowl, and offered burnt offerings on the altar.—Genesis viii.20. (2300 B.C.)

In another episode, a fire disaster is dramatically described:

Then the Lord rained upon Sodom and Gomorrah brimstone and fire from the lord out of Heaven. . . . And Abraham looked toward Sodom and behold the smoke of the country went up as the smoke of a furnace.—Genesis xix.24. (1300 B.C.)

In his fury, King Nebuchadnezzar, who seems to have been a very impulsive man, described one temperature as being seven times greater than another. The expression looks quantitative but it was largely rhetorical:

Then was Nebuchadnezzar full of fury. Therefore he spake and commanded that they heat the furnace seven times more than it was wont to be heated. And he commanded the most mighty men that were in his army to bind Shadrach, Meshach, and Abednego, and to cast them into the burning fiery furnace. Therefore, because the king's command was urgent and the furnace exceeding hot, the flame of the fire slew those men that took up Shadrach, Meshach, and Abednego.—Daniel iii.19.   (580 B.C.—in the days of Thales.)

This then is the first recorded attempt to express a quantitative relation between two temperatures.

The scriptures also refer to baking:

And Moses said unto them This is that which the Lord hath said, To-morrow is the rest of the holy Sabbath unto the Lord: bake that which ye will bake to-day and that which remaineth over lay up for you to be kept until the morning.— Exodus xvi.23.   (1500 B.C.)

and to the melting and refining of metals:

For he is like a refiner's fire. And he shall sit as a refiner and purifier of silver. He shall purge them as gold and silver.—Malachi iii.3. (400 B.C.)

That is to say, until he sees his own image in the clear surface of the molten metal.

The point is made that an optical shadow is also a thermal shadow:

And there shall be a tabernacle for a shadow in the day-time from the heat. Behold a king shall reign in righteousness as the shadow of a great rock in a weary land.—Isaiah iv.6.   (750 B.C.)

PROMETHEUS BOUND

Again we come into the Greek cycle of culture and this time into what is called its late summer period. Previous mention has been made of Prometheus as the central figure in the great poem *Prometheus Bound*, composed by the Athenian dramatist Aeschylus. The work of this poet, always forceful and lofty, has earned for him the appellation "father of the Greek drama." The tragedy of Prometheus describes in majestic verses the bringing of the gift of fire to man.

Hephaestus (or Vulcan) was the god of fire; Aether we have met before:

> (*Enter* HEPHAESTUS, STRENGTH, AND FORCE, *leading*
> PROMETHEUS *in chains.*)

STRENGTH. And now, Hephaestus, it is thine to do the hests the Father gave thee, to these lofty crags, to bind this crafty trickster fast in chains of adamantine bonds that none can break; for he, thy choice flower stealing, the bright glory of fire that all arts spring from, hath bestowed it on mortal man. And so for fault like this, he now must pay the Gods due penalty, that he may learn to bear the sovereign rule of Zeus, and cease from his philanthropy.

HEPHAESTUS. And many groans and wailings profitless thy lips shall utter; for the mind of Zeus remains inexorable. Who holds a power but newly gained is ever stern of mood.

PROMETHEUS. But I needs must bear my destiny as best I may, knowing well the might resistless of Necessity. . . . For I, poor I, though giving great gifts to mortal man, am prisoner made in these fast fetters: yea, in fennel stalk, I snatched the hidden spring of stolen fire which is to men a teacher of all arts, their chief resource.

> (*Flashes of lightning and peals of thunder.*)

PROMETHEUS. Yea, now in very deed, no more in word alone, the earth shakes to and fro, and the loud thunder's voice bellows hard by and blaze the flashing levin-fires. . . . Such is the storm from Zeus that comes as working fear, in terror manifest. O, Mother venerable, O, Aether! rolling round the common light of all, see'st thou what wrongs I bear?[2]

MOLECULAR MOTION

With the westward spread of civilization and science from Mesopotamia and Egypt into Asia Minor and Greece, the study of heat assumed a new aspect. It took on a more pronounced attitude of speculation, as did other sciences, although in this science the advancements were comparatively few and rather sporadic. The work of Democritus symbolizes this advent of hypothesis, which was the characteristic contribution of the imaginative Greek mind to science. Previously, thermal problems had been practical: how to make and control fire; how to use fire for warming a cave, for hardening wooden spear-points, for keeping dangerous animals at bay, for placating gods and devils by burnt offerings, for drying clothes, for cooking meat and baking bread, for purifying silver and gold. In Mesopotamia, fire was used to harden clay in making bricks, pottery, and tablets of written records. Sometimes it was used to cremate the dead. By the bloody war-lord, Ashur-nazir-pal, fire was misused in burning cities and torturing prisoners. There is no record, however, of anyone during that long period asking, "What is this thing called heat?", nor of anyone imagining a mechanical picture or model of what is going on in a body while it

[2]Trans. E. P. Plumptre, *Harvard Classics.*

changes temperature, or while two bodies mutually alter each other's temperatures. Democritus speaks of vibrating particles and, vaguely, of heat and temperature, but the connection between them is not specified clearly.

From Diogenes Laërtius:

The atoms move about everywhere in the universe describing whirling motions. It is of them that the qualities, fire, water, air and earth, are formed.

From Simplicius' *De caelo:*

The atoms vibrate and move about in the void by reason of their specific differences. Democritus says that the atoms are by reason of their own nature inert; they are in motion because of their impacts. As they encounter each other abruptly they come into collison.

From Galen, *De elementis:*

The atoms are supposed to be unalterable. They cannot undergo modification: so that no atom, they say, can be hot or cold, nor can it be dry or moist nor white or black.

How Democritus reached the idea of atoms (or molecules) vibrating and colliding is interesting to speculate. Lucretius hints that it came from the dance of dust-motes in a sunbeam. As the story unfolds we shall see two sets of ideas develop separately concerning motion of molecules and the temperatures of bodies. Then about A.D. 1800, by means of the concept that when a body warms its particles vibrate faster and farther, they are brought together in a unified theory of heat and particularly in the kinetic theory of gases. Similarly, in mechanics we saw the study of planetary motion and of falling bodies develop separately until Newton fused them in the law of universal gravitation. A further example will be the welding of magnetism and electricity by Oersted and Maxwell.

THE FOUR ELEMENTS

The Greek philosopher and democrat Empedocles (490-430 B.C.), a contemporary of Democritus who lived at Akragas in Sicily, supposed that the universe is made from four "elements"—the "roots of all things." These were, *fire,* air, water, and earth. His hypothesis, then, is an example of the general tendency to advance from the Many to the Few. Plato adopted Empedocles' assumption and Aristotle discussed it in his treatise *On Generation and Corruption.* Although it has long since been discarded as untenable, Aristotle's comment is quoted here because the hypothesis was universally employed for twenty centuries. Its influence can be seen in the selections we have just read from Galen (A.D. 130-200) and Diogenes Laërtius (fl. A.D. 230).

Now there are four elementary qualities, hot and cold and moist and dry, and the number of combinations of four things, two at a time, is six; opposites, however, cannot be combined; for it is impossible that the same body should be both hot and cold or again, moist and dry. Hence it is evident that there will be four combinations of the elementary qualities, namely, hot and dry, hot and moist, and also cold and moist and cold and dry. Now these couples have attached themselves to the four elements, fire, air, water and earth, in accordance with theory; for fire is hot and dry, air is hot and moist (air being a sort of aqueous vapour), water is cold and moist and earth is cold and dry.

Throughout this classification it is evident that the temperature terms, hot and cold, have a priority of importance which places them always first.

CONVECTION

There are methods of heat transfer other than conduction. Aristotle hints at one that is characteristic of fluids. It is called *convection*. Currents form in the fluid and carry heat from one point to another. The rising of smoke above a camp-fire was doubtless observed in prehistoric times, and from it early man gained some practical knowledge of convection currents in a gas. The tendency of smoke flames and sparks to move upward was called the property of levity in contrast with weight, the tendency to move downward. This led to the idea that far above the earth was a zone of fire, the empyrean.

Let us state once more that there is no void existing separately as some maintain. If indeed each of the elements has a natural locomotion, e.g. fire upwards and earth downwards toward the centre of the universe [at the centre of the earth], it is obvious that vacuum cannot be the cause of locomotion.[3]

At one place in his *De Caelo*, Aristotle shows that he has a knowledge of the tremendous increase of volume that occurs when water changes to steam, and the consequent expansive force capable of producing explosion. He repeats the error of the previous selection: that evaporated water is air. The word air was often used to mean gas.

When water evaporates and is transformed into air [i.e. into a gas, more properly called steam], the vessel containing this matter is ruptured because the space within is no longer sufficient.

CHANGE OF PHYSICAL STATE

Lucretius of Rome made some thermal observations in *Concerning the Nature of Things* that are worth noting, such as his reference to the change of physical state called *melting:*

The ice of brass [i.e. its solidity] yields to the flame and becomes liquid.

[3]Aristotle *Physics*.

Here is a poet's way of proposing the theory of the *motion of molecules* which Lucretius, of course, calls atoms:

For look closely whenever rays are let in and pour the sun's light through the dark places in houses: for you will see many tiny motes mingle in many ways all through the empty space right in the light of the rays; and as though in some never ceasing strife, wage war and battle, struggling troop against troop, nor ever giving pause, harried by numberless impacts and recoils. From this you may picture to yourself how the atoms are perpetually tossed about in the great void.

Until about A.D. 1800, heat was considered by many as a kind of substance called caloric (Lat. *calor*, heat). That it is not a substance in the ordinary sense was recognized, for it was known that warming a body or cooling it does not alter its weight appreciably. Since the addition of heat to a body does not increase its weight, heat must be weightless. It was given the paradoxical name of "imponderable substance," and because heat is assumed to flow from one body to another, it was called an *imponderable fluid*. Lucretius argued in favour of regarding heat as a substance:

Because the ultimate components of bodies cannot be discerned by the eyes, let me tell you of other bodies which you, yourself, must needs confess are material bodies and yet cannot be seen. We do not behold warm heat nor can we espy cold with the eyes, yet these must needs be composed of matter inasmuch as they can stimulate our senses: for nothing which is not composed of matter can touch or be touched. Furthermore, garments hung up upon the shore, where the waves break, grow damp; and again, spread in the sun, they dry. Yet never has anyone seen the moisture of the water stealing into them, nor again, when it was fleeing before the heat.

THERMAL EXPANSION OF GASES

As we have seen, that ingenious artificer Hero (or Heron) of Alexandria, like Leonardo da Vinci and von Guericke, had a fondness for devices that astonish and mystify the beholder. His "magic" temple-doors were after his own heart. They show that he had a practical knowledge of the fact that air when warming, expands and, when cooling, shrinks. Fig. 214 is Hero's own diagram labelled, of course, in Greek.

To construct a chapel whose doors open of their own accord in response to an altar fire and, when the fire is extinguished, close again. The said chapel stands on a base $\alpha\beta\gamma\delta$ on which a small altar $\epsilon\delta$ is also placed. Through the altar, passes a tube, $\xi\eta$, whose mouth, ξ is inside the altar while η is surrounded by a ball ϑ, and is not far from its centre. The tube $\xi\eta$ is soldered into the ball which contains a curved siphon $\kappa\lambda\mu$. The door-posts are extended downwards and turn freely on small pivots situated within the base. Attached to the rollers are small chains which join to form a single chain and pass over a pulley to support the hollow vessel ν ξ. Other small chains that are wound around the door-posts in the opposite sense to the former ones, likewise join to form a single chain and, passing over a pulley, are there fastened to a lead counterpoise, by the descent of which the temple-doors are closed.

Lead the outer leg of the siphon κλμ into the hanging vessel. The sphere is half-filled with water through an aperture, π, which is thereupon closed. If the fire is lit, the result is that the air in the altar warms, expands and tends to occupy greater space. Air passes through the tube ξη into the sphere and drives

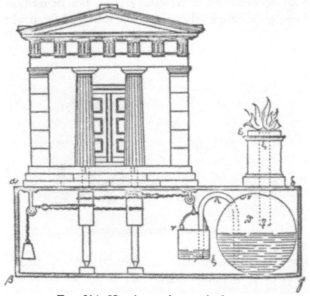

Fig. 214. Hero's magic temple-doors.

some of the water in it through the siphon κλμ into the hanging vessel. This now sinks down, pulls the chains and thus opens the doors. When the fire is extinguished . . . .(etc.)[4]

This philosophic toy is the prototype of the modern steam engine with cylinder and piston. It also foreshadows the air-thermometer.

THE AEOLIPILE

Hero invented also the first steam-turbine (A.D. 50?) and called it the "aeolipile" or eolipile, literally, the wind-ball (Gk. *Aeolus*, god of the winds, and *pila*, ball). He described the device in his *Pneumatica*, and fig. 215 is his original diagram.

Let αβ be a kettle containing water and subjected to heat. Its mouth γδ is closed by a lid through which a bent tube εξη, passes. The end of this tube passes

[4]Hero *Pneumatica*, p. 174.

into the sphere ϑκ by an air-tight joint. The end η, is diametrically opposite to the pivot λμ. The sphere is equipped with two small tubes, diametrically opposite to each other and bent at their free ends as shown, in opposite directions. If now the kettle is heated, the result is that steam is forced into the sphere through εξη and shoots out through the small bent tubes, and causes the sphere to rotate.

The aeolipile has some modern descendants, the common lawn-sprinkler, the pinwheel or Catharine wheel, and the electric whirl, which are all applications of Newton's third law of motion. The facts about the aeolipile were likely part of the material from which Newton

FIG. 215. Hero's aeolipile.

induced the law. In the modern turbine, of course, the steam is commonly led into the machine instead of outwards as in Herò's device.

Although Hero's "steam-turbine" did no useful *work*, yet it held that possibility. It was, therefore, a harbinger of a new epoch in human advancement. Henceforth, more and more work would be done by inanimate machines and engines of various kinds, and less and less by domesticated animals and human slaves. (In our own day, we have seen the automobile replacing the horse-driven vehicle.) Hero's aeolipile was, therefore, an earnest of the growth of freedom in the world.

THE HYGROMETER

Of the few contributions to thermics that were made during fourteen centuries after Hero, only three items will be mentioned here. The first is the invention, sometimes ascribed to Roger Bacon (1260), of gunpowder,[5] a substance that has had a pronounced influence in the history of the world. The burning of gunpowder is an example of a type of process called *explosive* which proved of fundamental importance

[5]In all probability a Chinese invention used for fire-works celebrations. Bacon records a recipe for preparation of gunpowder but does not claim invention.

later in the story of thermics and especially in the advancement of chemistry.

The next two contributions are from the manuscripts of da Vinci and have to do with the atmosphere. In describing the weather, we, like Aristotle, frequently combine the idea of moist or dry with that of hot or cold as, for instance, in the word "sultry." The condition of the air as to water vapour content is its *humidity*, and an instrument for indicating humidity is a *hygroscope* (Gk. *hugros*, moist). If the instrument performs this function quantitatively, it is called a *hygrometer* (Gk. *metron*, measure). Some desert plants (xerophytes) are natural hygroscopes, for they coil up in dry weather to conserve their moisture, and uncoil in wet weather. Nicolaus von Cusa (1401-63) was probably the first inventor of a hygroscope.[6] That invented by da Vinci about 1508 was an application of his study of the lever. In it a ball of wool was balanced on a small lever by a non-hygroscopic counterpoise of metal. As change in the humidity of the air made the wool heavier with moisture (or lighter) the lever moved and the reading of its position on a graduated circle was taken as a measure of the humidity of the air.

Da Vinci also pointed out the existence of a great "water-cycle" that keeps water flowing through the rivers:

Or do you not believe that the Nile has discharged more water into the sea than is at present contained in all the watery element? So, therefore, one may conclude that the water passes from the rivers to the sea and from the sea back to the rivers [by evaporation and condensation] ever making the selfsame round; and that all the sea and the rivers have passed through the mouth of the Nile an infinite number of times.[7]

Similarly da Vinci anticipated Harvey by suggesting the circulation of the blood in human beings.

[6]Gerland and Traumüller, *Geschichte der Physikalischen Experimentierkunst*, pp. 83, 108.
[7]E. McCurdy, *Science Note-Books of Leonardo da Vinci*, p. 96.

CHAPTER FIFTEEN

The Invention
of the Thermometer

BY an experiment, known as Galileo's experiment, and performed by that great physicist at Padua about 1592, the study of heat was raised to a stage worthy of the name of science. For Galileo invented the thermometer, which expresses temperatures numerically, and thereby brought the subject to a quantitative stage. There is no report of the invention in his extant writings, but a statement was made by his old student Viviani:

> After the commencement of his professorship at Padua, toward the end of the year 1592, Galilei invented the thermometer. It was made of glass, containing air and water [fig. 216], which serve to indicate changes and differences of temperature. This invention was later improved by the reigning grand-duke, Ferdinand II of Tuscany.[1]

This experiment is still performed in elementary courses, and the tube with bulb is called a Galileo-tube. P. Castelli, who saw Galileo perform the experiment about the year 1600, wrote an account of it to a friend:

> He took a glass bulb about the size of a hen's egg with a tube about two spans in length and the width of a straw-stem, warmed the glass with his hands, and turned it so that the end of the tube dipped into a tumbler placed underneath. When the air in the bulb cooled, the water rose to more than a span above the surface of the liquid in the tumbler. Galileo used this fact to prepare an instrument for determining degrees of heat and cold.[2]

That Galileo sometimes used wine or alcohol in his thermometers instead of water is seen in the following from a fragment of one of his letters:

> But if the air around the bulb is cooled by surrounding it with a cooler body, . . . the air (in the bulb) will become colder than before and, according

---

[1] F. Burckhardt, *Die Erfindung des Thermometers* (1867). A copy of this book may be seen in the Library of Congress.
[2] Ibid.

to the principle previously stated, will contract and occupy less space. That there be no vacuum, the wine rises to occupy the space vacated by the air.[3]

In this statement, Galileo uses the old "horror of the void," whereas in view of the work of his old students, Torricelli and Viviani, we would now replace his expression by the words, "to equalize the pressure."

Since the Florentine Academy's membership included former students of Galileo, it is not surprising to find that the society carried

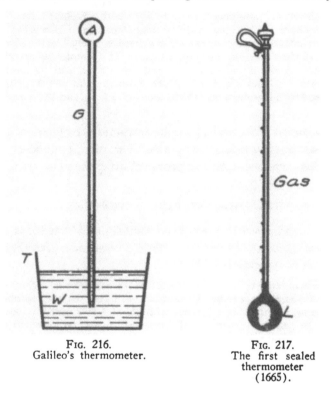

FIG. 216.
Galileo's thermometer.

FIG. 217.
The first sealed
thermometer
(1665).

on pioneer work in the study of heat. In one passage of their report, the secretary, Magalotti, described the first *sealed thermometer* (fig. 217). Viviani's account, which we read, suggests that Ferdinand of Tuscany, a patron of the society and an old student of Galileo, was in charge of this research. The sealed thermometer was an advance on its predecessor in that it was independent of atmospheric pressure; but it was less sensitive, since liquids expand thermally less than gases.

To measure the degrees of Heat and Cold in the Air. . . . Let the Lamp-blower then, make the Ball of this instrument of such Capacity, and joyn

[3]Galilei, *Opere*, tom. III, *Pensieri varj.*, **444.**

thereto a Cane of such bore, that by filling it to a certain mark in the Neck with Spirits of Wine [alcohol], the simple cold of Snow or Ice, Externally Applyed, may not be able to condense [contract] it below the 20 deg. of the Cane; nor on the contrary, the greatest vigour of the Sun's Rays at Midsummer to Rarefie it above 80 deg. . . .: which Instrument may be thus fill'd, viz., by heating the Ball very hot, and suddenly plunging the open end of the Cane in the Spirit of Wine, which will gradually mount up, being suck'd in as the Vessel Cools. But because 'tis hard, if not altogether impossible, to evacuate the Ball of all the Air by Rarefaction: and the Ball will want so much of being fill'd as there was Air left in it: we may thus quite fill it with a Glass Funnel, having a very slender shank, which may easily be made when the Glass is red hot and ready to run; for then it may be drawn into exceeding small hollow Threads. . . . Divide the whole Tube into Ten equal Parts with Compasses, marking each of them with a knob of white Enamel, and you may mark the intermediate Divisions with . . . black Enamel. . . . This done . . . the mouth of the Tube must be closed with Hermes' Seal at the flame of a Lamp, and the Thermometer is finish'd.[4]

The alchemist while sealing a tube incanted the charm of Hermes, praying that the glass would not break afterwards. The modern chemist composes his own incantation impromptu after the glass cracks.

THE BALL AND RING EXPERIMENT

The Academicians obtained experimental proof that brass expands when warming and shrinks while cooling. The surmise was that in this matter brass is typical of solids.

There was ordered to be cast a Ring of Brass, and by turning, it was fitted exactly to a Cilinder of the same Metal: this was put in the fire for a short space, and then being put upon the Cilinder while hot, it was sensibly loose; being dilated by the heat . . .: when it had remained some time upon the Cilinder, and had communicated its heat thereto, between the increasing of that and the shrinking of the Ring, by little and little as it cooled; they not onely came to fit as at first, but were so firmly united, that before they were quite cold, a considerable force was requisite to separate them. The contrary in all respects happened when we intensely froze [cooled] the Ring.

This experiment was probably inspired by the blacksmith's ancient method of putting an iron rim on a wooden waggon-wheel. The form of this apparatus, called the Ball and Ring (fig. 218b), which is used nowadays, was devised about 1730 by s'Gravesande as a modification of the Florentine experiment. Now and again, an examination candidate inadvertently refers to this experiment as the "Ball and Chain" experiment.

[4]*Saggi di Naturali Esperienze fatte nell' Accademia del Cimento*, p. 2. Also, *Essayes of Natural Experiments, &c.* (1684), trans. R. S. Waller, S.R.S., which may be seen in the library of the University of Chicago.

THE FLASK AND TUBE EXPERIMENT

The same little band of experimenters were first to perform the so-called Flask-and-Tube experiment (fig. 219). Their report on this research is replete with new facts. They showed that water when cooling contracts more than glass. They observed the slight initial rise of the level in the tube, due to the glass contracting first. They discovered that water when near the freezing-point is exceptional (anomalous), for it then expands when cooling. Consequently, at a certain temper-

FIG. 218*a*.
Florentine experiment
on thermal expansion.

FIG. 218*b*.
Ball and Ring apparatus.

FIG. 219.
Flask-and-Tube
apparatus.

ature near the freezing-point, water has a minimum volume and, therefore, maximum density. In the same set of experiments they witnessed the pronounced expansion (10 per cent) that occurs when water freezes. Finally, they discovered that the melting-point of ice is constant. The "bolt-head" they used was a spherical glass vessel with a narrow neck, and resembled, therefore, a Galileo-tube.

TOUCHING THE PROCEDURE OF ARTIFICIAL FREEZINGS WITH THEIR
WONDERFUL ACCIDENTS.

The first Vessel we made use of in these Experiments was a Bolt-head of Glass 2$\frac{8}{10}$ Inches in Diameter, with a Neck about 34 Inches long, slender and divided; into this, we poured fair water to a Sixth Part of the Neck, then setting the Bolt-head in the freezing-mixture, we attentively observed the Motion thereof. . . . You must know That upon the first immersion of the Ball, as soon as ever it touch'd the freezing mixture, we observed the Water in the Neck, a

little rising, but very quick, which, soon subsiding, fell in the Neck, with a
Motion regular enough and a moderate Velocity retiring to the Ball, till arriving
at a certain degree, it stopp'd for some time, as far as our Eye could judge,
immovable. Then by little and little, it remounted, but with a very slow Motion.
... and then of a sudden ... it flew up with a furious Spring. At last it happened ... that we let the water ... freeze in the Neck first ... and so brake
our Vessel. ... Whereupon, we were forced to make another; ... we made the
Neck longer to the height of $45\frac{1}{10}$ Inches(2 cubits). This we filled to 160 deg.
and set it to freeze in Ice; very diligently heeding it, we found, at first, that all
the accidents of rising, ... swiftly running up and stopping again, were the
very same, i.e. happened always when the Level of the Water was at the same
Mark or Degree in the Neck: for upon putting it in the Ice, we observed it was
reduced to the same Degree as in the former Tryal: that is to say at the same
Temperament of Heat and Cold: taking the whole Instrument for a nice Thermometer by reason of the largeness of the Ball and the proper proportion of
the Neck.

The old word, accident, meant occurrence (Lat. *accidere*, happen).
Since Waller's time, this word, like many others, has changed to a
worse meaning. *Facilis descensus Averni*—easy the road downward.
"Temperament" in this passage is an early approach to the term
temperature. The use of a "freezing-mixture" in this experiment emphasizes the fact that man was learning how to cool bodies as well as warm
them.

Another contribution that the Florentines made to thermics was
their invention of a primitive dew-point hygrometer (fig. 220):

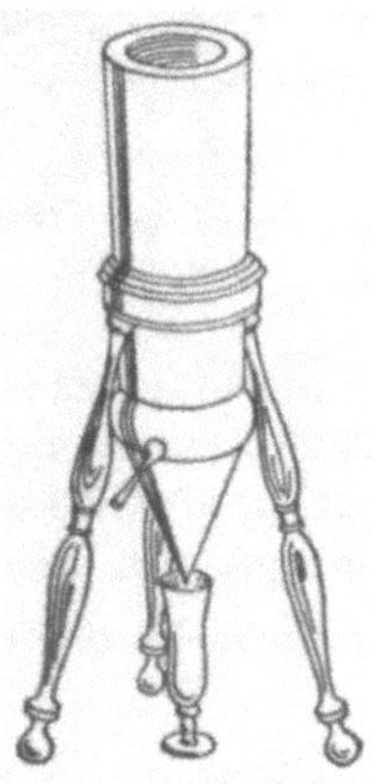

FIG. 220.
Primitive dew-
hygrometer.

To DISCOVER THE DIFFERENCE OF MOISTURE IN THE AIR.

It is part of a Cone of Cork, hollow within and pitched;
and covered on the outside over with Tin: at the smaller
end, it is inserted into a Vessel of Glass with a Conical
Point shaped as in the Figure and closed Hermetically:
The Vessel ... is to be filled with Snow or small beaten
Ice: the water whereof as it melts, shall have its issue by
the Pipe made in the upper part of the Glass. The Use of
it is this, The subtil Moisture carryed about by the Air,
adheres by little and little to the sides of the Vessel, covering it at first but with a dew or mist, till by the coming
of more Moisture, it gathers into great drops, and at last,
stealing down the sides of the Conical Glass, drops into
a tall Cup in the shape of a Mumglass, divided into equal
degrees.

"Mum" was a strong sweet beer; hence the name
mumglass.

The same book records the fact that when some
substances are dissolving in water, the system warms, whereas, with
other solutes, cooling occurs. The old name of sulphuric acid was *oil
of vitriol*:

Vitriol, the Spirit being drawn off, remains like a Tartar, . . . of a lively Fire Colour, which with a long and continued Fire distils a blackish Oyl almost like Inke, highly corrosive [impure sulphuric acid]. This being mixt with Water . . . produces an immediate Heat, which increases without raising any Bubbles, or perceivable Smoak, till the Glass wherein the mixture is contained can scarce be endured in the hand [60°C.]: . . . On the contrary, 'tis a known Experiment, That Nitre dissolved in Water, chills it: also sal ammoniac.

A process which gives out heat is called *exothermic*, and one which takes in heat *endothermic*.

THERMAL EXPANSION

In 1665, Robert Boyle published a book, *Experiments Touching Cold*, from which the next three excerpts are taken. The first reports that the flask-and-tube experiment was performed with turpentine as the liquid, and from the measurements taken, the coefficient of volume expansion for that liquid determined:

We found that rectifi'd Oyl of Turpentine of a moderate temper being expos'd to such a degree of Cold as would freez common water, did by shrinking lose about a ninty-fourth part of its Bulk.

Assuming that "moderate temper" for Boyle's laboratory was English room temperature, 15°C., the coefficient of thermal expansion for turpentine from Boyle's data was $c = \dfrac{e}{V.Dt} = \dfrac{1}{94} \times \dfrac{1}{15} = 0.00071$. A modern value is 0.00097.[5]

Boyle modified Galileo's experiment by using as an *indicator* a globule of water inside the tube instead of a whole tumblerful mostly outside it. In so doing, he came within striking distance of finding the coefficient of thermal expansion for air; but he let the opportunity slip, doubtless because he did not see it, and a century later the work was done by Citizen Jacques Charles of Paris, after whom the law he discovered is named. The apparatus that Boyle devised. is still used in elementary courses and is known as a Charles' law tube (fig. 221). A globule of mercury is generally used instead of water.

I took a thin glass-Egge, blown at a Lamp about the bigness of a Walnut, with a stem coming out of it about the bigness of a large Pigeons Quill four or five inches long and open at the Top; this slender pipe being dipp'd in water, admitted into its cavity a little Cylinder of Water, of half an Inch long or some- what more, which (the Glass being erected) subsided by its own weight[?], or the Temper of the Air in the Egge (in reference to the outward Air) till it fell to the lower part of the Pipe, where it comes out of the Egge, and thereabout it would rest.

[5]W. J. R. Calvert, *Heat* (Arnold, 1938), p. 56, or *The Handbook of Physics and Chemistry* (Chemical Rubber Publishing Co.).

The word "temper" in this quotation and the previous one, is another approach to the term *temperature*. Quite probably Boyle warmed the bulb with his hand before he dipped the tube in water.

Boyle was two years in advance of the Academicians in publishing the fact that the melting-point of ice is constant; but they probably were prior to him in performing the experiment. He proposed using that temperature as a fixed point in graduating thermometers. He also gave a qualitative demonstration of the principle known as Newton's law of cooling. The thermometers he used were Florentine.

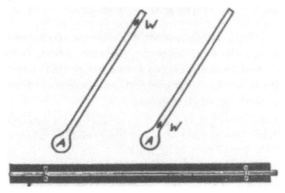

Fig. 221. Charles' law tube.

The seal'd Weather-glass [thermometer] being kept in the water till it began to freez, descended to 5½; Being immediately remov'd into the same snow and salt, that made the water begin to freez, it descended at the beginning very fast, and afterwards more slowly, till it came to the very bottom of the stem, where it expands it self into the Ball; then being remov'd into the same glass of water, whence it was taken, and which was well stor'd with loose Pieces of Ice, it did nevertheless hastily ascend at the beginning, and was soon after impell'd to the former Height of five Divisions and a half, or thereabouts. . . .

They that have a mind to prosecute Experiments of this kind and others that relate to the Degree of Cold may perchance be somewhat assisted even by these Relations and especially by those Passages that mention the use of the seal'd Weather-glass, furnish'd with spirit of Wine . . . , these being not subject to the Alterations of the Atmospheres Gravitation nor to be frozen, . . . the Instrument being accompanied with a memorial of the Degree it stood at, when exposed to such a Cold as made the Water begin to freeze.

Boyle's thermometer, then, had only one fixed point, namely, the temperature of melting ice. The expression, "at the beginning, very fast and afterwards, more slowly," indicates that the *rate at which heat flows from one body to another decreases as the difference between their temperatures grows less* (law of cooling); but apparently Boyle was not aware of this law.

BOILING-POINT OF WATER

Newton published anonymously in 1701 a paper entitled *A Scale of Degrees of Heat and Cold.*[6] Omitting his signature was a vain attempt to avoid controversy—his pet aversion. But nobody had much difficulty in naming the author; as John Bernoulli said on another occasion, "As by the claws, we know the lion; ergo, Newton wrote it."[7] In this paper, Newton was first to use as a second fixed point in a thermometer-scale the temperature of boiling water, as Huygens had suggested in 1665.

0°	—	—the temperature of winter air at which water begins to freeze.
12	—	—maximum reading that thermometer gives by contact with [healthy] human body.
$20\frac{2}{11}$	—	—freezing-point of wax ...
34	—	—temperature at which water boils vigorously.
48	—	—melting-point of mixture of equal parts tin and bismuth.
96	—	—melting-point of lead.
192	—	—temperature of coals in small kitchen fire, fanned to glowing with a bellows.

I found with a thermometer made with linseed oil, that when it was placed in melting snow, if the oil had a volume of 10,000 units, the same oil, expanded by the heat of the human body, occupied a space of 10256 units, and by the heat of water ... boiling vigorously, 10725 units. ... The expansion of air with an equal change of temperature, is ten times as great as the expansion of oil. ... By setting the temperatures proportional to the expansion of the oil and writing 12° for human body temperature, I obtained for the temperature of water ... when it boils vigorously, 34 degrees.

The calculation is $12 \times \dfrac{725}{256} = 34$. From the same paragraph, the expansion coefficients of linseed oil and of air can be calculated. Those who would like to read the discussion of Newton's law of cooling in the original words of the discoverer will find them in the same paper. Perchance, before they have finished the article, they will remark that John Bernoulli was right—there is not enough wool to hide the claws.

THE MERCURY THERMOMETER

Gabriel D. Fahrenheit (1686-1736) was an instrument maker of Amsterdam who contributed a number of valuable items to the science of heat. By inventing a new method of cleaning mercury he improved the mercury thermometer and did much to bring it into common use. He also discovered that the boiling-point of water varies with atmos-

---

[6]I. Newton, *Opuscula* (1701), "*Scala graduum caloris et frigoris.*"

[7]*Tamquam ungue leonem,* from the well-known fable about the lion in sheep's clothing.

pheric pressure. His papers appeared in the *Philosophical Transactions* of the Royal Society of London, of which he was a Fellow.

As I read in the History of the Sciences of the Royal Society of Paris about ten years ago, how the celebrated physicist Amontons discovered by means of a thermometer that he had invented that water boils at a fixed temperature, I was immediately inflamed with a burning desire to make such a thermometer for myself and produce this beautiful phenomenon . . . before my own eyes and satisfy myself as to the facts of the experiment. . . . But in spite of many attempts, all my efforts were in vain for lack of experience in this technique. . . . It occurred to me then that that keen observer of Nature had written about the correction of the Barometer and had found that the height of the mercury column in the barometer is affected to an observable degree by the temperature of the mercury. Hence I concluded that a thermometer might be made with mercury. In the construction of such a thermometer, I ultimately reached success, for, in spite of its defects, it satisfied my wishes. With the greatest excitement and joy, I now beheld the actual process. . . . Three years had now passed by . . . since I first determined to investigate whether other liquids have not also fixed boiling-points. The results are given . . . in the following table:

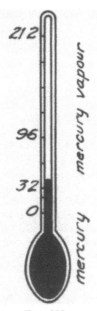

Fig. 222.
Fahrenheit's
thermometer.

Liquid	Specific Gravity	Boiling-Point
alcohol...............	0.8260	176
rainwater............	1.0000	212
caustic potash solution..	1.5634	240
sulphuric acid.........	1.8775	546

Volatile oils also begin to boil at fixed temperatures, but during the boiling their temperatures rise, probably because the more volatile parts are driven off and the resinous constituents remain.

Later, however, I discovered by various experiments and observations that the boiling-point of water is fixed only if the atmospheric pressure remains the same; but with variations of atmospheric pressure, the boiling point can change in either sense.

Fahrenheit found that the boiling-point went up with the barometer reading and down with it; increase of pressure raises the boiling-point of water.

Before I proceed with the experiments, . . . it will be necessary to mention briefly the thermometer I have made, also the graduation of its scale and, furthermore, to describe how the instrument was freed of air. . . . I make two kinds of thermometers, one kind containing alcohol and the other quicksilver. The scale is based on the determination of three fixed points. . . . The lowest lies at the beginning of the scale and is found by means of a mixture of ice, water, and sal ammoniac, or one may use sea-salt. When the thermometer is inserted in this mixture, the fluid level sinks to the point marked zero. . . . The second point is reached when water and ice are mixed without any salt. In this mixture, the thermometer registers 32 degrees. The third point is at 96°, and the alcohol expands to that level when the instrument is placed in the mouth or armpit of a healthy person and kept there until it has completely reached the temperature of the body.[8]

Tradition says that Fahrenheit considered the zero of his scale the coldest temperature possible.

SUPERCOOLING

It is possible to cool water below its freezing-point without freezing it; it is then said to be supercooled. In this state it is not stable; it is liable to congeal suddenly. As it does so, the system warms rapidly to its freezing- (or melting-) point. An interesting case of supercooling was reported to the Royal Society of London in 1731 by M. Triewald of Stockholm, Sweden, who was a Fellow of the Society:

SIR HANS SLOANE, P.R.S.   *Stockholm, 1731.*
    LONDON

Most Honoured Sir,

I have at present to communicate to you a somewhat strange Accident, which produced as unexpected a Phaenomenon. The 15th of December last, coming into the Hall where my Apparatus is placed in the Palace of the Nobility at Stockholm, the Weather being very cold, I feared that the Glass for shewing the Experiment with the Cartesian Devils (or those glass Figures in Water, which by the Pressure of the Air on the Surface of the Water, are made to change their Places, and to sink to the Bottom of the Glass) [fig. 85], would be in danger, if the Water should freeze in the same. I took it down from the Shelf, and was well pleased to see the Water in a fluid State; but before I would empty the Glass, as some Friends that were present, had not seen that Experiment, I placed my Hand on the Bladder tied on the Top of this Cylindrical Glass, which was of a pretty large Size, sixteen Inches high, and three Inches and a half Diameter, containing three Glass Figures: In that very Instant, and in the Space of a Second of Time, I found all the Water changed into Ice; when in that Time, two of the Figures had reached very near the Bottom, but the third, as well as they, fixed in the Middle of the Glass, surrounded with Ice as transparent as the Water, itself, before it congealed. This is in a few Words, the Matter of Fact; but the Reason why the Whole Body of Water, in such a short Space of Time, should turn into Ice, is, in my humble Opinion, not so easily to be accounted for.

Yours respectfully

M. TRIEWALD, F.R.S.[9]

[8] G. D. Fahrenheit, in *Phil. Trans. Roy. Soc. Lond.* (1727), XXXVII, 179. A copy of this volume is in the library of the University of Pennsylvania.
[9] *Phil. Trans. Roy. Soc. Lond.* (1731), XXXVII, 79.

THE CELSIUS OR CENTIGRADE SCALE

The professor of astronomy at Uppsala, Sweden, Anders Celsius, published in 1742 a proposal for a new thermometer scale. This scale has superseded the Fahrenheit scale in scientific work all over the world, and is in general use in many progressive countries. But Great Britain, which has been in many important matters a paragon of progressiveness, still clings to the old scale. Similarly, a century after the introduction of decimal currency and the convenient metric system, she still does it the hard way with cumbersome and inconvenient tables of weights and measures. The Celsius scale is also called the Centigrade (Lat. *centum*, 100 and *gradus*, degree) because the stem between melting- and boiling-points of ice and water is divided into 100 equal parts as suggested in a paper by Celsius:

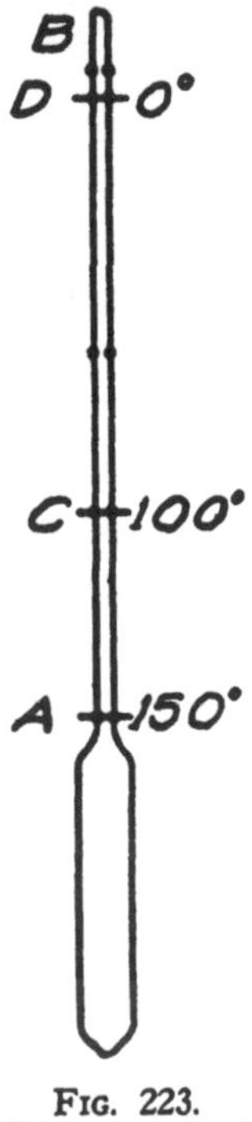

> Thermometers are in fairly general use now in Sweden, partly for the bath but mostly to hang on the wall in order to see how much the temperature of a room gains or loses. The commonest are the so-called Florentines which come to us in Sweden from Germany and are all "no good," because they give no definite measure of the degree of heat and cold [temperature] and furthermore, they do not all give the same reading for the same temperature.
>
> The best way to mark the degrees on a thermometer is as follows, for by it, one is certain that different thermometers of this kind, in any given air, will always give the same reading; and that, for example, a thermometer made in Paris, will stand at the same degree as that registered by one made at Uppsala.
>
> (1) Place the bulb of the Thermometer *AB*, fig. [223] in snow that will pack [melting] and observe accurately the point for freezing water *C*, which should be as high above the bulb at *A* as about half the distance between the points *C* and *D*, the freezing and boiling-points of water. . . . (2)

Fig. 223.
Celsius' scale of
temperatures.

The interval *CD* is divided into one hundred equal parts or degrees so that zero (0°) is at *D* and 100° at *C*. Extend the same series of degrees below *C* to *A* and the thermometer is ready.[10]

It is said that while Celsius was planning his new scale he called on his old friend Linnaeus (Linné), professor of botany at Uppsala. Linnaeus was a fairly shrewd man and when he had examined the new scale, he said "Do you know, Anders, old man, if I were you, I think I would reverse the numbers at *C* and *D*." Celsius said he would carefully consider his friend's advice. So he went home and arranged

[10]A. Celsius, *Abhandlungen der Schwedischen Akademie* (1742), IV, 197. Also, W. Ostwald, *Klassiker*, LVII, 117.

the scale exactly as he had it before. Eight years later, his colleague, M. Stromer, published a paper using the scale modified as Linné had suggested; and in the modern Celsius scale, C is marked 0° and D is 100°. It is well to note that Celsius' choice was not wrong but merely uncustomary and less convenient.

It is not an overstatement to say that the thermometer has been excelled by few instruments in the progress of science.

AN IMPROVED SOURCE OF HEAT

Benjamin Franklin of Philadelphia (1706-90), who styled himself simply "printer" even in his epitaph, was an outstanding citizen in many ways—journalist, businessman, statesman, diplomat, and scientist. He was a genius in both theory and practice. His autobiography is one of the greatest in literature. The United States does itself honour to place his portrait on its postage stamps. In the following excerpt from his writings, we see his practical bent in his invention of a new stove (1744), which made the heating of a room somewhat less like the warming of a cave with a camp-fire:

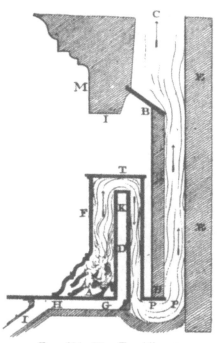

FIG. 224. The Franklin stove.

In these northern colonies, the inhabitants keep fires to sit by generally seven months in the year; .... Wood, our common fuel, . . . must now be fetched near one hundred miles. . . . As therefore so much of the comfort and conveniency of our lives, . . . depends on the article of fire; since fuel is become so expensive, . . . any new proposal for saving the wood and for lessening the charge and augmenting the benefit of fire . . . may at least be thought worth consideration. . . .

The new fire-places are a late invention to that purpose. . . . A fire being made, . . . in any chimney, the air over the fire is rarefied by the heat, becomes lighter and therefore, . . . rises in the funnel and goes out: the other air in the room . . . supplies its place. . . . Fire . . . throws out light, heat and smoke or fume. The large open fire-place, used in the days of our grandfathers, and still greatly in the country, . . . has generally the conveniency of two warm seats, one in each corner; but they are sometimes too hot to abide in, and, other

times, incommoded with the smoke.... They almost always smoke, if the door be.
not left open . . . so that the door can seldom be shut; and the cold so nips the
backs and heels of those that sit before the fire, that they have no comfort. . . .
In short, it is next to impossible to warm a room with such a fire-place; and I
suppose our ancestors never thought of warming rooms to sit in; . . . . Many
of the diseases proceeding from colds, . . . may be ascribed to strong drawing
chimneys, whereby in severe weather, a man is scorched before while he is froze
behind. To avoid the several inconveniences and at the same time retain all
the advantages of other fire-places, was contrived the Pennsylvania Fire-place
now to be described.

The fire being made at A [fig. 224], the flame and smoke will ascend and
strike the top T, which will thereby receive a considerable heat. The smoke
finding no passage upwards, turns over the top of the air-box, and descends
between it and the back plate. . . . The air of the room, warmed behind the
back plate, is obliged to rise, . . . is forced out into the room, and spreads all
over the top of the room . . . and the whole room becomes in a short time,
equally warmed. . . . A square opening for a trap-door should be left in the
chimney for the sweeper to go up.

COOLING BY EVAPORATION

In a letter dated New York, April 14, 1757, Franklin made the
first reference in the literature to the important principle of cooling
by evaporation.

Professor Simson of Glasgow lately communicated to me some curious
experiments of a physician of his acquaintance, by which it appeared that an
extraordinary degree of cold, even to freezing, might be produced by evaporation. . . . Wet the ball of a thermometer by a feather dipped in spirits of wine,
which has been kept in the same room, and has, of course, the same degree of
heat or cold. The mercury sinks presently three or four degrees, and the quicker,
if during the evaporation, you blow on the ball with bellows: a second wetting
and blowing, when the mercury is down, carries it yet lower.

Franklin's letters also contain an interesting discussion of the
conduction of heat. He uses the caloric theory of those days: instead
of the word *caloric*, he uses the term, *fire:*

Allowing common fire . . . to be a fluid, capable of permeating other bodies
and seeking an equilibrium, I imagine some bodies are better fitted by nature
to be conductors of the fluid than others; and that, generally, those which are
the best conductors of the electrical fluid, are also the best conductors of this;
. . . . Thus if a silver tea-pot had a handle of the same metal, it would conduct
the heat from the water to the hand, and become too hot to be used: we, therefore give to a metal tea-pot a handle of wood. . . . But a china tea-pot . . . may
have a handle of the same stuff. Thus, also, a damp moist air shall make a man
more sensible of cold . . . than a dry air that is colder because a moist air is
fitter to conduct . . . away the heat of the body. . . . My desk . . . and its lock
are both exposed to the same temperature of the air, and have, therefore, the
same degree of heat or cold: yet if I lay my hand successively on the wood and
on the metal, the latter feels much the coldest, not that it is really so, but, being

a better conductor, it more readily than the wood, . . . draws into itself, the fire that was in my skin. . . . How a living animal obtains its quantity of this fluid, called fire, is a curious question. . . . I imagine that animal heat arises . . . from a kind of fermentation in the juices of the body, in the same manner as heat arises in the liquors preparing for distillation.

We might well underline, in this paragraph, the words *temperature* and *equilibrium*.

CHAPTER SIXTEEN

The Fruits of the Thermometer

T HE professor of chemistry at Glasgow (1760), Joseph Black, M.D., introduced some useful terms—thermal equilibrium, quantity of heat, thermal capacity, and latent heat, also the idea of specific heat of a substance. Black was a great scientist and a great teacher. One of his most important services was the training of a number of men who became eminent scientists: Thomas Young, Watt, Hope, and Leslie. Black did not receive much formal training in physics and in his work one can see a freedom from any bias that schooling might give, such as that of the caloric theory.[1] Our quotations are taken mostly from transcriptions of his lectures, for he published little.

THERMAL EQUILIBRIUM

I remarked formerly that even without the help of thermometers, we can perceive a tendency of heat to diffuse itself from any hotter body to the cooler ones around, until it be distributed among them in such a manner that none of them is disposed to take heat from the rest. . . . When all mutual action has ceased, a thermometer applied to any one of the bodies, acquires the same degree of expansion: Therefore, the temperature of them all is the same. . . . The heat is thus brought into a state of equilibrium. . . . I call it the equilibrium of heat.

Black made prominent use of the "method of mixture" in which two or more bodies of different temperatures are brought together and the resulting processes examined with the aid of thermometers and the balance. He introduced the term *thermal capacity:*

I perceived that . . . the quantities of heat which different kinds of matter must receive, to reduce them to equilibrium, or to raise their temperatures by an equal number of degrees are not proportional to the quantity of matter in each [i.e. mass or weight] . . . . Dr. Boerhaave . . . tells us that Fahrenheit agitated together quicksilver and water unequally heated. . . . He adds that it was necessary to take three measures of quicksilver to two of water in order to produce the same middle temperature that is produced by mixing equal mea-

---

[1]E. Mach, *Die Principien der Wärmelehre* (Leipzig, 1923), p. 1.

sures of hot and cold water. . . . Let us suppose the water to be at the 100th degree of heat, and that an equal measure of warm quicksilver at the 150th degree is suddenly mixed with it. . . . The temperature of the mixture turns out 120 degrees. The quicksilver, therefore, is become less warm by 30 degrees, while the water has become warmer by 20 degrees only; and yet the quantity of heat which the water has gained is the very same quantity which the quicksilver has lost. . . . Here it is manifest that the same quantity of the matter of heat which makes two measures of water warmer by 25 degrees is sufficient for making three measures of quicksilver warmer by the same number of degrees.

FIG. 225. Joseph Black.

Quicksilver, therefore, has less Capacity for the matter of heat (if I may be allowed to use this expression) than water; it requires a smaller quantity of it to raise its temperature by the same number of degrees.[2]

These stimulating paragraphs suggest the choice of some convenient units. The unit *quantity of heat*, the *calorie*, is the heat gained by one gram of water (1 cc.) warming one centigrade degree, or lost when it cools through the same temperature interval. The kilocalorie (= 1000 cal.) is the unit commonly used in measuring the fuel value of foods. The *thermal capacity* of any body is the quantity of heat it loses when cooling through 1 centigrade degree or gains while warming 1C. deg. The *specific heat* of any substance is the thermal capacity

[2]J. Black, *Lectures on the Elements of Chemistry* (Edinburgh, 1803). Quoted in Magie, *Source Book of Physics*.

of one gram of it. The calorie may also be defined as the thermal capacity of one gram of water.

When a hot iron touches a block of ice, the iron cools but the ice does not warm: some of it melts. The final temperature of the system (iron, ice, and water) is the same as the initial temperature of the ice (0°C.). It seems that the heat which the iron lost has vanished. Black said it had been used in melting ice and to express the idea that it had not caused any rise of temperature, he introduced the term *latent heat* (Lat. *latesco*, hide) for any isothermal change of physical state:

> When ice . . . is changed into a fluid by heat, I am of opinion that it receives a much greater quantity of heat than what is perceptible in it immediately after, by the thermometer . . . and this quantity of heat, absorbed and as it were concealed in the composition of fluids, is the most necessary cause . . . of their fluidity.
>
> I put a lump . . . of ice into an equal quantity [weight] of water, heated to the temperature 176, and the result was the fluid was no hotter than water just ready to freeze. . . . In the process of freezing . . . water, the extrication . . . of the latent heat, if I may be allowed to use these terms, is performed . . . with such a smooth progress, that many may find difficulty in apprehending it. . . . Water may be cooled to six . . . or eight degrees below the freezing-point without being frozen; but if it be then disturbed, . . . there is a sudden congelation . . . and while this happens (and it happens in a moment of time) this mixture of ice and water suddenly becomes warmer and makes a thermometer immersed in it rise to the freezing-point.
>
> I can easily show in the same manner as in the case of fluidity that a very great quantity of heat is necessary to the production of vapour. The vapour . . . is found to be . . . of the same temperature as the boiling water from which it arose. . . . During the boiling, . . . heat is absorbed by the water and enters into the composition of the vapour . . . in the same manner as it is absorbed by ice in melting. . . . As in this case, . . . so in the case of boiling, the heat absorbed does not warm surrounding bodies but converts the water into vapour. In both cases, . . . we do not perceive its presence: it is concealed, or latent, and I give it the name of *latent heat*.[3]

Black is noted for his canniness in drawing a conclusion and for his painstaking accuracy in making measurements, as may be seen from his measurement of latent heat. The facts of his experiment are represented in fig. 226, which is called a thermometer diagram:

$$176°F. = (176\text{-}32)\frac{5}{9} = 80°C.$$

Let w be the weight of the lump of ice and of the water.

Let L be the latent heat of 1 gram of ice.

The quantity of heat lost by the water cooling =
$$w(80 - 0) = 80w \text{ cal.}$$

[3]*Life and Letters of Joseph Black* (ed. Ramsay).

The quantity of heat gained by the ice melting $= wL$ cal.

$$\therefore \quad wL = 80w$$
$$\therefore \quad L = 80 \text{ cal.}$$

This is the value for the latent heat of ice still commonly used in such calculations, one hundred and eighty years after Black made his determination.

Tradition has credited Black with using a special form of calorimeter (fig. 227), which was made from a block of ice. His unit quantity of heat was the heat required to melt one gram of ice at 0°C., i.e. 80 calories.

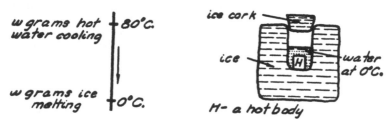

FIG. 226.
A thermometer diagram.

FIG. 227. An ice calorimeter.

THE STEAM ENGINE

There have been few men whose work was so revolutionary as to bring new ages into the world in their own lifetimes. In prehistoric days there may have been a few—for instance, the first maker of a stone fist-hatchet, the discoverer of fire, the first to extract copper from its ores, or the first to obtain iron; but the names of these inaugurators (if names they had) were not recorded. In historic times there have been two: Faraday, whose discovery of induction ushered in the electrical age, and James Watt, who brought the steam age into the world by his improvement and development of the steam engine. One has only to think of the effects of Watt's steam engine in the production of commodities, in transportation and travel, in the construction of buildings, and in almost every walk of life, to realize why it has been called the "king of machines." At least eighty years of endeavour preceded Watt's experiments: Papin, in 1690, was first to use a *cylinder and piston*, if we disregard the pump, which was in use before Aristotle's day. Newcomen's steam engine pumped water from Cornish mines in 1712, and advances were made by Savery, Potter, and Smeaton. In 1763, a Newcomen engine was brought to Dr. Joseph Black to "make it work." He handed the problem over to an instrument maker of

Glasgow who was an old pupil of his, James Watt. The outcome of Watt's work was the steam age. A share of the credit goes to Black. Watt's patent papers of 1769 give some description of the improved engine:

My method of lessening the consumption of steam and consequently fuel in fire-engines [i.e. steam-engines] consists of the following principles:

I. That vessel in which the powers of steam are to be employed to work the engine, which is called the cylinder . . . and which I call the steam-vessel, must, during the whole time the engine is at work, be kept as hot as the steam that enters it; first, by enclosing it in a case of wood, or any other materials that transmit heat slowly: secondly, by surrounding it with steam or other heated bodies: and, thirdly, by suffering neither water nor any other substance colder than the steam, to enter or touch it during that time.

II. In engines that are to be worked wholly or partially by condensation of steam, the steam is to be condensed in vessels distinct from the steam-vessels or cylinders, although occasionally communicating with them: these vessels I call condensers: and while the engines are working, these condensers ought at least to be kept as cold as the air in the neighbourhood of the engines, by application of water or other cold bodies.

III. Thirdly, whatever air or other elastic vapour [gas] is not condensed by the cold of the condenser and may impede the working of the engines, is to be drawn out of the steam-vessels or condensers by means of pumps, wrought by the engines themselves or otherwise.

IV. Fourthly, I intend in many cases to employ the expansive force of steam to press on the pistons, . . . in the same manner in which the pressure of the atmosphere is now employed in common fire-engines. In cases where cold water cannot be had in plenty, the engines may be wrought by this force of steam only, by discarding the steam into the air after it has done its office. . . .

VI. Sixthly, I intend in some cases to apply a degree of cold not capable of reducing the steam to water, but of contracting it considerably, so that the engines shall be worked by the alternate expansion and contraction of steam. . . .

Lastly, instead of using water to render the pistons and other parts of the engine air and steam-tight, I employ oils, wax, resinous bodies [i.e. substances], fat of animals, quicksilver and other metals in their fluid states.

This is certainly an important document we have just read; think of its results in the world for good and for evil. As in the case of Ashur-nazir-pal, so here also, certain unscrupulous persons misused the steam-engine by making it part of a scheme to employ labour and even child labour at criminally low wages. It is largely in this connection that not only the steam-engine but the machine in general has been styled a Frankenstein and accused of bringing more harm than good.

In the steam-engine, we have an example of one science, thermics, helping another, mechanics. The condenser which Watt has just described was a thermal contrivance, but his device for regulating the flow of steam into the cylinder, known as a Watt steam-engine governor, is an application of the centrifuge and belongs to mechanics.

FIG. 228. Lavoisier in his laboratory, with Mme. Lavoisier keeping his "science notes."

SPECIFIC HEAT

Lavoisier, the great French chemist, and Laplace, "The Newton of France," both of Paris, published a treatise, *Sur la Chaleur*, in 1780. They used Black's ice calorimeter and invented a modification of it. They redefined his term *specific heat*, and determined the specific heats of a number of substances.

Consider two bodies of equal mass and at the same temperature: the quantity of heat necessary to warm their temperatures one degree may not be the same for the two bodies. If we take as unity that quantity of heat which can raise the temperature of a litre [1000 cc.] of water one degree . . . we may conveniently consider all other quantities of heat gained or lost by different bodies as expressed in terms of this unit. In what follows, we shall mean . . . by specific heats, the ratios of the quantities of heat necessary to raise the temperatures of equal masses through the same number of degrees.

In their discussion of the method of mixture, i.e. placing two bodies of different temperatures together and observing the final temperature of the system, Lavoisier and Laplace derived a useful equation. The salient facts of their argument are given in the thermometer diagram of fig. 230, p. 341.

Hence one can derive a very simple general rule for determining the specific heats of substances by the method of mixture. Call the mass of the warmer body m' in pounds, . . . its temperature a', and q' the heat necessary to warm a pound of this substance one degree. If we designate by m'', a'', q'', the same quantities for the cooler body, and finally, if b is the temperature of the system when it has become uniform; it is evident that the heat lost by m' ... is $m'q'(a' - b)$. .... By the same argument the quantity of heat gained by m'' is $m''q''(b - a'')$. But since we assume that the quantity of heat is the same after the mixing as before, the heat lost by the body m must be equal to that gained by m'. Hence, we conclude that

$$m'q'(a' - b) = m''q''(b - a'').$$

This equation does not determine q' or q'' but gives their ratio

$$\frac{q'}{q''} = \frac{m''(b-a'')}{m'(a'-b)} \ .$$

Fig. 229. S. Laplace.

Consequently, if we compare various substances with one and the same substance, for example, with water, ... we shall be able to determine the specific heats of these substances.

It is plain to see that the study of heat was becoming more mathematical and the tendency has continued.

The modification which Lavoisier and Laplace made of Black's ice-calorimeter took a step toward the modern instrument. It has the double-wall feature previously mentioned (fig. 231).

If we wish to know the specific heat of a solid body, we raise its temperature through a certain number of degrees and then place it inside the ice block just mentioned [Black's ice calorimeter]. Leave it there until its temperature becomes zero, then collect the water produced by its cooling. This quantity of water, divided by the product of the mass of the body and the number of degrees its initial temperature was above zero, will be proportional to its specific heat. . . .

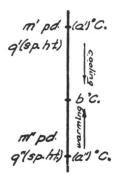

FIG. 230.
A thermometer diagram.

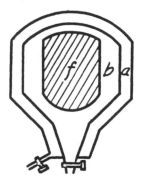

FIG. 231.
Double-wall calorimeter.

We have devised the following instrument. Its interior is divided into three parts. Its inner chamber, f, is made of iron wire netting. In this vessel we place the body under examination. The middle compartment b is intended to hold the ice which is to surround the inner vessel and to absorb the heat of the body under investigation. . . . Finally, the outer compartment a is intended to contain ice, which is to exclude the thermal interference of the external air and surrounding bodies.

Although the idea of caloric, the imponderable fluid, was present in the writings of Democritus and Lucretius, nevertheless it did not receive its name until about 1780. We saw that Franklin called it fire. The name caloric was introduced by Lavoisier.

RUMFORD'S EXPERIMENTS

Count Rumford's *Essays on Heat* make entertaining and instructive reading. He compared the *thermal conductivities* of a number of substances. One experiment he described is used in elementary courses at present (fig. 232):

To show the relative conducting power of the different metals, Doctor Ingenhouz contrived a very pretty experiment. He took equal cylinders of the different metals (being straight pieces of stout wire, drawn through the same hole and of the same length) and dipping them into melted wax, covered them with a thin coating of wax. He then held one end of each of these cylinders in boiling water, and observed how far the coating of wax was melted by the Heat communicated through the metal, and with what celerity the Heat passed.

Among all the various substances of which coverings may be formed for confining Heat, none can be employed with greater advantage than common atmospheric air. The warmth of the wool and fur of beasts and of the feathers of birds, is undoubtedly owing to the air in their interstices. . . . And in the same manner, the air in snow serves to preserve the Heat of the earth in winter. Double windows have been in use many years in most of the northern parts of Europe, and their great utility . . . is universally acknowledged. It is the confined air shut up between the two windows . . . that renders the passage of Heat through them so difficult.[4]

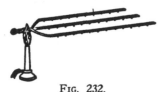

FIG. 232.
Thermal conductivity apparatus.

Rumford devised a beautiful indicator similar to one used by D. Bernoulli (1754) in studying streamlines. Using his new indicator, Rumford was the first to "see" convection currents in liquids (fig. 233):

I chose a glass tube . . . with long cylindrical neck . . . and putting into it about half a teaspoonful of yellow amber in the form of a coarse powder, . . I poured upon it . . . distilled water. . . . The amber remained at the bottom. . . . I increased the specific gravity of the liquid by adding alkaline solution . . . till the small pieces of amber . . . were just buoyed up and suspended in the Fluid [s.g. 1.05].

On inclining the tube to an angle of about 45 degrees, . . . and holding the middle of it over the flame of a candle, at the distance of three or four inches above the point of the flame; the motion of the Fluid in the upper part of the tube became excessively rapid, while that in the lower end of it . . . remained almost perfectly at rest. I even found that I could make the Fluid in the upper part of the tube actually boil without that in the lower part of it appearing to the hand to be sensibly warmed. . . . The motions in opposite directions in the liquid afforded a very entertaining sight; but, to a scientific observer, they were much more than amusing. They detected Nature as if it were in the very act, in one of her most hidden operations, and rendered motions visible in the midst of an invisible medium, which never had been observed before, and which most

FIG. 233.
Convection currents.

probably had never been suspected. . . . Heat cannot pass in that Fluid except when it is *carried* by its particles, which, being put in motion by the change it occasions in their specific gravities, *transport* it from place to place. . . . What has been called the gulph stream, in the Atlantic ocean, is no other than one of these currents.

[4]Count Rumford, *Essays, Political, Economical and Philosophical* (London, 1800).

The most famous of all Rumford's contributions to science was the series of experiments in which he showed a relation between heat and work. It sounded the knell of the caloric theory and ushered in a new period in the story of thermics. It led to the work of Joule, the law of conservation of energy, and the science of thermodynamics.

Being engaged lately in superintending the boring of cannon in the workshops of the military arsenal at Munich, I was struck with the very considerable degree of Heat which a brass gun acquires in a short time in being bored. . . . A thorough investigation . . . seemed even to bid fair to give further light into the hidden nature of Heat: and to enable us to form some reasonable conjectures respecting the existence or non-existence of an igneous fluid [caloric]. From whence comes the Heat, . . . produced in the . . . operation above mentioned? Taking a cannon, a brass six-pounder, cast solid, and fixing it horizontally in the machine, . . . I caused its extremity to be cut off . . . and by turning down the metal, . . . a solid cylinder was formed. This cylinder being designed for . . . generating Heat by friction, by having a blunt borer forced against its solid bottom at the same time that it should be turned around its axis by the force of horses, . . . a small hole . . . for the purpose of introducing a small mercurial thermometer was made in it. . . . A quadrangular oblong deal box, water-tight . . . being provided, . . . on pouring water into the box, . . . the cylinder would be completely covered. The box was filled with water at 60°F. and the machine was put in motion. . . . At the end of one hour, . . . its temperature had been raised no less than 47 degrees . . . and at 2 hours 30 minutes it actually boiled. It would be difficult to describe the surprize and astonishment expressed in the countenances of the by-standers, on seeing so large a quantity of cold water . . . actually made to boil without any fire. . . .

What is Heat? Is there any such thing as an igneous fluid? Is there any thing that can with propriety be called caloric? . . . We must not forget . . . that most remarkable circumstance, that the source of the Heat, generated by friction in these Experiments, appeared evidently to be inexhaustible. It is hardly necessary to add that anything which any insulated body, or system of bodies, can furnish without limitation, cannot possibly be a material substance; and it appears to me . . . extremely difficult if not quite impossible to form any distinct idea of anything capable of being excited and communicated in the manner the Heat was excited and communicated in these Experiments except it be MOTION.

RADIANT ENERGY

Rumford's *Essays on the Management of Fires* contain some remarks about radiation that are to the point:

The heat generated in the combustion of Fuel manifests itself in two ways: namely, in the hot vapour which rises from the Fire, . . . and in the calorific rays which are thrown off from the Fire in all directions. These rays may with greater propriety be said to be *calorific* or *capable of generating heat* in any body by which they are *stopped*, than to be called hot: for when they pass freely through any medium (as through a mass of air, for instance) they are not found to communicate any heat whatever to such medium; neither do they appear to excite any considerable degree of heat in bodies from whose surfaces they are reflected; and in these respects, they bear a manifest resemblance to the rays emitted by the sun. What proportion this *radiant heat* (if I may be allowed to use so inaccurate an expression) bears to that which goes off . . . in the smoke and heated vapour, is not exactly known.

CHARLES' LAW

Joseph Gay-Lussac was a very active and brilliant scientist, as witness the hundred and fifty papers he has to his credit. His early researches were in physics but the chemists have not failed to point out that as he grew in mental stature he became more and more a chemist. The article quoted here is from his first paper. It derives the volume coefficient of thermal expansion of gases and the law that is named after Professor J. Charles of Paris, whose priority it acknowledges. Apparently Guy-Lussac was not aware that John Dalton of Manchester had published the law one year previously (1801).

I must preface my report by saying that although I had frequently recognized that the gases, oxygen, nitrogen, hydrogen, carbon dioxide and atmospheric air, expand equally between 0° and 80° [freezing-point and boiling-point of water on the Réaumur scale], Citizen Charles had discovered the same property of gases fifteen years ago; but he never published his results and it was by merest chance that I happened to hear of them.

A large hollow glass sphere [fig. 234], is furnished with an iron tap to which is attached a curved tube *ID*. The handle of the tap bears a lever *LL* pierced at its ends to receive two cords by means of which one can open the tap under water. To introduce the gases into the sphere, I use a glass bell-jar *M* [fig. 235]. Next I warm the bath. . . . After 15 or 20 minutes of boiling, . . . I remove the end of *ID* from the mercury *K*, to establish pressure equilibrium between the gas in *B* and the outside air. . . . After cooling the bath with ice or water, . . . I plunge *B* full depth into the bath of known temperature. . . . On opening the tap, a volume of water enters *B* which is exactly equal to that of the gas which was expelled by the heat. When the tap is closed, I take out the sphere, dry its surface carefully and weigh it in this state. Then I weigh it full of water and also empty. . . . From these data, I have the capacity of the sphere . . . and the volume of the water corresponding to that of the gas expelled by the heating. It is then easy to determine the ratio between the initial volume and the volume after dilation.

With atmospheric air, I found the following six results, from the temperature of melting ice to that of boiling water, equal volumes of atmospheric air represented by 100 became 137.40, 137.61, 137.44, 137.55, 137.48 and 137.57, the average of which is 137.50.

I am therefore justified in drawing from what has just been said, the following conclusions:

I. All gases, whatever their densities, and the quantities of water they contain in solution, and all vapours, expand equally in changing through the same temperature interval.

II. For permanent gases, the increase in volume which each undergoes between the temperature of melting ice and that of boiling water, equals . . .

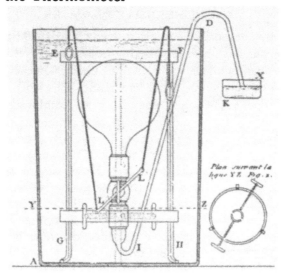

FIG. 234. Thermal expansion of gases.

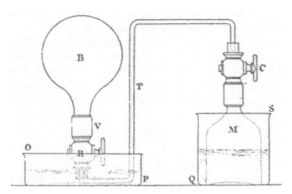

FIG. 235. Filling the flask with gas.

$\dfrac{100}{266.66}$ of the initial volume (at 0°C.) for each degree of the centigrade thermometer. (Charles' Law.)[5]

The value now generally accepted for this coefficient of thermal expansion (and contraction) is $\dfrac{1}{273}$ or 0.00366. Gay-Lussac's result was therefore about 2.5 per cent high.

[5]J. Gay-Lussac in *Annales de Chimie et de Physique* (1802), XLIII, 137.

MAXIMUM DENSITY OF WATER

Although the Florentines had discovered about 1660 that water has "anomalous expansion" near its freezing-point and therefore minimum volume and maximum density, it was not until 1804 that an accurate determination of that temperature was made. This work was done by Thomas C. Hope, professor of chemistry at Edinburgh University,

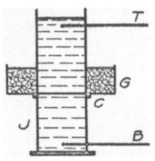

FIG. 236.
Hope's apparatus.

another of Black's old students. One of his experiments is regularly performed in elementary courses; it is known as Hope's experiment and the apparatus as Hope's apparatus (fig. 236). His report appeared in the *Transactions* of the Royal Society of Edinburgh. The accompanying excerpts from that paper describe one of his experiments.

I took a glass jar, *J*, 17.8 inches deep and 4.5 in diameter. . . . I provided also a cylindrical bason of tinned iron, *G*, 4.8 inches deep and 10 inches in diameter with a circular hole in the middle of the bottom, large enough to receive the top of the jar. By means of a collar and cement, *C*, I secured this bason so that it encircled the upper part of the jar. . . . I filled the jar with water at 32°F. [0°C.]. The air of the room was about 34°. I suspended two thermometers in the fluid, nearly in the axis of the jar, one with its ball about half an inch from the bottom *B*, the other at the same distance below the surface *T*. I then poured water of temperature 68°F. into the bason. The following readings were obtained.

Time	B	T
0 min.	32°F.	32°F.
10 "	35	32
15 "	36	32
20 "	36	32
25 "	37	33
30 "	38	33
38 "	38	33
45 "	39	33
50 "	39	44
55 "	39	45
60 "	39	48

Nothing can be more decisive with regard to the question in dispute than the particulars of this experiment. Heat is applied to the middle of a column of ice-cold water. The thermometer evinces that the warm current sets downwards. The inference is plain that the cold water at 32°F. is contracted by heat. No sooner did the inferior portion attain the temperature of 39° than the heated fluid altered its course, and, by ascending, carried the increase of temperature very rapidly to the surface, so that it soon surpassed the bottom and continued to rise while the other remained stationary.[6]

Hence water has its maximum density at 39°F. or 3.9°C.

THE DIFFERENTIAL THERMOSCOPE

We have met three of Black's old students who rose to eminence or fame, Young, Watt, and Hope. Now we are to consider the work of a fourth, Sir John Leslie (1760-1832), professor of natural philosophy at Edinburgh, three of whose contributions to thermics are commonly included in elementary courses. The differential thermoscope (or thermometer), which he invented and used to good purpose, is described in his treatise *Meteorology* (1810). It is an application of the Galileo-tube (fig. 238, p. 349).

A glass tube, terminated by a ball containing air, is joined hermetically . . . to another long tube, terminated by a similar ball containing air also but including a small portion of some coloured liquid. The tubes are then bent into a recurved or double stem, like the letter U. . . . If both balls have the same temperature, the liquid must evidently remain stationary: but if the ball of the shorter tube be warmed, the air expanding and exerting more elasticity, will depress the liquid in the stem: or if this ball be cooled, the air, by its contraction, allows the liquid to ascend, from the superior elasticity of the air contained in the opposite ball. The fall or rise of the liquid will, therefore, mark the excess of heat or cold in the adjacent ball, and the space through which it moves will measure the precise difference of temperature.[7]

At first, Leslie employed this instrument as a wet-and-dry-bulb-hygrometer, for he covered one of its bulbs with cambric, which he moistened with water, as he said, "as often as occasion required." This was an application of the principle of "cooling by evaporation" mentioned by Franklin and explained by Black, from his studies of latent heat.

There is another of Leslie's experiments which applies cooling by evaporation. It is a classic and is named after its inventor. Many find it very intriguing for several reasons; for instance, it comes as a surprise to many to see the same body of water both boiling and freezing at the same time (fig. 237):

The mere evaporation of some very volatile liquids is sufficient to produce intense cold. . . . This effect is augmented under the receiver of an air-pump. . . . We have now to relate a discovery which will enable human skill to command the refrigerating powers of nature and, by the help of an adequate machinery, to create cold and produce ice on a large scale at all seasons and in the hottest climates of the globe. In 1810, . . . having introduced a surface of sulphuric acid

[6]*Trans. Roy. Soc. Edin.* (1805), v, 387.
[7]J. Leslie, *Treatises in Natural and Chemical Philosophy* (1838).

under the receiver of an air-pump, Mr. Leslie perceived with pleasure that . . . the sentient ball of the hygrometer, which had been covered with several folds of wetted tissue paper, was observed . . . suddenly to lose its blue tint and assume a dull white, while the coloured liquor sprung upwards in the stem. . . . The act of congelation had, therefore, at this moment, taken place and the paper remained frozen several minutes. . . . The hygrometer was removed and a watch-glass filled with water substituted. By a few strokes of the pump, the whole was converted into a solid cake of ice. . . . If common water be used, it will evolve air bubbles copiously as the exhaustion proceeds; in a few moments . . . the ice spicules will shoot beautifully through the liquid mass and entwine it with a reticulate contexture. The appearance presented is extremely various.[8]

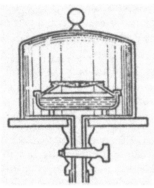

FIG. 237. Leslie's experiment on freezing and boiling of water.

The last sentence is no overstatement. A certain teacher who has demonstrated this experiment scores of times has seldom if ever seen it happen in exactly the same way twice. There are considerable differences produced by slight differences in the conditions, which are sometimes unintentional and sometimes unknown. Some time after the air-bubbles have ceased to appear, bubbles of water vapour form in the liquid and rise to the surface; this event satisfies the definition of boiling; at the same time freezing occurs with fully as broad a repertoire of variations as the Jack Frost patterns on windows.

Another important contribution of Leslie to heat was his study of so-called radiant energy, or radiation, for which Rumford, with apologies, used the term, "radiant heat." Leslie used as a radiator a tin vessel which is now called a *Leslie cube* (or cylinder).

I prepared hollow cubes of block-tin. One side of the cube was constantly kept clean and bright; the opposite one was painted over with a coat of lampblack and size [mucilage]. I had also two similar cylindrical vessels. The instrument most essential to this research was the differential thermometer.

The experiment is commonly set up nowadays as in fig. 238. Let one bulb F be blackened with candle-soot and the other G, coated with aluminium paint or tinfoil: also let both sides of the Leslie cylinder be dull black. If hot water is poured into the cylinder, the liquid sinks at L and rises at G. Thus Leslie found that a dull black surface is a better absorber of radiant energy than a bright shiny one.

Let both bulbs F and G be blackened to make them as efficient absorbers as possible and equal in that regard. Let the B-side of the

[8]Ibid.

Leslie cylinder be dull black and the W-side bright and shiny. When hot water is poured into the cylinder, the liquid at L sinks and at H rises. Leslie showed in this way that a dull black surface is a better radiator than a bright shiny one.

In another of his experiments, he filled a Leslie cube with ice and his report contains an early use of the term *energy* which seems to come close to the modern meaning of the word:

> If the canister be filled with ice, the ball F will be chilled and the coloured liquor (H) will consequently sink. It appears unquestionable that some hot or cold matter flows to the ball F. We have now to investigate what circumstances are capable of altering the Energy of that transmission.[9]

In the expression "matter flows to the ball," Leslie was still speaking in terms of the caloric theory, which continued to be used for some time after Rumford's epochal paper (1798). The first use in the literature

Fig. 238. Leslie differential thermoscope and Leslie cylinder.

of the term *energy* in its modern sense was made by Dr. Thomas Young in a lecture on the hydraulics of the blood-stream (1807).[10] Since both these men were old students of Dr. Joseph Black, it may well be that they heard the germ of this idea in his remarkable lectures.

THE DAVY SAFETY LAMP

When Count Rumford established the Royal Institution of London, he selected for its Director and Professor of Chemistry a young scientist who later became Sir Humphry Davy. Two of Davy's contributions to thermics will be considered here, the Davy Safety Lamp and the melting of ice by friction.

Some terrible coal-mine disasters turned Davy's mind to inventing a lamp that would not ignite marsh-gas. At the first glance, this project looked impossible, but Davy found a brilliant solution. The device he invented was completely efficient, yet simple, and further, it turned a danger into an actual benefit. This success was very gratifying to his humane temperament for it meant the saving of hundreds of lives as well as much property. There is reason to surmise that part of the credit for this brilliant invention should go to Davy's assistant, young Michael Faraday, for the instrument bears the earmarks of his handiwork.

[9]J. Leslie, *An Experimental Inquiry into the Nature and Propagation of Heat* (Edinburgh, 1804).
[10]T. Young, *Lectures on Natural Philosophy* (1807), VIII.

If a piece of wire gauze sieve is held over a flame of a lamp or of coal gas, it prevents the flame from passing it. . . . The air passing through it is very hot for it will convert paper into charcoal; and it is an explosive mixture, for it will inflame if a lighted taper be presented to it; but it is cooled below the explosive point by passing through the wires, even red hot, and being mixed with a considerable quantity of comparatively cold air.

FIG. 239. Sir Humphry Davy.

In this connection, I shall describe a light [fig. 240] that will burn in an explosive mixture of fire-damp and the light of which arises from the combustion of the fire-damp itself [marsh gas or methane].

The invention consists in covering or surrounding a flame of a candle by a wire sieve; the coarsest that I have tried with perfect safety contained 625 apertures in a square inch, and the wire was one-seventieth of an inch in thickness, the finest, 6400 apertures in a square inch, and the wire was one-two-hundredth of an inch in diameter. . . . When a lighted lamp or candle, screwed into a ring soldered to a cylinder of wire gauze, having no apertures except those of the gauze, . . . is introduced into the most explosive mixtures of carburetted hydrogen [marsh-gas] and air, the cylinder becomes filled with a bright flame and this flame continues to burn as long as the mixture is explosive. . . . In all these experiments, there was a noise like that produced by the burning of hydrogen gas in open tubes. . . . These extraordinary and unexpected results led to many enquiries respecting the nature . . . of flame, but my object at present is only to point out their application to the use of the collier. All that he requires to ensure security are small wire cages to surround his candle or his lamp. which may be made for a few pence. . . . The application of this discovery will not only preserve him from the fire-damp, but enable him to apply it to use, and to destroy it at the same time that it gives him a useful light.[11]

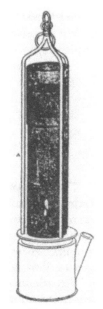

Sir Humphry refused to patent his invention, in spite of the advice of friends, on the ground that the state paid him his salary to do such work. Later he saw that this view was wrong, for presently some money-grabbers made and sold inferior lamps, with the result that once more there were mine disasters. Because Davy had no patent, he could not restrain those unscrupulous persons from making money at the cost of the lives of others. They had infringed no patent. They were within their *rights*!

Davy performed a famous experiment which clinched Rumford's argument in showing the inadequacy of the caloric theory. His report of the experiment was an important pronouncement:

Matter is possessed of the power of attraction. By this power, the particles of bodies tend to approximate and to exist in a state of contiguity. The specific gravity of all bodies can be increased by diminishing their temperatures. Consequently, on the supposition of the impenetrability of matter, the particles of bodies are not in actual contact. There must, then, act on the corpuscles . . . some other power which prevents their actual contact: this may be called repulsion. . . .

Fig. 240.
Davy safety lamp.

EXPERIMENT. I produced two parallelopipedons of ice, of the temperature of 29° (i.e. about −1.6°C.). . .; they were fastened by wires to two bars of iron. By a mechanism, . . . their surfaces were placed in contact and kept in a continued and violent friction for some minutes. They were almost entirely converted into water, which water was collected and its temperature ascertained to be 35°(1.6°C.). . . . It is a well-known fact that the capacity of water is much greater than that of ice: and ice must have heat . . . added to it before it can be converted to water. Friction, consequently, does not diminish the capacities of bodies for heat.

Since bodies become expanded by friction, it is evident that their corpuscles must move or separate from each other. Now a motion or vibration of the corpuscles of bodies must be necessarily generated by friction and percussion. . . . Heat, then, . . . may be defined as a peculiar motion, probably a vibration, of the corpuscles of bodies tending to separate them.

[11]*Phil. Trans. Roy. Soc. Lond.* (1818).

The caloric explanation of Rumford's experiment (or any change of substance or state) was that brass filings had lower specific heat or capacity for caloric than solid brass. In the change to filings it was claimed there was an evolution of caloric which warmed the water. Rumford measured the specific heat of filings and found it the same as that of solid brass. In spite of this devastating evidence, the caloric theory remained in vogue for five decades. Davy's experiment presented even more crippling opposition. Black showed that to melt ice required heat or caloric. In the process ice→water, therefore, the caloric theorist was forced to assume that in Davy's experiment the requisite caloric was obtained by a decrease of specific heat or capacity for caloric. Hence the specific heat of water must be less than that of ice. But by measurement it was shown that the specific heat of water is twice that of ice.

In the last paragraph one sees the ideas of molecular motion and temperature being brought together to form a kinetic theory of heat.

THE DEW-POINT HYGROMETER

The professor of physics at the Collège de France, Paris, in 1845, was Henri V. Régnault (1810-1878). He was a master of painstaking and accurate measurement and experimental technique. In modern tables, many of his values of expansion coefficients and specific heats are quoted, but few determinations by any of his contemporaries. He devised a dew-point hygrometer whose ancestor we found in the records of l'Accademia del Cimento. It is called the Régnault hygrometer or dew-point apparatus (fig. 241).

I believe that all the disadvantages [of other hygrometers] are eliminated in the instrument which I propose to physicists under the name condensing hygrometer. . . . This apparatus is composed of a thin . . . polished silver thimble *abc*, . . . ground to fit accurately a glass tube *cd*. . . . The tube carries a small lateral tube *T'*. . . . Some ether is poured into the tube as far as *MN* and . . . *T'* is connected . . . with an aspirator. . . . The air flows through the tube *gf*, and bubbles through the ether which becomes cooled by evaporation. . . . In less than a minute, the temperature is lowered sufficiently to produce an abundant deposit of dew. At this moment, the thermometer is observed by means of a telescope.[12]

THE KELVIN OR ABSOLUTE SCALE

One of Régnault's students was Lord Kelvin of Glasgow (1824-1907), whose absolute scale of temperatures is met in elementary courses in heat, frequently in connection with the study of Charles' law (fig. 242):

[12]H. V. Régnault, "Studies on Hygrometry," *Comptes Rendus* (1845), xx, i, 1127.

The particular kind of thermometer which is least liable to uncertain variations is that founded on the expansion of air, . . . and this is, therefore, generally adopted as the standard. . . .

The characteristic property of the scale which I now propose is that all degrees have the same value: that is, that a unit of heat descending from a body, A, at the temperature $T°$, of this scale, to a body, B, at the temperature $(T\text{-}1)°$, would give out the same mechanical effect, whatever be the number T. This may justly be termed an absolute scale, since its characteristic is quite independent of the physical properties of any specific substance.

The best result . . . which Mr. Joule and I have yet been able to obtain is that the temperature of freezing water is 273.7 on the absolute scale: that of the boiling-point being consequently 373.7.

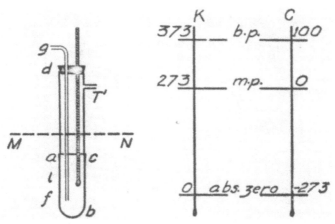

FIG. 241. Régnault's dew-
point hygrometer.

FIG. 242. Kelvin's absolute scale
of temperatures.

The absolute scale is often called the Kelvin scale. Thus $32°F = 0°C = 273°K$. One immediate convenience of the Kelvin or absolute scale is the simplification it effects in the statement of Charles' law which takes the form: "The volume of a gaseous body at constant pressure is directly proportional to its absolute temperature."

Kelvin stood in sharp contrast to his teacher, Régnault, with respect to deftness. What a trial a clumsy student must be to a master of technical skill. Yet Kelvin was by far the greater scientist because of his preeminence in theory and the magnitude of his contribution.

MECHANICAL EQUIVALENT OF HEAT

In the previous article, Kelvin mentioned his collaborator, James P. Joule (1818-89), of Manchester, one of Dalton's pupils. The most famous of Joule's papers was probably that in which he measured the relation between heat and work, which is fundamental in thermodynamics:

Read before the Royal Society, June 21st, 1849

I have the honour to present . . . the results of experiments I have made to determine the mechanical equivalent of heat. . . .

Count Rumford, . . . that justly celebrated philosopher, demonstrated . . . that the very great quantity of heat excited by the boring of a cannon could not be ascribed to a change taking place in the calorific capacity of the metal: and he . . . concluded that the motion of the borer was communicated to the particles of the metal, thus producing the phenomenon of heat. . . . There were many facts such as . . . the warmth of the sea after a few days of stormy weather, which had long been attributed to fluid friction. Nevertheless, the scientific world, preoccupied with the hypothesis that heat is a substance [caloric] . . . has almost unanimously denied the possibility of generating heat in that way. . . . In 1845, . . . I employed a paddle-wheel [fig. 243] to produce the fluid

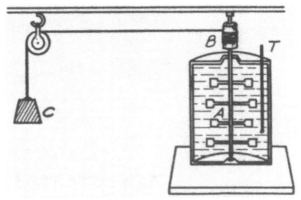

FIG. 243. Joule's apparatus for determining the mechanical equivalent.

friction and obtained the equivalents, 781.5, 782.1, 787.6, respectively, from the agitation of water, sperm-oil and mercury. . . . Results so closely coinciding with one another, . . . left no doubt in my mind as to the existence of an equivalent relation between force [work] and heat, but still it appeared of the highest importance to obtain that relation with still greater accuracy.

Joule then proceeded to thrash out every last detail in a difficult and complicated series of experiments which ultimately brought him fame. Though these details are omitted here, the student is recommended to thrash them out for himself as they are the very stuff of which science is made.[13] It was for his handling of the details here omitted that Science gave Joule her applause and named an important unit after him. His paper closes with these sentences:

I will, therefore, conclude by considering it as demonstrated, . . .

(1), that the quantity of heat produced by the friction of bodies, whether

[13]A. Wood, *Joule and the Study of Energy* (Classics of Scientific Method) (Bell, 1934).

liquid or solid, is always proportional to the quantity of force [work] expended: and

(2) that the quantity of heat capable of increasing the temperature of a pound of water . . . by 1° Fahr., requires for its evolution the expenditure of a mechanical force [energy] represented by the fall of 772 pounds through a space of one foot [772 ft. pd.].[14]

The thermal unit used by Joule is called the British Thermal Unit (B.T.U.). The symbol commonly used for the mechanical equivalent of heat is J, Joule's initial, and the value now generally accepted is $J = 778$ ft.pd. per B.T.U., which is about 1 per cent higher than that found by Joule [4.2 joules (or 4.18) per cal.].

THE BUNSEN BURNER

The advent of a new source of heat is an event of importance industrially, academically, and in the lives of the people. In fact, the discovery of any new source of energy has regularly inaugurated a new epoch in human advancement. After the discovery of fire in prehistoric times, until about 1900, advances in applied thermics were mostly with regard to fuel and fuel devices. The story can be sketched in three words: solid, liquid, and gas. Solid fuels are wood, fats, resins, peat, coal, coke; liquids are kerosene, gasoline, oil (1850); gases are coal gas, water-gas and producer gas.

Coal was first described by Theophrastus (370 B.C.) in his treatise *On Stones*, and called anthracite. Our name, coal, however, is an old British word meaning black. In A.D. 852, the Abbot Ceobred burned 12 loads of "coalle" per annum.

The advancement in sources of light followed a parallel path often preceding that of heat, for example, in the case of oil and gas. Oil was used for lamps in

Fig. 244.
The Bunsen burner.

prehistory, whereas the oil furnace is recent. The first illuminating gas was used by William Murdock (1795), who was Watt's chief assistant. Gas heating came later when a suitable burner had been invented by Bunsen.

[14]J. P. Joule, *Collected Papers*, Pt. I, p. 298. Also *Phil. Trans. Roy. Soc. Lond.* (1850).

The inventor of the spectroscope was also the inventor of the Bunsen burner. Here is his description of the latter device, which has certainly been of tremendous utility in laboratories, in homes, and in industry. Bunsen used it, in this instance, to obtain coloured flames for spectroscopic analysis (fig. 244):

In these investigations, we made use of a type of burner which one of us designed and introduced in this laboratory [Heidelberg, 1855]. . . . It serves better than any other device for obtaining steady flames of various power, colour, and form. The principle of this burner rests simply on the fact that the gas is allowed to issue under conditions in which, by reason of its own motion, it can draw in and mix with the proper amount of air. . . . In the figure, *a* is an ordinary short-nosed burner; from such a burner, the gas streams out in three diverging flames, whose edges fall on the vertices of an equilateral triangle. The burner rises in the centre of the cylindrical chamber *b*, . . . which communicates with the outer air. . . . If we screw the tube *e* on *b*, . . . it draws through the aperture *d* enough air to cause it to burn at the mouth of the tube *e* with a colourless and altogether sootless flame. The brightness of the flame thus produced by mixture with air hardly exceeds that of a hydrogen flame. If we close the openings *d*, the ordinary luminous sooty gas flame immediately appears again.[15]

WATER EQUIVALENT

Clerk Maxwell published in 1875 an excellent text, *Theory of Heat*, which had his characteristic terseness and precision of language. In one passage he discusses the idea expressed by the term *water equivalent:*

The instrument by which quantities of heat are measured is called a calorimeter [fig. 245], probably because it was invented at a time when heat was called caloric. . . . Weigh out a pound of water in your vessel (*I*) and ascertain its temperature. . . . Lift the roll of lead out of the boiling water . . . and immerse it . . . in the cold water. . . . It was assumed . . . that all the heat which escapes from the lead enters the water . . . and remains in the water. . . . The latter part of this assumption cannot be quite true, for the water must be contained in a vessel of some kind and must communicate some of its heat to the vessel. . . . If we could form the vessel of a perfect non-conductor, . . . this loss . . . would not occur; but no substance can be considered . . . a non-conductor. If we use a vessel which is merely a slow conductor, . . . it is very difficult to determine . . . how much heat is taken up by the vessel. A better plan is to use a vessel which is a very good conductor . . . but of which the capacity for heat is small, such as a thin copper vessel. . . . We may speak of the capacity for heat of . . . such a copper vessel, in which case, the capacity depends on the weight as well as on the kind of matter [substance]. The capacity of a particular thing

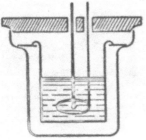

FIG. 245. A calorimeter.

[15]R. Bunsen and H. Roscoe, in *Poggendorff Annalen*, C, 84.

is often expressed by stating the quantity of water which has the same capacity [i.e. its water equivalent].

KINETIC THEORY OF HEAT

Although the term "water equivalent" is technically important and Maxwell's discussion an example of clear exposition, nevertheless it does not represent his greatest service to thermics. As in optics so here also, applying his great mathematical abilities, he gathered together the threads of the subject and wove them into a pattern of unity and continuity. Such organization is of the very essence of science. Outstanding in Maxwell's work was his mathematical development of the kinetic theory of heat with special emphasis on the gaseous state of matter. An intimation of this work is given in the following excerpt from one of his papers. Here again we meet Democritus' ideas of vibrating and colliding molecules but now they are closely related to the ideas of temperature and heat.

So many of the properties of matter, especially when in the gaseous form, can be deduced from the hypothesis that their minute parts are in rapid motion, the velocity increasing with the temperature, that the precise nature of this motion becomes a subject of rational curiosity. Daniel Bernoulli, Herapath, Joule, Krönig, Clausius, and others have shown that the relations between pressure, temperature, and density in a perfect gas can be explained by supposing the particles to move with uniform velocity in straight lines, striking against the sides of the containing vessel and thus producing pressure. It is not necessary to suppose each particle to travel to any great distance in the same straight line; for the effect in producing pressure will be the same if the particles strike against each other. . . . M. Clausius has determined the mean length of path in terms of the average distance of the particles and the distance between the centres of two particles when collision takes place. . . . Certain phaenomena, such as . . . the conduction of heat through a gas, . . . and the diffusion of one gas through another, seem to indicate the possibility of determining accurately the mean length of path which a particle describes between two successive collisions. In order to lay the foundation of such investigations on strict mechanical principles, I shall demonstrate the laws of motion of an indefinite number of small, hard, and perfectly elastic spheres acting on one another only during impact.

If the properties of such a system of bodies are found to correspond to those of gases, an important physical analogy will be established. . . . If experiments on gases are inconsistent with the hypothesis of these propositions, then our theory, though consistent with itself, is proved to be incapable of explaining the phenomena of gases.[16]

THE CROOKES RADIOMETER

The little whirligig that a jeweller sometimes displays in his window is called a radiometer. It was invented and named by Sir William

[16]J. Clerk Maxwell, "Illustrations of the Dynamical Theory of Gases," *Phil. Mag.* (1860), Ser. 4, xix, 19. Quoted more fully in Magie, *Source Book of Physics*.

Crookes (1832-1919), of London, and is one of his best known contributions to thermics. He described the device in his paper, *Repulsion from Radiation*, which was published in the *Transactions* of the Royal Society for 1875:

> Since the experiments mentioned in the foregoing abstract were concluded, the author has examined more fully the action of radiation on black and white surfaces. . . . The luminous rays repel the black surface more energetically than they do the white surface. Taking advantage of this fact, the author has constructed an instrument which he calls a radiometer. This consists of four arms suspended on a steel point, resting on a cup, so that it is capable of revolving horizontally. To the extremity of each arm, is fastened a thin disc of pith, with lamp black on one side, the black surfaces facing the same way. The whole is enclosed in a glass globe, which is then exhausted . . . and hermetically sealed. The author finds that this instrument revolves under the influence of radiation, the rapidity of revolution being in proportion to the intensity of the incident rays.

This device caused a great sensation throughout the intellectual world; and the jeweller and the science-teacher find that it still intrigues most people. At the first glance it seems to be an example of perpetual motion; but the dream is dashed when a cloud veils the sun.

LE CHATELIER'S PRINCIPLE

To explain a new fact is to show its relation to facts previously known. The less obvious the relation, the more brilliant is the derivation of the principle that discloses the relation. That ice floats in water and that a snowball can be made by squeezing a handful of snow crystals may seem unrelated facts, but Le Chatelier's principle shows that either can be predicted from the other. Le Chatelier's treatise on chemical equilibrium, published at Paris in 1888, states the principle in these words:

> Every system at chemical equilibrium experiences by reason of any change in one of its conditions of equilibrium, a transformation in such a sense that if it occurred alone, it would cause, in the factor considered, a change of the opposite sign. . . . Every elevation of the temperature produces in a system of chemicals at equilibrium, a transformation in the sense which corresponds to a reduction of temperature, if it occurred by itself. This law has already been stated by M. van t'Hoff. . . . All the reversible phenomena of fusion or of volatilization which are accompanied by absorption of heat, are produced by an elevation of temperature.[17]

Dr. E. S. Ferry, in his *General Physics*,[18] gives the statement thus:

> When a system at equilibrium is subject to a constraint by which the equilibrium is altered, a reaction takes place which opposes the constraint.

[17] *Recherches Expérimentales et Théoriques sur les Équilibres Chimiques* (Paris, 1888), p. 48.
[18] Wiley, 1925.

If we increase the load L (fig. 246), thus increasing the pressure of the ice-water system I-W, in the cylinder PC, that change will occur which itself would decrease the pressure (Le Chatelier's law), namely, a shrinkage. Since ice floats in water, its density must be less, or, in other words, water takes less room than the same weight of ice, and melting of ice involves a shrinkage. Hence, when snow is squeezed, crystals melt where they touch and the water drops thus formed coalesce (fig. 247). When the pressure is released, the water refreezes, thus,

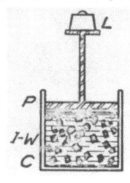

Fig. 246.
Ice-water system in cylinder.

Fig. 247. Two ice crystals.

as it were, "soldering" the crystals together, and a coherent lump is formed.

Le Chatelier's law explains many common and rather mysterious facts: why a snowball cannot be formed on a very cold day; how the ice on a pavement may pull off a man's rubbers when he is running to catch a streetcar; that a giant with steel gauntlets could not form an "iron snowball" from hot iron filings since iron sinks in its melt, and so forth.

THE DEWAR FLASK OR THERMOS BOTTLE

Sir James Dewar (1842-1923) is a member of the long series of celebrated scientists who have graced the staff of the Royal Institution of London. One incidental outcome of his researches on the liquefaction of gases was the thermos bottle, which Dewar invented and described in a paper on liquid air (1893).

The prosecution of research at temperatures approaching the zero of absolute temperature [−273°C.] is attended with difficulties of no ordinary kind. . . . The necessity of devising some new kind of vessel for storing and manipulating volatile fluids like oxygen [boiling-point, −180°C.] and liquid air [b.p., −190°C.] became apparent. . . . It naturally occurred to me that the use of high vacua

surrounding the vessels containing liquefied gases, would be advantageous. . . .
A high vacuum diminishes the rate of evaporation to one-fifth part of what it is
when the substance is surrounded with air at atmospheric pressure, or, in other
words, liquid oxygen lasts five times
longer when surrounded with vacuous
space. The next step was to construct a
series of glass vessels surrounded by a
vacuous space, suitable for various experiments. The vacuum vessels described
equally retard the loss as well as the gain
of heat and are admirably adapted for
all kinds of calorimetric observation.
The future of these vessels in thermal
observations will add greatly to the accuracy and ease of conducting investigations.

    If the inner vessel is coated with a
bright deposit of silver, then the radiation is diminished also with the result
that the rate of evaporation is further
reduced by more than half. In such
vessels, liquid oxygen [−180°] or liquid
air can be kept for hours and the economy and ease of manipulation greatly
improved.

FIG. 248. Sir James Dewar.

    Some notion of the temperature of liquid air is given by running on to the
surface some absolute alcohol, which, after rolling about in the spheroidal state,

suddenly solidifies into a hard transparent ice,
which rattles on the sides of the vacuum test-
tube like a marble. On lifting the solid alcohol
out by means of a looped wire, the application
of the flame of a Bunsen burner will not ignite
it. After a time, the solid melts and falls from
the looped wire like a thick syrup.[19]

    It is said that it was after the publica-
tion of his paper that Sir James realized,
in conversation with some friends, that a
workman might like to have hot tea with
his luncheon, and a picnicker would probably enjoy cold lemonade on a hot after-

FIG. 249. Thermos bottle.

noon. The commercial exploiter, however, won the race, and the first
thermos bottles on the market were not of English manufacture.

    Thus we have witnessed the discovery of some elementary prin-
ciples in thermics and a few of their applications. To follow the story
further would be just as interesting and profitable—but decidedly more
mathematical.

[19]*Proceedings of the Royal Institution*, XIV, 1. Also, *Collected Works of Sir James
Dewar* (Cambridge, 1927). I, 354.

THE STORY OF
Electricity
AND
Magnetism

CHAPTER SEVENTEEN

Amber and the Magnet

THE history of the science of electricity differs markedly from that of any other branch of physics whose story we have traced. One difference is in age. Electricity is younger than its sister sciences, for it had little or no preliminary practical stage in prehistoric times as they had. The dates of their births are unknown, but the birth of electricity was recorded about 600 B.C. by Thales of Miletus. It was christened in A.D. 1646 by Sir Thomas Browne of London when he introduced the name electricity into the literature of science. The early stages of the development of this science are chronicled in detail. In the other sciences, in general, we are able to refer to the inventors of the first instruments only by such mythical names as Jubal, Pan, Hermes, and Prometheus; but not so in electricity. The first electrical instrument was the *versorium*, invented by Dr. William Gilbert of London about 1600, and the first application of electrical knowledge which was of economic importance was the *lightning-rod*, invented in 1744 by Benjamin Franklin of Philadelphia. In this case, therefore, we have the unusual opportunity of following closely the early fact-gathering stage of a science, that stage which always precedes the theoretical period in which valid hypotheses are formulated.

So spectacular has been the progress of this science, however, that already it has given its name to the age in which we live. It shares this distinction with mechanics; acoustics, optics, and thermics have never reached such prominence. One reason for the late debut of the science of electricity was that it is not related directly to any of our special sense-organs, as acoustics and optics are to the ear and the eye, and thermics to the thermal nerves. Electrical phenomena are sufficiently rare and unimportant in the zoological world that animals, in general, have not developed special sense-organs to perceive them. This places a student of electricity somewhat in the position of a deaf person studying acoustics; but it brings also a peculiar mystery; and mystery

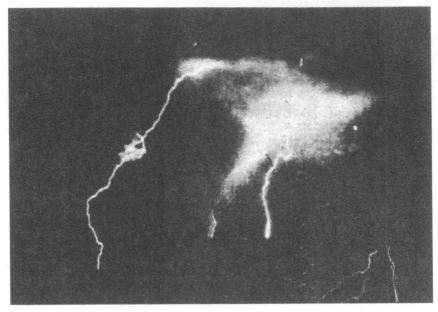

FIG. 250. Lightning.

intrigues. No other science has attracted to itself investigators from so many other sciences and from so many other walks of life: Thales, a geometrician; Faraday, a chemist; Dufay, a botanist; Gilbert, a physician; Gray, a classicist;Franklin, a printer; Bennet, a churchman; Wesley, an evangelist; Canton, a schoolmaster; Geissler, a glass-blower; Marat, a revolutionist; Morse, a painter, and so on.

PREHISTORIC OBSERVATIONS

It is true, however, that in prehistoric times, some observations were made which had to do with electricity. Lightning, that most awe-inspiring spectacle of nature, was known, of course, and feared from earliest times (fig. 250). Sometimes a tribe witnessed the death of a man by lightning. In mythology, the thunderbolt was the attribute of the supreme ruler of the heavens, whether called Zeus, Jupiter, or Thor. Thunder was interpreted as the roar of the mighty chariot of Thor, after whom is named the fifth day of our week. Classical representations of Jove's flaming thunder-bolt showed the torpedo-shaped bolt, four wings, and also coruscations emanating in all directions—suggesting (remarkable to relate) transverse waves (fig. 252). The familiar design used on a telephone directory is a modern treatment of the same motif. Lightning and thunder, however, are really optical and acoustical phenomena, and it was only in comparatively recent times that Gray and Franklin showed their origin to be electrical. Lightning, as its name implies, is perceived by the eye, and thunder

FIG. 251. Lead sling-shot, with thunderbolt design. Roman, 3rd century B.C.

FIG. 252.
Jove's thunderbolt.

by the ear, but electricity, in the original sense of the term, can be neither seen nor heard.

The eerie Northern Lights, the Aurora Borealis (fig. 253), were observed by primitive man. He also saw, occasionally, St. Elmo's Fire (fig. 255). Probably a few persons in prehistoric times experienced the shock of the torpedo or electric eel. A few may have seen sparks when they rubbed a tiger's fur in the dark, and may have lived to tell the tale. The priests of Bubastes would know of the sparks obtainable from

FIG. 253. Aurora Borealis.

a cat's fur, and, undoubtedly, some cave-men saw and heard sparks when, at long intervals, they pulled off their fur garments over their hair in the dark.

Early man, however, did not turn any of these discoveries to practical account. He went little further than attempting to avoid the effects which he dreaded. He learned that sheltering deep within a cave was safer from lightning than standing in the open or under a tree. Now and again, one hears advice that has been handed down through generations, such as the warning not to stand at the mouth of one's cave during a thunderstorm, that is to say, in modern terms, at an open door or window. There are still households in which it is the custom, at the approach of a thunderstorm, to hide the cutlery from the sight of the angry gods by placing it under the table-cloth!

One of the earliest descriptions of lightning and thunder occurs in the Babylonian story of the Flood, in which Pernapistim corresponds to Noah of the Biblical account:

> Then spake Pernapistim to him, even to Gilgamesh. The city of Shurippak . . . as thou knowest, is situated on the bank of the river Euphrates. That city was already ancient when the gods . . . set their hearts to bring a deluge. . . . On the fifth day, I drew the vessel's design. One hundred and twenty cubits high were its sides. One hundred and twenty cubits measured the length of its beam. . . . I selected a mast and added what was wanting. Three sars of pitch I poured out on the inside . . . so as to make everything water-tight. . . . The ship sank into the water two thirds of its height. . . . There arose a dark cloud. The weather-god, Ramman, thundered in its midst. The gods, the Annutaki, lighted on high their torches. . . . Six days and seven nights continued the storm. Raged cyclone and tempest.[1]

AMBER

Faraday's discovery in 1831 of the method of obtaining induced currents led to the invention of the *generator*, which is the fundamental machine of the Electrical Age. This great advance originated, however, not from contemplation of the sublime grandeur of lightning, but from the study of an early observation which seems by comparison insignificant, and, at the first glance, even trivial. "For the Lord was not in the wind that rent the mountains, nor in the earthquake that brake in pieces the rocks, nor in the fire, but in the still small voice." It was probably before the invention of writing that someone discovered that if the gem amber is rubbed on fur or cloth it acquires temporarily the mysterious power of attracting little bits of dust or chaff (fig. 254). That this observation was first made in prehistoric times is indicated by the discovery of amber in the excavations of ancient lake dwellings

[1] *Library of Original Sources* (Milwaukee University, 1907), I, 17.

in Europe.[2] Later it was discovered that tourmaline when warmed by a camp-fire exhibits the same ability. This is recorded in its Dutch name *Aaschentrikker*, which evidently means "ash attractor." The ability of amber (or tourmaline) under these circumstances to attract dust particles may be expressed briefly by saying that it gives the "dust-test."

This property of amber seemed so devoid of practical possibilities that it received almost no attention, except in idle amusement, until about three centuries ago. Even Gilbert studied it only because he thought that a subordinate chapter on this type of attraction might properly be included in his great book about the magnet, *De Magnete*, which deals with magnetic attraction. Insignificant as the dust-test of amber, may seem, yet it holds in embryo the whole electrical age. The Greek name for amber was *electron*, and from this came the name electricity.

FIG. 254.
Amber dust-test.

The Hebrew scriptures make several references to electrical phenomena. In one of these, Job's servant announces the destruction of his flocks and their shepherds by lightning (1200 B.C.):

The fire of God is fallen from heaven, and hath burnt up the sheep, and the servants, and consumed them.—Job i.16.

In another passage, Job's notorious comforters attempt to defend the justice of an omnipotent deity who allows dire calamity to devastate "a perfect and an upright man":

Then the Lord answered Job out of the whirlwind, and said, Who is this that darkeneth counsel by words without knowledge? Gird up now thy loins, like a man; for I will demand of thee, and answer thou me. Who hath divided a way for the lightning of thunder? Canst thou send lightnings, that they may go, and say unto thee, Here we are?—Job xxxviii.1 and 35.

It has been said that Samuel Morse, the inventor of the electric telegraph, was the first person who could answer that last question affirmatively.

Moses once saw an intriguing display of St. Elmo's Fire:

And the angel of the Lord appeared unto Moses in a flame of fire out of the midst of a bush: and he looked, and, behold, the bush burned with fire [was covered with flames] and the bush was not consumed. And Moses said I will now turn aside and see this great sight, why the bush is not burnt.—Exodus iii.2-3 (1500 B.C.)

[2]Park Benjamin, *A History of Electricity* (Wiley, 1898), p. 12.

It is interesting to note here the attempt to express the new phenomenon in terms of the known facts of fire and, failing that, to resort to the supernatural. Under certain conditions the flames of St. Elmo's Fire, a rather rare electrical display, are seen on spires, on masts (fig. 255), always at *points*, sometimes even perched on the finger-tips of

FIG. 255. St. Elmo's Fire.

an outstretched hand. Pliny observed them on the spearheads in a Roman camp in A.D. 70; Columbus saw them on the masts of his ships in 1493 during his second voyage, and Darwin described them in his *Voyage of the Beagle* (1832). An explanation of these facts was reached by a study of "brush discharges" of electricity in 1729 by Stephen Gray and other "electricians" of the same period. Recently, St. Elmo flames have been reported by a number of aeroplane pilots.

One writer is inclined to consider that Solomon or his architect Hiram had some practical knowledge of electricity: for the historian Josephus states that the roof of the Temple at Jerusalem had a forest of gold *points* which were connected by metal pipes with caverns in a neighbouring hill. During a thousand years, not once was that temple struck by lightning,[3] for it really was equipped with lightning-rods. The Etruscans had a temple that was similarly ornamented and protected.

The *ligure*, which was probably amber, is listed in the scriptures as the seventh jewel in the famous breast-plate of the high priest.

The first row shall be a ruby, a topaz, and a carbuncle. The second row shall be an emerald, a sapphire, and a diamond. The third row, a ligure, an agate, and an amethyst. And the fourth row a beryl, and an onyx and a jasper.— Exodus xxviii.17.

[3] P. F. Mottelay, *Bibliographical History of Electricity and Magnetism* (Lippincott, 1922), p. 9. Also, Park Benjamin, *Hist. Elec.*, p. 564.

Homer's *Iliad* and *Odyssey* both contain references to the thunderbolt of Zeus, the king of gods and men.

Therefore, dreadfully thundering, he sent forth his gleaming thunderbolt, and cast it into the earth before the steeds of Diomede. There arose a terrible flame of burning sulphur and the two frightened steeds crouched trembling beneath the chariot.[4]

These words illustrate the primitive method of "explaining" natural occurrences by attributing them to the whims and caprices of superhuman personalities with superhuman powers, the gods and goddesses. In such terms as force and electricity, we do somewhat the same thing, except that the agencies whose existence we assume are impersonal and their actions more predictable and much more general in scope.

THE MARINER'S COMPASS

The science of electricity is closely linked with that of magnetism, one of the chief applications of which is the mariner's compass. Some have claimed that the following passage from Homer (Smyrna, 900 B.C.) refers to that instrument:

> In wondrous ships instinct with mind
> No helm secures their course, no pilot guides:
> Like man intelligent, they plough the sea,
> Though clouds and darkness veil th' encumbered sky,
> Fearless thro' darkness and thro' clouds, they fly.[5]

Some of these words apply to the compass, but as a description of the instrument they are vague.

The Phoenician sage Sanconiathon lived at Tyre about the time of Homer. Among his sayings there is one that refers to the magnet or compass:

It was the God, Ouranos, who devised Betulae, contriving stones that moved as having life.[6]

INDICATORS

As has been mentioned, Thales of Miletus was first to bring into scientific discussion the fact that amber, if rubbed on fur, acquires *temporarily* the power of attracting to itself little bits of *dust*, chaff, cork, and, in fact, any sufficiently light particles (fig. 254). From the report of Diogenes Laërtius, it seems that Thales spoke of the ability of amber to give this "dust-test" as a fact previously known:

Herodotus and Democritus are agreed that Thales belonged to the Thelidae, who are Phoenicians. After engaging in politics, he became a student of Nature.

[4]Iliad, VIII.
[5]Odyssey, VIII, trans. Alexander Pope.
[6]Park Benjamin, *Hist. Elec.*, p. 56.

As Plato testifies, he was one of the Seven Sages. . . . He held that there was no difference between life and death. "Why, then," asked one, "do you not die?" "Because there is no difference," he replied. . . . According to some, he left no writings. Aristotle and Hippias affirm that he attributed a soul or life even to inanimate objects, arguing from the magnet and amber.[7]

Since amber beads have been found in the lake dwellings of Europe, this property of amber was probably known thousands of years before Thales.

Each dust-particle in Thales' experiment is an *indicator*. If any body, *I*, is introduced into a system containing the body *B*, so that by the behaviour of *I*, we can detect, examine, or measure a condition or change of condition in *B*, then *I* is an indicator, e.g. a pressure-gauge, a piece of litmus-paper, an ammeter, and so on. Because of the lack of electrical sense-organs in human beings, there will necessarily be at least one indicator present in every experiment throughout the whole story of electricity. Furthermore, the series of indicators introduced as the science advances exhibits an increase in sensitivity and reaches, for instance, the complexity of a radio-set. The contribution of more than one scientist has been a new indicator or the improvement of an old one. The same need of indicators exists in the science of magnetism as in that of electricity, for we have no special sense-organ to perceive magnetic phenomena.

MAGNETIZATION

Socrates, in order to illustrate a point in his dialogue with the poet Ion, referred to the fact that a magnet can magnetize a piece of iron or steel by contact:

> For as I was saying just now, Ion, this is not an art in you whereby you speak well on Homer, but a divine power, which moves you like that in the Stone which Euripides named the magnet but which most people call the Heraclean stone. For this stone not only attracts iron rings, but also induces in them a power whereby they in turn are able to do the very same thing as the stone and attract other rings: so that sometimes there is formed quite a long chain of bits of iron and rings suspended one from another: and they all depend for this power on that one stone. In the same manner, also, the Muse inspires men herself, and then by means of these inspired persons, the inspiration spreads to others and holds them in a connected chain.[8]

In the works of Lucretius, we shall find an echo of this passage with an interesting addition. Socrates studied physics under the best physicists in Athens, Anaxagoras and others; but his consuming interest was in human character. He was once asked why he did not go outside

[7]A. M. Still, *Soul of Amber* (Murray Hill, 1944), gives an interesting history of electricity.
[8]Plato *Ion*, trans. Lamb (Heinemann), III.

the city gates and study the rocks, trees, birds and clouds. He replied
that these had little to teach Socrates but that the market-place had
much. It is a safe wager that Socrates would favour the humanistic
approach in the study of science.

THE TORPEDO

The only contribution of Aristotle to electricity which will be considered here is his description of the paralysing shock which the fish
called the torpedo (literally "the stunner") can administer by contact.
The excerpt is from his famous treatise on zoology:

> In marine creatures also, one may observe many ingenious devices which
> are adaptations to their environments. . . . The torpedo narcotizes the fishes
> that it intends to capture, stunning them by the power of shock which its body
> possesses; then it feeds on them. It hides in the sand and mud and bags all the
> poor fish that swim near enough to be paralysed by its action. All these occur-
> rences have been observed by the writer himself. . . . The torpedo is also known
> to be capable of paralysing a man.
> The so-called sea-scolopendra, after swallowing the hook, turns itself inside-
> out until it spits out the hook; then it turns itself right-side-out again.

The last paragraph indicates that in 350 B.C. a good fish-yarn was
allowable in a zoology text. Aristotle's pupil Alexander, the Macedonian conqueror, used to send home to his old professor consignments
of specimens from his marauding expeditions in many countries, and
with the specimens, no doubt, he sent some tall and lively accounts.
In speaking of the torpedo, however, Aristotle declared that he was
reporting his own observations. That the lightning-bolt, the behaviour
of amber, and the shock of the torpedo are closely related, was apparently not suspected by Aristotle, and who could blame him? The torpedo has intrigued many scientists into a study of its shocking habits.
Cavendish, for instance, built a model torpedo and succeeded in imitating the shock. His test object in these experiments was Cavendish.

THE LYNCURIUM

Part of the plunder shipped home by Alexander to his honoured
professor at Athens was a gorgeous collection of jewels which greatly
enriched the cabinets of the Lyceum. Aristotle's best student, whom
he named Theophrastus because of his godlike power of using apt
phrases, studied the collection and wrote a famous treatise *Concerning
Stones*, in which he described a second mineral that resembles amber
in giving the dust-test. He called it *lyncurium*, and may have meant
tourmaline or the hyacinth, a variety of zircon. The name means "lynx-
water," and refers to a strange myth about the origin of the gem. A

controversy raged for centuries among the schoolmen of the Middle Ages as to which gem is the lyncurium. As Roger and Francis Bacon both contended, if the energy that was wasted in disputing about vague passages in old books had been used in performing new experiments and in discussing them, progress would have been faster. This discovery by Theophrastus of a second substance which, like amber, gives the dust-test, represented the total advance in electricity during three centuries—and for all that, it was a 100 per cent advance on Thales. It showed that amber was not unique in this respect, and suggested that there might be still more substances with the same property.

It requires expert workmanship to bring out the lustre of the emerald, for in its rough state, its lustre is rather dull. It has excellent qualities, however, and so has tourmaline which is also used for making engraved seals. Its hardness is very great as is the case with many precious stones. It has also the power of attraction, characteristic of amber, and is said to attract not only straws and small bits of bark, but also, according to Diocles, even copper and iron if they are beaten into thin flakes. Tourmaline is exceedingly clear and its colour is an elegant red. Amber is also a gem. It is dug out of the earth in Liguria and has the same property of attraction as the stone just mentioned. But the strongest and most evident power of attraction occurs with that stone which attracts iron.

Thus Theophrastus pointed out that the magnet is *selective*, as it attracts only iron (and a few other substances), whereas amber is utterly promiscuous. It is worth emphasizing the fact that it is a property of *all* matter to be attracted by rubbed amber.

Theophrastus succeeded Aristotle as Dean of the Lyceum and taught to the age of eighty-four. At this ripe age he retired and shortly afterwards died. His parting words were: "When we are just beginning to live, lo! we die. Farewell, and may you be happy."

MAGNETIC REPULSION

Lucretius was first to record magnetic *repulsion* and also the fact that magnetic action can take place *through brass*. His words should be compared with Socrates' reference to magnetic action.

I shall now proceed to tell by what law of nature it comes to pass that iron can be attracted by the stone which the Greeks call the Magnet from the name of its native place; for it was first found occurring naturally within the boundaries of the land of the Magnetes. . . . At this mineral men marvel, indeed, it often forms a chain of little rings all attached to itself; for sometimes you may see five or more hanging in a festoon and swaying in the light breezes when one hangs from another, clinging to it beneath, and each from its neighbour experiences the attractive force of the stone; in such penetrating fashion does its force prevail.

It also happens that at times iron is repelled by this stone and is wont to flee and follow alternately. For I have seen Samothracian iron rings even leap up and at the same time iron filings move in a frenzy inside brass bowls, when this Magnesian mineral was placed beneath; so eagerly is the iron seen to desire to flee from the stone.

This is the only instance of Lucretius performing an experiment, and, even on this occasion, it would seem that somebody carried out the manipulations while Lucretius Carus looked on. From his report of repulsion, we can infer that the Samothracian rings were of steel or hardened iron; for soft iron has not sufficient *retentivity* to exhibit repulsion as Lucretius describes it.

At Samothrace there was a temple and an iron mine. The priests at the temple and the chemists at the mine were one and the same and this situation held also in Egypt; but since those times there has been a division of labour. The priests made iron finger-rings which they sold to pilgrims as souvenirs. It became customary to give one of those rings as a pledge to close an agreement or contract. The custom spread ultimately to include the most important of all contracts—the marriage contract. Hence arose the custom of the bridegroom giving his bride a ring at their wedding, or, in the more beautiful ceremony, the exchanging of rings. Increase of wealth led to the use of gold and platinum rings, but the iron ring is still used and may be seen, for example, on the finger of an engineer as a symbol of his initiation into the engineering fraternity.

Thus far and for the next dozen centuries man's knowledge of magnetism and his knowledge of electricity developed separately. Then for about four centuries progress was greater in the young science of magnetism because of the commercial importance of the compass. During the next two centuries the science of electricity made the greater progress and there arose a growing notion that the two sciences were related. Finally during the nineteenth century, in the work of Oersted, Maxwell, and others, we shall see the two sciences joined in a brilliant marriage.

FLOW OF ELECTRICITY

A famous treatise, *Natural History*, which was published about A.D. 70 by the Roman historian, Pliny the Elder, contains some mention of electrical facts. It adds three names to the "amber-list," or list of substances which resemble amber in giving the dust-test, namely ruby, garnet, and amethyst. This jumped the list from two to five at one swoop, and represented a 150 per cent growth in the science in about four centuries. Pliny's *Natural History* also records a display of Aurora Borealis and gives a description of St. Elmo's Fire:

I have seen during the night-watches of the soldiers a luminous appearance like a flame surmounting the spears on the ramparts. They are also seen sometimes on the yard-arms and on other parts of sailing-ships. They perch there and make a sort of rustling sound like that of birds flitting about. When these flames occur singly they bring disaster, even sinking the ships; and if they strike the bottom of the keel [!] setting them on fire. When these flames are twinned, however, they bring good luck and foretell a prosperous voyage; for by their arrival, the dire and menacing influence of the single flame, called Helena, is driven away. So the sailors call them Castor and Pollux and invoke them as gods upon the sea. They also shine sometimes as a halo around the head of a man in the evening hours and thus constitute an important omen. The whole phenomenon is shrouded in mystery and concealed in the majesty of nature.

Pliny's book contains the first reference to the property of electricity of flowing along certain bodies which are therefore called *conductors*. In this case, the conductor was a moist fishing-spear:

Would it not have been quite sufficient merely to cite the example of the torpedo, another inhabitant of the sea, as a manifestation of the mighty powers of Nature? From a considerable distance even, and if only touched with the end of a spear or staff, this fish has the power of benumbing even the most vigorous arm and of paralysing the feet of a runner, however swift he may be in the race.

This observation foreshadows the second epoch in the story of electricity, which is characterized by the study and application of conduction. To tell the truth, there have already been two examples of conduction of electricity mentioned in this story, but it is left for the reader to detect them.

JET

A century and a half after Pliny, the Roman compiler Julius Solinus (A.D. 220) published a book, *Concerning Marvels*. One of its chapters, in describing Britain, mentions the jet mines and, incidentally, states that jet belongs to the "amber-list." The old English translation used here was made in 1587 by A. Golding:

Moreover, to the intent to passe the large aboundance of sundry mettals (whereof Britaine hath many rich mynes on all sides), Here is store of the stone called Geate, and y⁰ best kind of it. If y⁰ demaund y⁰ beautie of it, it is a black Jewell:... if the nature, it burneth in water, and goeth out in Oyle; if the power, rubbe it till it be warme, and it holdeth such things as are laide to it; as Amber doth. The Realme is partlie inhabited of barbarous people, who even fro theyr childhoode haue shapes of divers beastes cunninglye impressed and incorporate in theyr bodyes.[9]

Necklaces of jet beads need frequent polishing to free them from dust particles which the beads attract to themselves after they have rubbed against clothing.

[9]Quoted in Silvanus Thompson, *Notes on Gilbert's De Magnete*, p. 37.

PENETRATION

St. Augustine, the bishop of Hippo, near Carthage, recorded an important fact about the magnet in his book, *The City of God*, which was published about A.D. 400. He showed that the action between a magnet and an iron filing is not impeded by the interposition of various bodies, as Lucretius had shown for brass:

> We know that the magnet draws iron strangely: and surely, when I observed it for the first time, it struck me with astonishment. . . . Yet stranger was that experiment with this stone which my brother bishop shewed me. . . . He told me that he had seen Bathanarius, when he feasted him once at his house, take the said stone and hold it under a silver plate upon which he laid a piece of iron: and as he moved the stone underneath the plate, so did the iron move about, the plate not moving at all; and exactly in the same motion with which his hand moved the stone, did the stone move the iron. This I saw. . . .
>
> There was an iron image hung in a certain temple. . . . It hung so poised in mid-air between two magnets, whereof one was placed in the roof of the temple and the other in the floor, that it did not touch anything at all. . . . I do not believe all the things that I have set down, so firmly that I make no doubt of them; but for that which I have tried . . . as the magnet's drawing of iron and not moving a straw, these I have seen and believe without any doubt at all.

Science does not boast many saints in its ranks. If one reads the *Confessions of St. Augustine*, one gathers that in his youth he was no saint; but according to his own testimony, the study of science and philosophy and the influence of his mother and of religion enabled him to reform his life.

Apparently St. Augustine's mind harboured some doubts concerning the iron image story. In fact the same myth took different forms, being told of the horse of Bellerophon on the island of Rhodes, of the effigy of Mausoleus in his mausoleum, and finally of Mohammed's coffin in a mosque.

CHAPTER EIGHTEEN

Poles and Charges

LAWS OF MAGNETIC POLES

ROGER BACON had a friend whom he held in high esteem, Peter Peregrinus. This name, which means Peter the Pilgrim, indicates that Peter was a returned Crusader, but apparently he was master of more than one trade. In 1269, he was an engineer in the French army besieging the town of Lucera, in Italy. From the trenches he sent home to his neighbour in Picardy, "by way of a letter" (as Boyle put it), the first treatise ever written on magnetism. This remarkable brochure, *Epistola de magnete*, records several of Peregrinus' contributions to magnetism, including his invention of the first European pivot-compass and his discovery of the law of magnet poles.

In Camp, at Lucera, Aug. 8, 1269.

Dearest Friend,

In response to your request, I am going to disclose to you in brief style a certain mysterious quality of the magnet stone, and write down in plain language some facts that are quite unknown to the majority of students. Such stone is found as a rule in Northern parts and is reported by sailors in all ports of the Northern Seas. But its Virtue is discerned through its strong attraction of iron. The best magnet is the one which can lift the greatest weight of iron.

Peregrinus was first to locate the poles of a magnet and to suggest the name "pole," which he borrowed from astronomy. The magnet he used was of spherical shape and was made of the mineral magnetite, a black oxide of iron. The letter continues:

You must know that this stone bears in itself the similitude of the heavens. . . . There are two points in the heavens more noteworthy than all others, because the celestial sphere turns about them as upon axes: One of these is called the Arctic or North Pole, whilst the other is named the Antarctic or Southern. Likewise, in this stone, there are two points, of which one is called the North, and the other, the South. . . . To locate these points, . . . have the magnet rounded with a tool for rounding crystals and other stones. . . . Then let a Needle . . . be placed over the stone; draw a line on the magnet marking the position of the needle. Move the needle to another place and mark its new position in the same way. . . . Do this at several positions and it will be found that all the lines thus drawn will without exception meet in two points, just as all the meridian circles of the earth meet in the two opposite poles of the world. . .

There is another method. . . . Observe the place on a rounded stone . . . where the end of a needle . . . adheres . . . more strongly. . . . Break off a little piece from the needle . . . and place it on the spot where the point has already been located . . . and if it stand upright on the stone, without doubt, the point sought for is there. But if not, move it until it does stand upright. . . . Mark these Poles with incisions on the stone.

The space surrounding a magnet or a magnet pole, in which its action can be detected by any indicator, is called its magnetic field or, briefly, its field. This passage is the earliest instance of exploring a magnetic field; the indicator was a needle or compass, and for part of the experiment, the point of the needle. The location of the pole as the point where the needle-point stands upright foreshadows the invention of the dipping-needle by Norman in 1581, and anticipates the location of the earth's Arctic magnetic pole by Sir James Ross in 1831, and the Antarctic magnetic pole in 1909 by Sir Ernest Shackleton, both of these by the use of a dipping-needle.

As we have seen, a magnet, in general, has two poles and may be described as bipolar. If a light and slender or needle-shaped bipolar magnet is supported so as to be free to turn in a horizontal plane, it comes to rest with its axis in a northerly-southerly direction. Any magnet thus set up to indicate directions is called a *compass*. In Peregrinus' day, the magnet was often called "the stone" and a compass "the needle." The mineral magnetite, because of its use in navigation, was called lodestone (Anglo Saxon *laedan*, lead or guide). Before Peregrinus invented the pivot suspension, the needle used aboard ship was floated on wood on water and was called a water compas. Peregrinus described the preparation of one of these:

Having performed the technique of finding the poles of this stone, . . . take a round wooden vessel in the shape of a cup or dish, and in it place the stone . . . then place that vessel with the stone in it . . . in another vessel, a large one filled with water, so that the stone may be in the first vessel like a sailor in a boat. But let the first vessel be in the second one with plenty of room. . . . This stone so placed will turn its small vessel, until the North pole stands in the direction of the northern point of the sky, and South pole in the direction of the southern point. And if this stone be moved aside a thousand times, a thousand times will it return to its position . . . by the will of God. . . . One part will turn toward the star called the nautical star [pole star or lodestar] because it is near the pole of the heaven. For the truth is that it does not move toward the said star but toward the pole of the heavens.

We have seen that in the Middle Ages many occurrences that were not understood were said to be caused by the stars. For example, the name of the disease *influenza*, which is an Italian word for influence, records the mediaeval belief that epidemics of that disease were caused by the malevolent influence of the stars. In assuming that it is the north star or the pole of the heavens which attracts one end of the compass, Peregrinus adopted an opinion prevalent in his day. No person is entirely independent of the received opinions of his times. That Peregrinus was an independent researcher and a sufficiently free thinker to be an innovator, cannot be doubted, but to be a reformer at every moment of one's life would be too utterly exhausting. Peregrinus introduced three improvements in the compass: pivot suspension, a sighting-arm, and a circle graduated in degrees.

Let a vessel be made of wood or any solid material, and let it be . . . in the fashion of a box [fig. 256] . . . and over it a lid of transparent substance such as

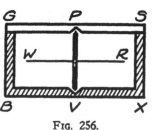

FIG. 256.
Peregrinus' compass.

glass or crystal. So let there be arranged in the middle of the box . . . a slender Axis [*PV*] of brass . . . fitting at its extremities to the two parts of the box. . . . Let there be two holes in the middle of the Axis, at right angles to each other and let an iron wire . . . be passed through one of these holes and through the other, . . . another wire of silver or brass. Then you shall bring near the crystal [hence, near the iron wire] whichever pole of the Magnet you wish . . . until the needle [iron wire] moves toward it and receives virtue from it.

In this account Peregrinus described incidentally the magnetization of an iron wire by bringing it near a magnet pole; Socrates and Lucretius had already described magnetization by contact. The needle Peregrinus used in locating the poles of a spherical magnet was similarly magnetized by one of these methods. The double pivot suspension here described is like that of a clock-wheel.

That Peregrinus was anticipated by the Chinese in his invention of the pivot compass is indicated by the accompanying picture of a Chinese war-chariot compass dated A.D. 450.[1] The little mandarin has the needle up his sleeve and the pivot shows beneath his sandals. The artist has shown a grand disregard for scale in making the compass as large as a cart-wheel. Pivot suspension allows the needle to turn in only one plane whereas the earlier Chinese method of tying a silk thread around its middle allowed the needle to turn in almost any direction. This method is called "free suspension." Chinese tradition says that the Emperor Hwang-ti (2500 B.C.) won a battle during a fog because he had on his chariot a magic needle that pointed always south. His opponent, lacking such a device, became confused and his army was

[1]Park Benjamin, *Hist. Elec.*, p. 73.

routed. It is interesting to note that the Chinese said their compass pointed south whereas we say ours points northward.

Peregrinus next performed for the first time an experiment which is still repeated annually in the schools. By it he discovered a law of magnetic poles:

Prepare two magnets as follows: . . . Place one in its vessel so that it may float like a sailor in his boat . . . but hold the other magnet in your hand. And bring the Northern part of the Stone which you are holding near the Southern part of the one floating in the vessel: for the floating stone will then follow the one that you are holding, as if wishing to adhere to it: And furthermore, if you present the Southern part of the stone in your hand to the Northern part of the floating one, the same thing will happen. . . . Know you, then, as the rule, that the Northern part of one stone attracts the Southern part in another and the Southern, the Northern. But if you do the opposite, namely, bring the Northern part near the Northern— the stone in your hand will seem to repel the floating one . . . and in fact, for this reason, that the Northern part seeks the Southern. . . . If the Southern part is held . . . towards the Southern part of the floating stone, you will see it repel immediately.

FIG. 257.
Chinese compass.

The law is generally stated: Like magnetic poles repel each other and unlike or opposite poles attract each other. The rule is easy to remember, for in certain important human affairs it is said also that "opposites attract each other."

Peregrinus worked out the rule for the arrangement of poles produced by any touch method of magnetization.

When an oblong piece of iron has touched a magnet and has been fastened to a light piece of wood . . . and is put on water, . . . the part of the iron which touched . . . the South of the stone, will turn toward the North of the heavens.

The magnetizing pole leaves opposite polarity where it last touches the body it magnetizes. The properties of a magnetic pole and of a magnet to which the pole belongs are referred to collectively as polarity. In the next passage, reversal of polarity is described.

But if violence be done, . . . namely, if the Southern part of the iron . . . be joined to the Southern part of the Stone, the virtue in the iron will be easily altered . . . and that will become the South which was the North. And the cause of this is the impression of that which acted last, confounding and altering the virtue of the first.

Very naturally, the old Crusader tried to prepare a magnet which would have only one pole and if obtained would be called monopolar.

His method, of course, was to cut a bipolar in two at its neutral equator (*EQ*, fig. 258). He reported his attempt, and failure:

Take one stone *NS* . . . in which *N* is the North, whilst *S* is the South point. And divide it into two parts. . . . Then put the stone which contains *N* on water that it may float; and you will see that *N* will turn toward the North as before. . . . And so that part . . . just at the break, which is *E*, must needs be the South.

At this juncture, the iron-filing test at *E* and an application of the law of poles, showing repulsion, would have improved the demonstration.

Thus you see that the two parts of the two Stones which before the separation were continuous [at *E*] in one stone [and neutral], are found after the separation, the one to be a Northern part, the other, Southern.

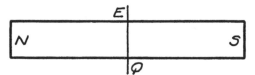

Fig. 258. Cutting a magnet in two.

This is an example of a failure or negative result that proved useful. Up to the present, in spite of many attempts, there is no accepted record of success in preparing a monopolar magnet. If a magnet has one pole, it regularly has also at least one more pole of opposite kind. Since the rule applies to the fragments *NE* and *QS*, and to their fragments and so on, by extrapolation, one can state Peregrinus' result briefly in terms of particles: Every magnet is a system of minute bipolar magnets.

Since the equator *EQ* was initially neutral by the iron-filing test, one naturally asks, "How was the south-seeking pole produced at *E* and the north-pointing pole at *Q*?" Since cutting is not a method of magnetization, the two poles must have been at *E* before the cut was made. When they were close together, therefore, the two opposite poles *E* and *Q* must have nullified or *neutralized* each other.

To close this interview at a mediaeval level, we shall read Peregrinus' instruction to his neighbour for making a perpetual motion machine from a spherical magnet:

In this we may suppose not only that the poles of the stone receive their influence and virtue from the poles of the World but also that the whole stone does so from the whole heavens. . . . Let a stone be rounded . . . and afterwards dispose it upon two sharp styles . . . so that it may turn on them without difficulty. . . . Arrange the stone on the meridian circle on its pivots . . . in such wise that the elevation of its poles may accord with the elevation of the poles of the heavens. . . . Now if the stone rotate in harmony with the motion of the heavens, rejoice that you have arrived at a secret marvel. . . . By means of this instrument . . . you will be relieved from every kind of clock for by it you will be able to know . . . at whatever hour you wish . . . all the dispositions of the heavens . . . after which astrologers seek. But if not, let it be ascribed rather to your own want of skill than to a defect of nature.

What a quaint escape from a false prediction! In this passage, Peregrinus shows that he may have suspected the earth of being a magnet, so that when Gilbert wrote his book, in 1600, to establish this idea, he had at least one forerunner.

DECLINATION

That very indomitable man, Columbus, on his first voyage, found to his dismay that Peregrinus and others were wrong in stating that all compasses point due north. He could hardly have made a discovery less welcome to himself for it threatened to ruin his gold-seeking expedition. His jail-bird sailors had mutinied and declared their wish to turn back before the ships fell over the edge of the big pan-cake into the bottomless abyss, even though the adventure offered them the sight of Atlas at work. Columbus had soothed and cajoled them by reference to their trusty friend the compass which served them loyally night and day regardless of cloud or storm or whatever might betide. Now this last comfort and argument was snatched away. Realizing that if his discovery became known among the ratings, it would mean the end of his expedition, Columbus resorted to devious artifice, by which he eluded the crisis. The story of his memorable voyages, from which our excerpts are taken, was written by Las Casas, a member of the expedition who drew his materials from his own experiences and largely from the Admiral's own log-book.

Friday, August 3, 1492. Set sail from the bar of the Saltes [near Cadiz] at 8 o'clock and proceeded with a strong breeze till sunset, sixty miles or fifteen leagues S, afterwards SW. and S. by W. [average speed, 6 mi. per hr. or about 5 knots]. . . .

Sunday, September 10. This day and night, sailed sixty leagues at the rate of ten miles an hour. Reckoned only forty-eight leagues, that the men might not be terrified if they should be long upon the voyage. . . .

Thursday, September 13. The needles varied to the NW. and the next morning about as much in the same direction.

This observation Columbus did not publish in the ship's newspaper, but the next entry tells how the bad news got abroad, and then the fat was in the fire:

Monday, September 17. Steered W. and sailed day and night above fifty leagues. . . . The pilots took the sun's amplitude and found that the needles varied to the NW. a whole point of the compass [about 11°]. The seamen were terrified and dismayed. The Admiral ordered them to take the amplitude the next morning.

Thus by casting doubt on the reading, the Admiral postponed the return voyage by at least twenty-four hours and gained opportunity, it is said, to "correct" the compass during the night by shifting the card (!) about 10°. The next entry states the outcome of his "experiment."

Tuesday, September 18.   Then they found the needles were true.

The Admiral soothed the crew with the following explanation:

The cause was that the pole-star moved from its place [!] while the needles remained stationary.

At any place on the earth, the angle between the axis of the compass needle at equilibrium and the true north line was called by some "variation," but now in general it receives the name *declination.* The source of the term is in the account of Columbus' third voyage:

Thursday, August 16, 1498.   When he left the Canaries, . . . having passed three hundred leagues to the west, at once, the needle turned to the NW. . . . and now on this voyage it had never declined to the NW. until last night, when it declined more than a quarter and a half, and some needles declined to the NW. . . . two quarters. . . . 2½ degrees east of the island of Corvo in the Azores, the magnetic declination changed and passed from NE. to NW.[2]

Lines drawn on charts, joining points that represent places with the same declination, are called *isogonic lines;* and the isogon for zero declination which Columbus found passing through Corvo is called an *agonic line.* Another agonic line passes through Lake Superior, Venezuela, and Uruguay.

In October, 1493, during his second voyage, Columbus wrote:

St. Elmo appeared on the top-gallant masts with seven lighted tapers.

INCLINATION

The first treatise on the earth's magnetism was written by Robert Norman, a maker of ships' compasses at London. According to the opinion of his collaborator, Dr. William Gilbert, he was "an ingenious artificer." Norman published his book, *The Newe Attractive,* in 1581. Like many books in those days, it was generously titled: *The Newe Attractive, containing a short discourse of the Magnes or Lodestone, and amongest other his virtues, of a newe discovered secret and subtill propertie concerning the declinyng of the Needle touched therewith, under the plaine of the Horizon.*

[2]N. M. Penzer, *The Voyages of Columbus* (Argonaut Press), p. 254.

Norman's book gives an account of how he came to make the discovery on which his fame rests:

Having made many and divers compasses and using alwaies to finish and end them before I touched the needle, I found continuallie that after I had touched the yrons with the stone, that presentlie the north point thereof woulde bend or decline downwards under the horizon in some quantitie: in so much that to the flie of the compass, which was before levell, I was still constrained to put some small piece of ware on the south point and make it equall againe.[3]

Applying his discovery, Norman invented a new instrument, the dipping-needle or dip-needle or inclinometer. He gave his directions for making a simple form of dip-needle:

Then you shall take a deepe Glasse, Bowle, Cuppe or other vessell, and fill it with fayre water, setting it in some place where it may rest quiet and out of the winde. This done, cut the Corke circumspectly, by little and little, untill the wyre with the Corke be so fitted, that it may remain under the superficies [surface] of the water two or three inches, both ends of the wyer lying levell with the superficies of the water, without ascending or descending, like to the beame of a payre of ballance beeing equalie poysed at both ends. Then take out of the same the wyer without mooving the Corke, and touch it with the Stone, the one end with the South of the Stone, and the other with the North, and then set it againe in the water, and you shall see it presentlie turne it selfe upon his own Center, shewing the aforesay'd Declining propertie, without descending to the bottome, as by reason it should, if there were any Attraction downwards, the lower part of the water being neerer that point then the superficies thereof.[4]

In this experiment, Norman magnetized an iron wire by a touch method. The wire before being magnetized is called a *magnetic body*, a term applied to any body which is believed capable of becoming a magnet. The substance of a magnet or of a magnetic body is called a *magnetic substance*, e.g. iron, nickel, cobalt, Heussler alloy, alnico and a few others.

At any place on the earth, the angle between the horizon and the axis of the dipping-needle at equilibrium in the magnetic meridian, is known as the dip or inclination for that place. Norman found it to be at London "71 degrees and 50 mynutes." He also drew the important conclusion that the point toward which compasses point is not in the sky but in the north of the earth and some distance beneath the surface; thus he brought the study of magnetism down from heaven to earth.

GILBERT'S *DE MAGNETE*

The scientist who is seen in fig. 259 demonstrating some experiments for Good Queen Bess is her court physician, Dr. William Gilbert,

[3]Weld, *Hist. Roy. Soc.* (1848), III, 432. Also, P. F. Mottelay, *Bibliog. Hist. Elec. and Mag.*, p. 75.
[4]Quoted in Silvanus Thompson, *Notes on De Magnete*, p. 60.

President of the London College of Physicians. He has been called the "Father of Experimental Science in England." In 1600, he published the famous book, *Concerning the Magnet*, from which the following excerpts are taken.[5] In repeating the experiments of Peregrinus with a spherical magnet, which Gilbert called a "terrella" or small earth, he introduced some useful terms such as *axis* and *equator* of a magnet, and *magnetic meridians* of the earth. He pointed out that the relation between the distance between two poles and their force of attraction

FIG. 259. Dr. William Gilbert at the court of Queen Elizabeth.

or repulsion is inverse. It was nearly two centuries later that Charles Coulomb proved that the force varies inversely as the square of the distance. Gilbert corroborated Peregrinus' observation that the equator of a bipolar magnet is neutral:

> Attraction is always more powerful when the poles are nearer. Places declining from the poles have attractive forces, but weaker . . . in the ratio of their distances: so that at length at the equator, they are utterly enervated.

The space surrounding a magnet in which its action can be detected Gilbert called its "Orbe of Virtue"; Priestley used the term "atmosphere." Now it is called magnetic field or just "field." The small compass that Gilbert used to explore a field, he called a "magnetic ver-

[5]*De Magnete*, edited by Silvanus Thompson.

sorium" (Lat. *verto*, turn). His ideas, like those of Peregrinus, approached Faraday's concept of *magnetic lines of force:*

A terrella sends out in an orbe its powers in proportion to its vigour. But when iron or any other magnetick of convenient magnitude comes within its orbe of virtue, it is allured. Rays of magnetick virtue spread out in every direction in an orbe.

Peregrinus measured the strength of a pole by its lifting strength— the weight of iron filings it could lift with contact. Gilbert showed that the heavier a magnet is the greater its strength can be. One passage in which he draws this conclusion is interesting also because it contains an early use of the important term *mass:*

Let there be given a loadstone of eight ounces weight which lifts twelve ounces of iron; if you cut off from that loadstone a certain portion from that loadstone which is then reduced to two ounces, such a loadstone lifts a piece of iron of three ounces, in proportion to its mass ("*pro molis proportione*").

Gilbert also considered the intensity or "vigour" of a pole's action at a distance:

Loadstones from the same mine are similar in strength. Nevertheless, one which is of greater size exhibits greater attractive power, for it lifts greater weights and it has a wider orbe of virtue.

RETENTIVITY

In studying the ability of a magnet to retain its magnetism, i.e. its *retentivity*, Gilbert invented the "keeper" (fig. 260), and called it an *armature* (Lat. *armatura*, equipment), a term which later became important in connection with devices such as the electric bell and buzzer, the telegraph sounder and relay, the induction coil, the motor and generator.

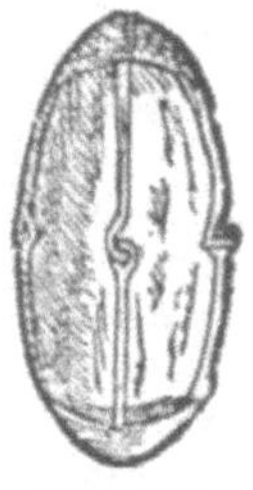

A loadstone loses some of its attractive virtue and, as it were, pines away with age, if exposed too long to the open air instead of being laid in a case with filings or scales of iron.

Steel fetches a much better price than mere [soft] iron. Owing to its superiority, . . . it acquires the virtues from the loadstone more quickly [?] and retains them longer at their full. . . .

If with a single small stone you touch a thousand bits of iron for the use of mariners, that loadstone attracts iron no less strongly than before.

Fig. 260.
Magnet with
Gilbert's
armatures.

Conceive . . . a piece of iron shaped like an acorn . . . to be attached to the loadstone. . . . The greatest force of a combining nature . . . is seen when two loadstones, armed with iron caps . . . are joined by their contrary ends so that they mutually attract each other. Iron unites to an armed loadstone more firmly than to a loadstone. . . . For by the near presence of the magnet, they are cemented together and since the armature conceives a magnetick vigour . . . they are firmly bound together.

MAGNETIC PENDULUM

Pursuing St. Augustine's experiment, Sir William showed that all substances except magnetic ones are "transparent" to magnetic rays or lines of force. He described the action of interposed iron, which is now called *shielding*. He also repeated Peregrinus' experiment with two magnets, one of them floating in a dish "like a sailor in a boat," and pointed out the similarity between its motions and those of a pendulum:

It turns in a circle until its austral pole points northward and its boreal, southward. . . . Although by the first rather vehement impulse, it swings past the poles, yet after oscillating several times, it finally comes to rest in the magnetic meridian, pointing to the poles.

In Coulomb's hands, these facts led to the invention of the magnetoscope and the magnetometer (1785).

THE EARTH A MAGNET

The main purpose of Dr. Gilbert's book was to show that the earth behaves like a huge magnet. The full title of the book was *Concerning the Magnet, Magnetick Bodies and the Huge Magnet the Earth* (*De magnete, magneticisque corporibus et de magno magnete tellure*).

Those experiments which have been demonstrated by means of the terrella, showing how magnetick bodies behave with respect to the terrella, are all or at least the principal ones and the most remarkable ones, exhibited by the body of the earth. . . . In the first place, as in the terrella, the equator, meridians, parallels, axis, and poles are characteristic features, . . . so also in the earth, these features are characteristic. . . . Just as at the periphery of a terrella, a magnet . . . turns its proper pole, so, at the earth's surface, the same characteristic deflections are manifested by magnets. Iron rods also, when placed for some time pointing toward the pole, become magnetized. . . . So all the variations of the versorium or mariner's compass everywhere by land or sea, vagaries which have sore perplexed men's minds, are discerned and recognized as due to the same causes. The magnetick dip, which is the wonderful turning of magnetized bodies towards the terrella, in systematic course, is seen in clearer light to be the same thing upon the earth. And that single experiment, by a wonderful indication, as with a finger, proclaims the grand magnetick nature of the earth. . . . There resides, therefore, a Magnetick force in the earth just as in the terrella.

Gilbert's proposition that the earth is a magnet is supported, as he showed, by the fact that an iron rod can be magnetized or demagnetized at will by *hammering or heating or both*, according to how it is held with respect to the earth during the operation. Then it must be in a magnetic field (fig. 261).

Let the blacksmith beat upon his anvil a glowing mass of iron of two or three ounces weight into an iron spike, the length of a span or nine inches. Let the smith be standing with his face to the north, his back to the south. Let him always, whilst he is striking the iron, direct the same point of it toward the north and let him lay down that end toward the north [to cool].

The rod is found to be magnetized with a north-pointing pole at the end which was kept toward the north. Gilbert added:

Those, however, which are pointed toward the eastern or western point, conceive hardly any verticity [polarity].

In fact, this procedure produces demagnetization, i.e. the poles lose some or all of their strength.

Fig. 261. Effect of hammering and heating on magnetization.

ELECTRICS

It is interesting to observe at this point how much advancement has been made in the study of magnetism, with no equivalent advancement in the study of amber. There were no important applications of the knowledge of amber, such as the compass was of magnetism. The only factor which maintained the study of amber was the curiosity of scientists who were not concerned with applications. Practical men laughed at them as children and dreamers, but without two centuries of their work, the electrical age could never have come into being.

Gilbert included in *De Magnete* what he considered a subsidiary chapter about the attraction of amber, although it was slightly aside from the theme of the book. It became, however, the most important part of the volume because of the impetus it gave to the study of electricity. It is referred to as the "famous second chapter" and has earned for its author the undisputed title of "Founder of the Science

of Electricity." "Thou knowest not which shall prosper, this or that."
Let us consider a few passages from Gilbert's second chapter:

It is not only amber and jet which entice small bodies; but Diamond,
Sapphire, Carbuncle, Opal, Amethyst, and Beryl do the same thing. Electrics
are those bodies which attract in the same manner as amber. Similar powers of
attraction are seen also to be possessed by Glass as also by false gems made of
glass. Rather hard Resin entices. Sulphur also attracts and mastick and hard
sealing-wax. All this one may see when the air in mid-winter is sharp and clear
and dry. These electrics attract all substances, not only straws and chaff but
also metals, woods, leaves, stones, earths, even water and oil, in fact, every
body which is perceived by our senses. But in order that you may be able to
apply convincing tests as to which bodies attract and how much attraction
occurs, make yourself a [non-magnetic] versorium of any metal you wish, three
or four digits long, lightly pivoted on a needle after the style of a magnetic
versorium. Bring near one of its ends an amber rod or a smooth polished gem
that has been rubbed; the versorium turns at once.

FIG. 262. Non-magnetic versorium.

Thus by the use of the new indicator, the versorium, Gilbert trebled
the amber-list, which now included seven-twelfths of the jewels in the
high priest's breast-plate. Gilbert considered that this group of bodies
was large enough to warrant a name. So he coined for them the name
electrics (Lat. *electrica*) from the Greek word for amber, *electron*, to
suggest their resemblance to amber in giving the dust-test when excited
by rubbing or contact with other bodies. He did not use the term
electricity. Those bodies from which he failed to obtain the dust-test
he classified as non-electrics.

On the other hand, many gems as well as other bodies do not allure and are
not excited by any amount of friction; thus emerald, agate, pearls, jasper,
porphyry, coral, touchstone, flint, blood-stone, do not acquire any power, nor
do metals, silver, gold, brass, iron nor any loadstone.

In the sequel, as more and more sensitive indicators were devised,
and a better understanding of electricity gained, the statements of the
last paragraph were disproved one by one.

To find whether gaseous bodies are attracted by electrics, Gilbert
brought rubbed amber near a candle-flame, a flame being an incandescent gaseous body. He observed no attraction. This observation
will call for further investigation. As it stands it classifies gases as non-
electrics.

ELECTRICITY

Half a century after the publication of Gilbert's *De Magnete*, two English scientists, Sir Thomas Browne (1646)[6] and Walter Charleton (1650)[7] in turn proposed the assumption that when an electric is in the condition to give the dust-test, there is on it a Something to which they assigned the name *electricity*, literally, "amber-stuff." This does not attempt to say what electricity is. Properties will have to be assigned to it in the light of experiments. Lord Kelvin once asked a class the question, "What is electricity?" and playfully dropped it on an inattentive student who tried to bluff through with the old reply, "I forget." "Ladies and gentlemen," remarked the professor, "a great disaster has just happened in our midst. The only man in the world who knew what electricity is, has just forgotten." It is not improbable that the word electricity was used for some time prior to 1646 in discussions about electrics.

The electricity on an electric or on a region of it is called a *charge*. The term is related to the word "cargo," and means a load or quantity. Similarly, a gun is said to be loaded with a charge of powder. From experiments with excited electrics, amber and others, it is plain that a charge of electricity can be neither seen nor heard; nor can we perceive it with any of our senses. If it were not for the behaviour of the dust-particles or other indicators we should be unaware of the existence of the charge on amber.

In spite of what was said previously about the comparative rarity of electric charges, nevertheless, it may be added paradoxically that as the story unfolds, it will appear that the whole physical universe is so replete with electrical charges and events that these are its main content. The composition and properties of all bodies and substances prove to be expressible in terms of electricity. Light is found to be electromagnetic. The whole science of optics may be viewed as a special department of electricity and the energy of the universe is for the most part electrical. The study of electricity truly leads to what Lucretius called "a new view of creation."

IRON-FILING INDICATOR

The originator of the well-known iron-filing method of exploring a magnetic field was Descartes, whose famous book, *Principles of Philosophy*, published in 1644, twelve years after it was written, is considered one of the clearest and most polished pieces of exposition in the whole

[6]Sir Thomas Browne, *Pseudodoxia Epidemica*, 1646.
[7]Walter Charleton, *Ternary Paradoxes*, 1650. Reference given in Boyle's *The Mechanical Origine of Electricity*, 1675 (Oxford, Old Ashmolean Reprints, VII).

literature of physics. On learning what the Inquisition had done to Galileo, Descartes, who had no hankering for a martyr's crown, decided to follow Horace's famous advice about putting his new book away in his desk for several years before publishing. Every now and again, however, during twelve years, he would bring out the manuscript and give it another polishing. Descartes was more philosopher and mathematician than physicist. In this respect he resembled Aristotle, whose mediaeval schools he criticized drastically. He performed comparatively few experiments and proceeded mostly by deduction from what

FIG. 263. Descartes' diagram of spherical magnet and field.

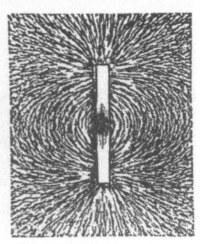

FIG. 264. Field of bar-magnet.

he considered fundamental principles. He entertained but scant regard for Galileo, who was a thoroughgoing experimenter and proceeded by induction to general principles. Furthermore, Galileo avoided many of Descartes' errors. One passage in Descartes' *Principia Philosophiae* refers to his iron-filing method:

If we examine carefully how iron filings arrange themselves around a magnet, we shall perceive from their behaviour, many facts that will confirm what I have already said. For in the first place, it can be observed that the filings are not massed together in a haphazard fashion but lying end to end, they form as it were certain filaments. . . . In order that the positions of these filaments may be rendered clearly visible to one's very eyes, let a little of the said iron filings be sprinkled on a flat surface that has an aperture into which a spherical magnet fits so that its poles, at opposite sides of the sphere, both touch the plane. . . . Then the filings will arrange themselves in lines which display to view the curved paths of the filaments around the magnet . . . which I described previously.

Fig. 263 is a copy of Descartes' original diagram. It is one of the last appearances of the spherical magnet in the literature. From now on it is replaced more and more by the bar magnet (fig. 264) and other shapes. One of the last to use a spherical magnet was Huygens (1680).

THE FLORENTINE EXPERIMENTS

Two experiments in electricity and magnetism performed by the Florentine Academy (1665) are of special interest. In the first, the conditions of Gilbert's versorium tests were reversed by pivoting the charged amber rod and bringing a neutral or uncharged body near the *charged versorium* which was thus prepared. Magalotti reports:

> It is commonly believed, That Amber attracts the little Bodies to it self; but the Action is indeed mutual, not more properly belonging to the Amber, than to the Bodies moved, by which it also it self is attracted; or rather, it applies it self to them: of this we made the Experiment; and found that the Amber being hung at liberty by a thread in the Air, or counterpois'd upon a Point like a Magnetical Needle, when it was rubb'd, . . . made a stoop to those little Bodies, which likewise proportionally presented themselves thereto, and readily obey'd its call.[8]

This discovery, in fact, could easily be predicted as a simple *deduction* from Newton's third law of motion; reaction equals action. Newton's *Principia*, however, was published in 1687, so the Florentines, experimenting in 1665, did not have that law at their disposal and could not reach the conclusion by deduction. Quite possibly, their result may have been part of the materials from which Newton *induced* the law. That an uncharged body attracts a charged one may be referred to as the Florentine theorem. An explanation of this phenomenon will be found in the third epoch of electricity of which this experiment was a harbinger. It will be found that what the Florentines thought to be an uncharged body was temporarily charged by a process called *influence*.

In the event that a charged body becomes neutral, it is said to lose its charge or to be *discharged*. The same experimenters found that when charged amber was dipped in mercury or in river water it lost charge, but not when dipped in oil. The significance of this observation was not realized even by its discoverers nor by anyone else until Stephen Gray found in 1729 that there are good conductors of electricity and slow or bad conductors:

> There are some Liquors wherewith the Amber being wetted, after rubbing, draws not; and there are others not producing this Effect: they that so act, are

[8] *Saggi d. Nat. Esper. fatte nell' Accad. d. Cim.*, p. 230.

generally Natural Waters, Distilled Waters, Wines, Vinegar, burning Waters, all Acids. . . . On the other hand, these are ineffectual:—Oyl of Flints, Sallet Oyl, Oyl of sweet Almonds, . . . Tallow, Fat; and Lastly, all Butter whether simple or perfumed with any Flowers, Ambergrice, or Muske. . . .

This experiment, then, foreshadows the second epoch in the story of electricity, characterized by the phenomenon of conduction. Any charge which is not being conducted along a body but is stationary on a charged body (fig. 266, p. 394) is called static (Lat. *sto*, stand). The discovery and study of *static charges* was characteristic of the first epoch of electricity, which began with Thales' amber experiment.

The Florentines tried the magnetic analogue of von Guericke's bell-jar experiment in acoustics, and Magalotti's report says:

> The needle is attracted by a magnet at the same distance in vacuo as in air.

VON GUERICKE'S ELECTRIC MACHINE

The fourth section of von Guericke's *De Vacuo Spatio* was devoted to electrical experiments. If his findings had been heeded by other scientists and followed up, the advance of the science could have been speeded half a century, for included were two experiments which later proved epoch-making. With his characteristic fondness for large-scale operations, von Guericke constructed the first *electric machine*. Instead of costly amber and pea-sized sapphires, he used a cheap electric of the size of a cabbage and thereby had at his disposal more copious charges than any of his predecessors (fig. 265).

> Secure one of the glass globes which are called phials, about the size of a youngster's head; fill it with sulphur, ground in a mortar and melted by application of a flame. After it refreezes, break the phial, take out the sulphur globe and keep it in a dry place, not a moist one. . . . Perforate it with a hole so that it can spin upon an iron axle . . . and thus the globe is prepared. . . . To demonstrate the power developed by this globe, place it with its axis on two supports . . . in the machine . . . a hand's breadth above the base-board, and spread under it various sorts of fragments, such as bits of leaves, gold-dust, silver filings, snips of paper, hairs, shavings etc. Apply a dry hand to the globe so that it is stroked or grazed two or three times or more. Now it attracts the fragments and as it turns on its axis, carries them around with it.

The ensuing "coda" should be taken *cum grano salis*.

> By this demonstration, we see displayed as it were, before our very eyes, the Terrestrial Globe which by attraction, holds all the animals and all other bodies that are at its surface and by its daily motion, carries them around with it once every twenty-four hours.

If by "attraction" Otto means electrical attraction, then it is to be feared that his imagination has overpowered his caution in this instance. Even his next observation should have been a warning to him against such an explanation of how we stick to the big centrifuge on which we live and have our being.

Electric repulsion was first reported in 1629 by the Italian Jesuit Nicolaus Cabaeus and again in 1672 by von Guericke in greater detail:

A repelling force is also to be plainly observed in this globe. For example, when it is taken in hand and rubbed or stroked with a dry palm, as described, it not only attracts but also, after different intervals of time, propels the same particles away from itself. Nor will it attract them again until some other body

Fig. 265. Von Guericke's sulphur-ball machine.

has touched them. This repulsion, however, is best observed with feathers which are soft and light and do not fall as quickly as other small bodies. When they are repelled upward, they can be held suspended in the electric field [*orbe virtutis*] of the globe for a fairly long time and in this manner may be carried about in the room with the globe wherever you wish.

If we are held to the earth by electric attraction, as von Guericke suggested, it is disquieting to think what would happen when the phase of repulsion set in, as with the sulphur sphere. From this account comes the classic pith-ball experiment which illustrates several important ideas in electricity (fig. 266). It is essentially Thales' dust-test, except that we use a single large particle, the pith-ball, as indicator, instead of many smaller ones such as a heap of cork particles. Any indicator used for examining a charge or its field is called an *electroscope*. Already we have met three: the particles used in the dust-test, Gilbert's versorium, and a charged versorium; now we add the pith-ball pendulum. In stage IV of this experiment, the ball is found (by the dust-test) to have gained a charge by contact with *E*, for it attracts

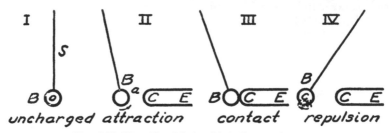

FIG. 266. Von Guericke's pith-ball experiment.

cork filings, whereas in stage I, it did not. Von Guericke discovered that *B* can be discharged by touching it and not until *B* is discharged can the experiment be repeated.

Von Guericke's next experiment demonstrated the important fact that a body can become charged by entering the field of a charge without touching the charged surface. The process is known as *influence* or *induction* (fig. 267).

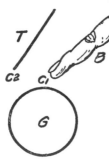

FIG. 267.
Influence charges.

If a linen thread, *T*, supported from above, is brought near the globe *G*, and you try to touch it with your finger or any other body *B*, the thread moves away, and it is difficult to bring the finger near the thread.

The laws of influence charges, discovered by Canton, Wilcke, and Aepinus in the years 1753 to 1758, afford an explanation of this observation just recorded by von Guericke, and also of his next observation:

The part of a feather ... which was attracted and afterwards repelled by the globe, always holds, with reference to it, the same position unaltered in the electric field, so that if the globe is brought above the feather, it reverses its position in mid-air and the same part of it always faces the globe. ... Similarly, the Moon always turns the same face to the Earth.

The study and application of influence charges is characteristic of the third epoch in the story of electricity, inaugurated by the work of Canton; but the germ of that epoch was present in these experiments of von Guericke. Coming events cast their shadows before them.

Von Guericke showed that an electric charge can flow along a thread.

A linen thread terminates in a sharp splint of wood and is firmly fastened to a table or stool and extends for a cubit or more so as to offer a space somewhat remote from the globe where a body can touch it from underneath. The globe is excited and the wooden point brought to it. . . . Then, it is plainly seen that the force has passed along the thread to its opposite end, for now it attracts.

In this experiment von Guericke anticipated the second epoch in electricity, which had to do with *conduction*. It is worth noting here that von Guericke terminated the thread in a *sharp point*.

Von Guericke discovered the method of *discharging* a charged body by touching it:

When a feather is in contact with the globe, and afterwards in the air, it puffs itself out and displays a sort of vivacity. Any nearby body it attracts or, if its force is not strong enough for that, it flies to the body. Thus it can fasten upon the projections of any bodies that approach it, for example, on somebody's nose. If a finger or any other body approaches a feather which has spread out its barbules, it flies to the finger and then back to the globe again and this performance is repeated time and again.

If one barbule is charged, in all probability its neighbour is also charged; and since the two charges were obtained in identical fashion, they may be described as *similar* or *like charges*. When the feather puffs out as von Guericke reports, like charges are seen to repel each other. The same thing was seen in stage IV of the pith-ball experiment. The charged feather flying to an uncharged wall or chair illustrates the Florentine theorem. Otto's prankishness is seen again in his phrase "on somebody's nose."

Von Guericke demonstrated discharge by proximity to flame:

If someone places a lighted candle on the table and brings the feather to within a hand's-breadth of the flame, the feather regularly darts back suddenly to the globe, and, as it were, seeks sanctuary there.

Thus the failure of amber to attract a flame in Gilbert's experiment was owing to the amber losing its charge. But a new question arises, "Why and how does flame discharge amber or any charged body?"

Von Guericke suggested that electrical energy may be transmuted into that of *sound* and *light:*

This globe can produce light . . . for if you take it into a darkened room and rub it with a dry hand, especially at night, it will give out light similar to that emitted by sugar when it is broken.

It can also produce sound; for when stroked by hand, . . . and brought near the ear, it is heard to emit a crackling sound.

Although von Guericke did not attempt to make the point, yet this is probably the first hint in the literature of the electrical origin of lightning and thunder. It must be added that the crackling sound may also be attributed at least partly to thermal causes: expansion of crystals due to warming and consequent separations. You can hear the sound by holding a piece of roll sulphur in your hand near your ear.

The closing words of von Guericke's chapter on electricity give us some good advice as they declare one of his own guiding principles:[9]

> Now, many other mysterious facts which are displayed by this globe, I shall pass by without mention. Nature often presents in very commonplace things, marvellous wonders, which are not discerned except by those who through insight and innate curiosity consult the oracle of experimentation.

BOYLE ON ELECTRICITY

The treatise on *The Mechanical Origine of Electricity*,[10] published at London in 1675 by Robert Boyle, was the first in any language to have the term *Electricity* in its title. It added some new items to the list of electrics and by the same token decreased the list of non-electrics. Gilbert, after showing that all solids and liquids are attracted by a charge on amber, attempted the demonstrations for gases but failed. Nor did Boyle succeed in this project, for smoke particles, which he used as indicators, are solid; his experiment, however, is an interesting modification of the dust-test.

> Having well lighted a Wax-taper which I preferr'd to a common Candle to avoid the stink of the snuff, I blew out the flame; and when the smoak ascended in a slender stream, held at a convenient distance from it . . . a chafed Diamond, which would manifestly make the ascending smoak deviate from its former line, and turn aside to beat, as it were, against the Electric, which, if it were vigorous, would act at a considerable distance, and seemed to smoak for a pretty while together. But as for flame, our Countryman, Gilbert, delivers as his Experiment, That an Electric though duly excited and applied, will not move the flame of the slenderèst Candle. . . .

It is amusing to note Boyle's reference to the trickiness of many demonstrations in static electricity:

> I think it will not be amiss . . . to give this Advertisement, That the event of Electrical Experiments is not always so certain as that of many others, being sometimes much varied by seemingly slight circumstances, and now and then, by some that are altogether overlook'd. . . . A late most learned Writer reciting the Electricks, reckon'd up by our industrious Countryman, Gilbert, and increasing their number by some observed by himself (to which I shall now add . . . white Saphyrs, and white English Amethysts . . .), denies Electricity to . . . the Cornelion and the Emrald. . . . I usually wear a Cornelian Ring, that is richly enough endowed with Electricity. . . . I proceeded to make trial with three or

---

[9]P. Lenard, *Great Men of Science*, trans. H. Hatfield (Bell, 1933), p. 54, gives a good biography of von Guericke.
[10]Boyle, *The Mechanical Origine of Electricity*.

four Emralds . . . and found them all somewhat . . . endow'd with Electricity, which I found to be yet more considerable in an Emrald of my own.

The list of electrics had now grown to include (for example) five-sixths of the gems in the high-priest's breast-plate—all but topaz and agate.

It was quite natural that Hooke and Boyle, the first men to perform the classic "Guinea and Feather" experiment, should also try to find whether electric action can occur in a vacuum:

> To try, whether Amber would draw a light body in a Glass whence the air had been pumpt out. . . . Having a vigorous piece of Amber which I had caused to be purposely turn'd and polish'd for Electrical Experiments, I found . . . that it would retain a manifest power of attracting for several minutes. . . . We suspended it, being first well chafed, in a Glass Receiver . . . just over a light Body; and making haste with our Air-Pump, . . . when the Air was withdrawn, we did by a Contrivance, let down the suspended Amber till it came very near the Straw, . . . and perceived . . . that in some Trials, upon the least Contact, it would lift it up: and in others, . . . the Amber would raise it without [first] touching it.

This result is analogous to the magnetic one obtained by the Florentines, to whose report Boyle had access, since it was presented to the Royal Society of London in which Boyle was a charter member. The technique of this experiment, as suggested by the nature of the "Contrivance," required considerable dexterity and ingenuity, and the plural pronoun in the expression, "We did by a Contrivance," suggests that Boyle's clever assistant, Robert Hooke, was called upon to operate, though, as usual, he received but scant mention.

Boyle outlined a number of hypotheses which had been advanced to explain electrical attraction (and repulsion). They were all corpuscular, assuming streams of particles or "effluvia":

> There are differing Hypotheses (and all of them Mechanical) propos'd by the Moderns to solve the Phaenomena of Electrical Attraction. Of these Opinions the First is that of the learned Jesuite, Cabaeus. . . . Another Hypothesis is that proposed by that Ingenious Gentleman, Sir Kenelm Digby, and embrac'd by the very Learned Dr. Browne (who seems to make our Gilbert himself to have been of it). . . . And according to this Hypothesis, the Amber or other Electrick, being chaf'd or heated, is made to emit certain Rayes or Files of unctuous Steams which, when they come to be a little cool'd by the external air, are somewhat condens'd, and having lost their former agitation, shrink back to the body whence they sallied out, and carry with them those light bodies, that their further ends happen to adhere to.

A person may wonder how this assumption will account for electrical repulsion as in stage IV of the pith-ball experiment.

> A third Hypothesis there is which was devis'd by the Acute Cartesius, who dislikes the Explications of others, chiefly because he thinks them not applicable

to Glass, which he supposes unfit to send forth Effluvia and which is yet an Electrical body.

Boyle also discussed *electrostatic shielding:*

Whereas the Magnetical Steams are so subtile, that they penetrate and perform their Operation through all kinds of Mediums hitherto known to us: Electrical Steams are . . . easily check'd in their progress, since 'tis affirm'd by Learned Writers, who say they speak upon particular Trial, that the interposition of the finest Linnen or Sarsnet is sufficient to hinder all the Operations of excited Amber upon a Straw or Feather plac'd never so little beyond it.

EXPERIMENTS WITH GLASS

The substance glass figured prominently in a number of Newton's well-known experiments, in his famous prism experiment, his reflecting telescope, in the Newton Rings experiment, and now in two experiments in electricity. The *Philosophic Transactions* of the Royal Society of London for 1676 describe an amusing little experiment performed by Newton. It is a modification of von Guericke's pith-ball experiment.

Having laid upon a table a round piece of glass about two inches broad, in a brass ring, so the glass might be about one-third of an inch from the table, and the air between them inclosed on all sides . . . then rubbing the glass briskly, till some little fragments of paper, laid on the table under the glass, began to be attracted and move nimbly to and fro; after he had done rubbing the glass, the papers would continue a pretty while in various motions; sometimes leaping up to the glass, and resting there a-while; then leaping down and again resting there. . . . Sometimes they skip in a bow from one part of the glass to another: . . . and sometimes hang by one corner, every paper with divers motion. . . .

December 16. Mr. Newton's experiment was tried but succeeded not. Ordered That the Secretary should write again to the said Mr. Newton.

Newton replied:

Cambridge, Dec. 21, 1675.

Sir,

Upon your letter etc. . . . I am apt to suspect the failure was in the manner of rubbing. . . . At one time, I rubbed the aforesaid . . . glass with a napkin, twice as much as I used to do with my gown, and nothing would stir.

ISAAC NEWTON.

Then in January, Newton came down to a meeting of the Society and showed the boys how the trick was done. To Newton the subject of electricity seemed a "pleasant divertissement," not one that merited the serious attention of a real hard-headed physicist and mathematician. Yet he made one valuable contribution to the improvement of the electric machine. He found that von Guericke did not need to use sulphur for an electric nor to chop up a perfectly good globe; for the glass globe itself would serve:

A Globe of Glass about 8 or 10 Inches in diameter, being put into a Frame
where it may be swiftly turn'd round its Axis, will, in turning, shine where it
rubs against the palm of ones Hand apply'd to it: And if at the same time, a
piece of white Paper or white Cloth, at the end of ones Finger be held at the
distance of about a quarter of an Inch, . . . from that part of the Glass where
it is most in motion, the electrick Vapour which is excited by the friction of the
Glass against the Hand will, by dashing against the white Paper, Cloth or
Finger, be put into such an agitation as to emit Light, and make the white
Paper, Cloth or Finger appear lucid like a Glow-worm.[11]

Modern electric machines such as the Wimshurst (fig. 268) frequently have glass discs but their action is by influence or induction.

Fig. 268. Wimshurst influence machine.

"THE TUBE"

In 1706, during the presidency of Sir Isaac Newton, the curator of
experiments in the Royal Society of London was Francis Hauksbee,
who succeeded Robert Hooke when the latter assumed the office of
secretary. Hauksbee merits our attention as a skilful and indefatigable
experimenter, a keen observer and an improver of machines; and his
quaint writings are full of enthusiasm and delight in his work. Though
his researches receive little mention nowadays, yet his contemporaries
and immediate successors speak highly of him and his labours were a
source of assistance and inspiration to such men as Gray, Dufay, and
Canton. Some of Hauksbee's fascinating experiments with pale, cada-
verous, shimmering lights, produced in darkened rooms and duly reported to the Royal Society,[12] if they had been performed a century
earlier would probably have brought him the grisly summons of the
witch-finder. The spirit of the times had improved so as to allow a

[11]Newton, *Opticks* (1704).
[12]*Phil. Trans. Roy. Soc. Lond.* (1706-1717).

scientist greater freedom in conducting his investigations. One of the strongest factors in overcoming superstition was the Royal Society, which encouraged Hauksbee's work and protected him under its aegis.

In 1675, Jean Picard of Paris had discovered a strange luminescence in the Torricellian vacuum of a barometer when he carried it through a room in the dark! John Bernoulli gave to this intriguing phenomenon the name "mercurial phosphorus" (Gk. *phos*, light; *phero*, bear). In some of Hauksbee's beautiful experiments on this topic his observations made him pioneer in the notion that *electric sparks* and *lightning* are similar in kind though different in degree. He exhausted the globe of Newton's electric machine and made many new observations, some of which anticipated the present epoch in electricity with its electron theory.

I Took a Glass Globe of about 9 inches diameter, and exhausted the Air out of it. . . . I fix'd it to a Machine, which gave it swift Motion . . . and then applying my naked Hand . . . to the surface of it . . . a considerable Light was produc'd [in it].

The significance of this observation will be more obvious when we come to study electric discharges in gases at low pressure. Suffice it to say that what we have just seen is the ancestor of all vacuum tubes.

Continuing his experiment Hauksbee made a number of new observations some of which were vivid advancements on von Guericke's discovery of charges produced by influence.

At the same time, the Room . . . became sensibly enlightned. . . . Now after . . . the Cock was turn'd, which gave liberty to the Air to enter the Globe, . . . while my Hand continued upon the Glass (the Glass being in Motion), if any Person approach'd his Fingers towards it . . . a Light would be seen to stick to 'em, at the distance of an inch or thereabouts, without their touching the Glass at all. And my Neckcloth, at the same time . . . appear'd of a fiery Colour, without any Communication of Light from the Globe.[13]

Hauksbee improved the received methods of electrifying glass by friction. This step made the subsequent work of Stephen Gray (1725) possible, for the latter's feeble finance could not support such luxuries as air-pumps and glass globes. The new apparatus was essentially an improved form of Thales' amber experiment.

Having procured a Tube . . . of fine Flint Glass, about one inch diameter, and thirty in length, I rubb'd it pretty vigorously with Paper in my Hand till it acquir'd some degree of heat. I then held it towards some pieces of Leaf-Brass which were no sooner within the Sphere of Activity of the Effluvia emitted by the Tube, but they began to be put into brisk motions, and yielded the following surprizing appearances. . . . They would leap towards the Tube at a very con-

[13]F. Hauksbee, *Physico-Mechanical Experiments* (Brugis, London, 1709). The volume was kindly loaned by the University of Pennsylvania from the John Fraser Memorial Library.

siderable distance from it . . . sometimes the distance of 12 or more inches. . . .
Sometimes, they would be thrown off from it. . . . And (which still adds to the
Wonderfulness of the phaenomenon) they would often repeat this alternate
rising and falling.

The tube described here was subsequently used by Gray, Dufay,
Canton, and others, and was often referred to as "the Tube." The term
Effluvia shows that Hauksbee's theory of electricity was corpuscular.

We are now to consider what occurr'd upon the like Attrition given it in
the Dark . . . when the Tube was *exhausted* of its Air. . . . The *attractive power
was very little discernible* . . . but a *much greater Light indeed did ensue*. . . . The
Light produc'd . . . appear'd to be wholly within it: whereas that which was
discover'd when the Tube was full of Air, seem'd to be altogether on its outside."

Fig. 269. Hauksbee's experiment with threads as indicators.

Here then was a second model of vacuum tube. In the original text, the
long "s" was used; it is interesting to note that it came to represent
the operation of integration in calculus, ∫ suggesting summation.
Hauksbee made considerable use of threads as indicators (fig. 269):

I contriv'd a Semi-circle of Wire, making it encompass the Glass. This Wire
had several pieces of Woollen Thread fasten'd to it, so as to hang down at pretty
nearly equal distances. When I came to apply my Hand to the Glass (so swiftly
whirl'd about), the Threads all harmoniously pointed to the center of the Circle,
in whose Plane the wire was plac'd. If the Wire with its loose Threads was
revers'd, so as to encompass the lower part of the Cylinder, the Threads[14] were
all erected into so many strait lines, still directing themselves to a center in the
Axis of the Glass.

This experiment foreshadows Faraday's term "lines of force." Hauksbee's work kept the study of electricity alive at a time when it had
reached a dull pause of stagnation such as sometimes occurs in a science
just prior to the advent of a new epoch in its progress.

[14]The author has described an inexpensive thread electroscope in "A Home-made
Electroscope," *The School* (June, 1937), p. 880.

CHAPTER NINETEEN

Conduction

A NEW chapter or epoch in the history of electricity was opened by the researches of Stephen Gray, who was a pensioner of Charterhouse Inn, the famous London poorhouse (and school). There have been few other scientists whose apparatus consisted so largely of common and cheap household articles—possibly the chemist Scheele and the entomologist Fabre worked under equivalent limitations. In spite of Gray's unusual restrictions he rendered yeoman service and his work must have been well esteemed by the Royal Society or he would not have been elected a Fellow. He made the far-reaching discovery that electricity can move or flow along some bodies which are therefore called *conductors* of electricity. The process is called *conduction* and it is characteristic of the second epoch in electricity. Gray and his successors found, for instance, that electricity can flow readily along a brass wire but not nearly as readily along a dry silk thread. If the silk thread is moistened with river water, it becomes a good conductor. Gray's results are given in the *Transactions* of the Royal Society for three or four years around 1729.[1]

I made several Attempts on the Metals to see whether they might not be made attractive by the same method as other Bodies . . . but without Success: I then resolved to procure me a large Flint-Glass Tube [*T*, fig. 270]. To each End I fitted a Cork C, to keep out the Dust. . . . upon holding a Down-Feather [*F*] over against the upper End of the Tube, I found that it would go to the Cork . . . at which I was much surprized, and concluded that there was certainly an attractive Vertue communicated to the Cork by the excited Tube. . . . Having by me an Ivory Ball, . . . this, I fixed upon a Fir-Stick [a stick of pinewood] about four Inches long [fig. 271] thrusting the other End in the Cork, and upon rubbing the Tube, found that the Ball attracted and repelled the Feather. . . . Then I made use of . . . Brass Wire to fix the Ball on [fig. 272] and found that the Attraction was the same . . . and that when the Feather was held over against any part of the Wire [*WR*], it was attracted by it. [Ther-

---

[1] The library of the School of Medicine and Dentistry of the University of Rochester has copies of these volumes.

fore, metals can be charged.] Upon suspending the Ball on the Tube by a Pack-thread [parcel-string, P] about three Feet long [fig. 273], it attracted . . . the Leaf-Brass [L], as did also a Ball of Lead. After I had found that the Metals were thus Electrical, . . . I next proceeded to try at what greater Distances the *Electrick Vertue Might be carried*. I made a Loop at each End of a Line [LN, fig. 274] and hanging it on a Nail [N] drove into a Beam [M], the other End hanging downwards; through the Loop at this End the Line [PHB] with the Ivory Ball was put; The other end of this Line was by a Loop hung on the Tube [T]. Then the Leaf-Brass [F], being laid under the Ball, the Tube rubbed, . . . not the least sign of Attraction was perceived. Upon this, I concluded that when the Electrick Vertue came to the Loop [L], . . . it went up . . . to the Beam . . . ;

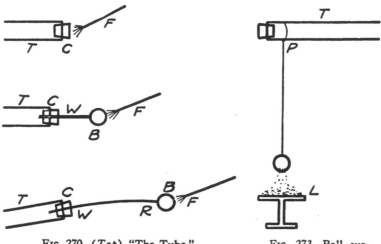

FIG. 270. (*Top*) "The Tube."
FIG. 271. (*Middle*) Tube with ivory ball.
FIG. 272. (*Bottom*) Ball supported by wire.

FIG. 273. Ball supported by moist parcel-string.

so that none or very little of it . . . came down to the Ball. . . . Mr. Wheeler then proposed a Silk Line to support the Line [of packthread] . . . The Line . . . being eighty Feet . . . in Length, was laid on Cross Silk Lines; when the Tube, being rubbed, the Ball attracted the Leaf-Brass. . . . The whole length of the Line being now 293 Feet, our Silk Lines broke. . . . We then took Brass Wire . . . but though the Tube was well rubbed, yet there was not the least . . . Attraction given by the Ball. By this we were now convinced that the Success we had before, depended upon the Lines that supported the Line of Communication [PH], being Silk [SK, fig. 274], the same Effect happening here [with Brass] as when the Line is supported by Pack-thread, viz., that when the Effluvia come to the Wire or Packthread . . . they pass through them to the Timber.[2]

Thus Gray discovered and demonstrated the conduction of electric charge along a brass wire. Once the distinction was made between good and bad conductors, the study of electricity took on new life. New light was shed on previous results, errors were corrected and advance-

[2]*Phil. Trans. Roy. Soc. Lond.* (1731), XXXVII, 18.

ments followed with accelerated pace. Some of Gray's contemporaries were slow to appreciate the consequences of his discovery and even Gray himself seemed not to realize its full significance. He seemed to be more intent on finding over how long a supported line he could "communicate the Electrick Vertue" than on its regrettable loss over a brass wire or a moist string. Reading between the lines, one can see that Mr. Wheeler, who was also an F.R.S., took an interest in the old pensioner and his ingenious poorhouse experiments. He invited Gray to come and spend a holiday on his estate. There Gray could have all

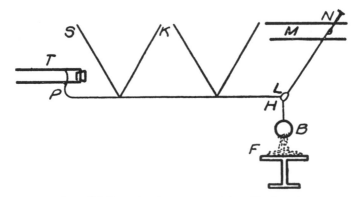

FIG. 274. Parcel-string supported by silk threads.

the Silk and Brass wire and other equipment he wished, and room to carry out experiments impossible in his own narrow quarters, for example, one with a conductor 293 feet long.

Some of Gray's experiments on what he called "brush discharge" made some advance toward an understanding of St. Elmo's Fire, and even the accompanying sounds that Pliny described as being "like the flitting of birds":

We caused to be made an Iron Rod, 4 Foot long, and about half an Inch Diameter, pointed at each End, . . . this being suspended on lines . . . At Night, . . . applying one End of the Tube to one End of the Rod, not only . . . that End had a Light upon it, but there proceeded a Light at the same Time from the other, extending in Form of a Cone, whose vertex was at the End of the Rod; and we could plainly see that it consisted of Threads, or Rays of Light, diverging from the Point of the Rod. . . . This Light is attended with a small hissing Noise . . . not to be heard without good Attention.

A Brass Plate, four Feet square, . . . was placed upon a Stand so that it stood perpendicular, the Stand being set on the Cylindrick Glass: then the Rod with its Stand and Glass was set so as that one Point of it was about an Inch from the Centre of the Plate; then the Tube being applied to the other End of the Rod, and . . . after striking it gently with my Finger on the back Side, a

Light appeared upon the Plate, and at the same Time, the Brush of Light came out from the Point of the Rod.

When Thomas Sutton endowed Charterhouse in 1611, he made provision for a hospital or almshouse to accommodate 80 pensioners. These were to be men of good education who had fallen on hard times. Having given them food and lodging, he then furnished them with occupation by making provision for 40 bright boys who could not otherwise afford education. The pensioners were their teachers. What a remarkable school, with a staff numbering twice the pupil enrolment!

Gray's boys were only too glad to assist him in his experiments. In one of these he insulated a boy by placing him on a board suspended from the ceiling by cords. He then charged the boy's body by contact with "the Tube" and now other boys drew sparks from their electrified accomplice by bringing their fingers near his body.

ALL BODIES ELECTRICS

One of Gray's warmest admirers was a vivacious and charming Frenchman, Charles Dufay, Director of the Royal Botanic Gardens of Paris. In 1733, Dufay sent to the Duke of Richmond a letter which is of considerable moment in the history of electricity.

My Lord,

I Flatter my self your Grace will not be displeased with an Account of some . . . Discoveries I have made in the Electricity of Bodies. . . . Having read in one of Mr. Gray's Letters that Water may be made Electrical by holding the excited Glass Tube near it, a Dish of Water being first fix'd to a Stand, and that set on a Plate of Glass, . . . I have found upon Tria! that the same thing happens to all Bodies without Exception, whether solid or fluid.[3]

This paragraph culminated twenty-three centuries of research. Thales noted that amber can give the dust-test; Theophrastus named a second such substance; Pliny described five; Gilbert found that there are many items in the "amber-list" and called them electrics. Boyle and others added to the list, and finally Dufay's work showed that all bodies can become electrified and are therefore, electrics. Hence Gilbert's term "non-electric" became meaningless and obsolete. The term "electric," on the other hand, became superfluous and thenceforth (after 1800) rarely used. Any quality that is common to all bodies is known as a *property of matter*: it was Dufay who showed that it is a property of matter to become electrified, and, in so doing, he foreshadowed the modern electron theory proposed in 1895 by Sir J. J. Thomson and a number of collaborators.

[3]*Phil. Trans. Roy. Soc. Lond.* (1733), xxxviii, 258.

INSULATORS

Dufay was an able experimenter and a prodigious worker, who made numerous contributions to electricity. One passage in his writings is the source of the term "insulator" and gives an early use of the term "charged" (fig. 275).

I was at Tremblay . . . with M. l'Abbé Nollet, who has been of infinite assistance to me in all these experiments and who even devised several that appear in this memoir. I took two pieces of cord about as thick as one's finger [*SA* and *CB*], resting each on two silk threads [*DE*, etc.] so that they could be placed at any desired distance from each other. Presenting the excited tube [*T*] to *S* when *AC* was one inch, the electricity was as perceptible on the Ball [*B*] as if the ropes had been in contact; also when *AC* was 3 inches; at 6 inches, a little less, and at 1 foot, about the same as after traversing 1256 feet of continuous cord. . . . This experiment shows how necessary it is that the cord one

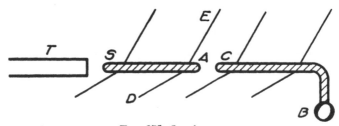

Fig. 275. Insulators.

uses for transmitting electricity to a distance should be isolated, i.e. supported only by bodies as little capable as possible of becoming themselves charged with electricity [i.e. supported by non-conductors or insulators].[4]

Dufay was one of those who helped to terminate the mediaeval custom of writing scientific articles in Latin. The last sentence of the passage above reads in the original:

Cette expérience prouve combien il est nécessaire que la corde, dont on se sert pour transmettre au loin l'électricité, soit isolée, ou ne soit soûtenue que de corps les moins propres qu'il est possible à se charger eux-mêmes de l'électricité.

Any conductor that is surrounded by insulators is called an *insulated conductor*. We shall see many forms of insulated conductors in the sequel. To understand an electric device, a good first step is to note which parts are conductors and which insulators.

Dufay's assistant, Abbé Nollet, did much to popularize the science of electricity: all fashionable Paris flocked to his brilliant and witty lectures. Fig. 276 shows him repeating one of Gray's experiments on electrifying a living human body insulated by silk cords. The lady

[4]*Mémoires de l'Académie Royale de Paris* (1733).

N. le Sueur Invenit   R. Brunet fecit

Fig. 276. Abbé Nollet electrifies a human body.

brings her finger-tip near the gentleman's nose and sparks occur. It is impossible to say which of them does the sparking.

The most famous passage in Dufay's writings is that in which he names *two kinds of electricity:*

Chance has thrown in my way another Principle . . . which casts . . . new Light on the Subject of Electricity. This Principle is that there are two distinct Electricities, very different from one another; one of which I call *vitreous* Electricity, and the other, *resinous.* . . . The first is that of Glass [Fr. *vitre*], Rock-Crystal, Precious Stones, Hair of Animals, Wool and many other Bodies [substances]: The second is that of Amber, Copal, Gum-Lack, Silk, . . . Paper and a vast Number of other Substances. The Characteristick of these two Electricities is, that a Body of the vitreous Electricity, for example, repels all such as are of the same Electricity; and on the contrary, attracts all those of the resinous Electricity; so that the Tube, made electrical, will repel Glass, Crystal, Hair of Animals &c; when render'd electrick, and will attract Silk, . . . Paper &c, though render'd electrical likewise. Amber on the contrary, will attract Glass, and other Substances of the same Class, and will repel Gum-Lac, Copal, Silk . . . &c. Two Silk Ribbons, rendered electrical, will repel each other; two Woollen Threads will do the like; but a Woollen Thread and a Silk Thread will mutually attract one another. This Principle very naturally explains, why the Ends of Threads, of Silk, or Wool, recede from one another in Form of a Pencil or Broom, when they have acquired an electrick Quality. From this Principle, one may with the same Ease deduce the Explanation of a great Number of other Phaenomena. And 'tis probable, that this Truth will lead us to the further Discovery of many other things.[5]

It will be instructive to see how scientists tested and criticized Dufay's statements in the light of new facts, discarding unsatisfactory parts and remodelling the retained parts so as to agree with all the facts both old and new. Dufay himself, of course, assisted in the remodelling. His laws of charges are stated thus: *Like electric charges repel each other and unlike or opposite charges attract each other.* These are reminiscent of Peregrinus' laws of magnet poles and suggested to more than one scientist that there may be a relation between magnetism and electricity.

The secretary of the Royal Society in Gray's time was Rev. J. Desaguliers, tutor of the Prince of Wales. Although he looked down his nose at Gray and avoided him because of the old pensioner's sharp tongue, yet his only important service to electricity was to emphasize Gray's discovery of conduction and insulators. In rendering this service, Desaguliers coined the term conductor:

In the following Account, I call Conductors those Strings to one End of which, the rubb'd Tube is applied; and Supporters (insulators), such Bodies as the Conductor rests upon.[6]

⁵ *Phil. Trans. Roy. Soc. Lond.* (1734), xxxviii.
⁶ *Phil. Trans. Roy. Soc. Lond.* (1739), xli, i, 193.

This sentence is of value for introducing a term, but as a definition of conductor and/or insulator, it is entirely inadequate. Desaguliers was fond of a meretricious theorem which he devised, "electrics are non-conductors and non-electrics are conductors." The whole statement collapses when we find that there are no non-electrics and, in the literal sense, no non-conductors. Yet the idea intrigued a considerable number of electricians and we shall find the term electric used to mean bad conductor or insulator as late as 1800 by Volta. Thus Desaguliers' "rule" was more a deterrent than an accelerator of progress.

THE LEYDEN JAR

During the first half of the eighteenth century, several German physicists, by improving the electric machine, obtained more powerful charges than had ever before been available for study. In 1745, Georg von Kleist, dean of the cathedral at Kamin, Pomerania, placed a nail in a tumbler containing water and charged the water by conduction from a powerful electric machine (fig. 277). The astonishing results obtained from this rather unpretentious apparatus he communicated to the Berlin Academy in a letter:

When a nail or a piece of thick brass wire is put into a small apothecary's phial and electrified, remarkable effects follow: but the phial must be very dry and warm: . . . . As soon as this phial and nail are removed from the electrifying glass, . . . if . . . I put my finger . . . to the nail, I receive a shock which stuns my arms up to the shoulders. If a little mercury . . . be put into [the phial] the experiment succeeds all the better [!] . . .

In the following year, the professor of physics at Leyden, Peter Musschenbroek, and his assistant Cunaeus, made the same discovery and because of the former's writings and renown, the new apparatus came to be known as the Leyden Phial or Jar, although in point of priority it should have been called a Pomeranian. Musschenbroek described the experiment in a letter to his friend Réaumur:

I wish to describe to you a new but dangerous experiment which I advise you not to attempt yourself. I was engaged in some researches on the power of electricity, and for that purpose, I had suspended, by two blue silk lines, a gun-barrel, which received the electricity of a glass globe rapidly turned on its axis and rubbed by applying the hand to it. At the end of the gun-barrel, away from the globe, there hung a brass wire, the end of which dipped into a round glass bottle, partly filled with water. I was holding the bottle with one hand, and with the other, I was trying to draw sparks from the gun-barrel, when, suddenly, the hand holding the bottle received such a violent shock that my whole frame was shaken as if by a lightning stroke. I thought that all was over with me, for my arms and my whole body were affected in a dreadful way, which I cannot describe. I would not take another such shock for the whole Kingdom of France.[7]

[7]Quoted in Ivor Hart, *Great Physicists*, p. 97.

Peter's expression, "as if by a lightning stroke," may have been meaningful to people who had been struck by lightning, but to Réaumur and himself it must have been largely imaginary. The attitude of the English investigator Cavendish (1785) was quite different for he frequently took the shock several times in a single experiment without wincing or crying aloud.

Few electrical inventions have made more stir in the world in their early days than the Leyden jar. Two Englishmen, Watson and Bevis, placed a sheet of tinfoil inside and another outside the jar, and others

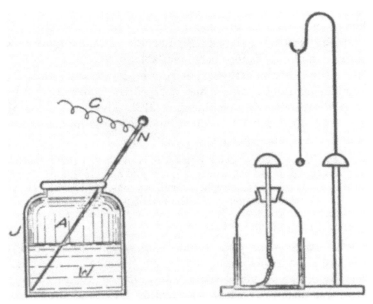

FIG. 277. Leyden jar.   FIG. 278. Electric chime.

introduced further improvements. Experiments with the Leyden jar became the rage. Men travelled Europe giving demonstrations, and shocking their audiences—for a consideration. Groups of people would join hands as for *Auld Lang Syne* and enjoy the shock together. The circle thus formed for the flow of electricity was called the circle or the *circuit*. A scientist named Gordon invented the first electric bell or chime by modifying the Leyden jar to the form shown in fig. 278, which has been dubbed by some students the "Leyden Jar with Bells On." The Leyden jar was much in evidence in the experiments of Benjamin Franklin, a consideration of whose work takes us, for the first time in the story of electricity, to America.

THE LIGHTNING ROD

Dr. William Wall, an English divine, first put on record, in 1708, the surmise that lightning and thunder are of electrical origin.[8]

> I found by gently rubbing a well polished piece of amber with my hand in the dark, that it produced a light: Whereupon I got a pretty large piece of amber, which I caused to be made long and taper. . . . upon drawing the amber . . . swiftly through the woollen cloth, . . . a prodigious number of little cracklings was heard, and every one of these produced a little flash of light. . . . By holding one's fingers at a little distance from the amber, a large crackling is produced, with a great flash of light succeeding it. And what to me is very surprising, upon its eruption, it strikes the finger very sensibly, wheresoever applied, with a push or a puff like wind. The crackling is full as loud as charcoal on a fire. . . . Now I make no question but upon using . . . a larger piece of amber both the cracklings and light would be much greater. . . . This light and crackling seems, in some degree, to represent thunder and lightning.

Stephen Gray showed that discharge from an insulated rod occurs most readily at its point, and the significance of sharp points in electrical phenomena has been referred to in the work of Pliny, von Guericke, Wall, and others. Benjamin Franklin, however, with his penchant for turning knowledge to practical account, pursued the idea until he achieved the invention of the lightning-rod, which has saved thousands of lives and property worth millions of dollars:

> Place an iron shot of three or four inches diameter on the mouth of a clean dry glass bottle [insulated conductor]. By a fine silken thread from the cieling,[9] right over the mouth of the bottle, suspend a small cork-ball, about the bigness of a marble: the thread of such a length as that the cork-ball may rest against the side of the shot. Electrify the shot and the ball will be repelled [stage IV, fig. 266] more or less according to the quantity of electricity. If you present to the shot, the point of a long slender sharp bodkin, at six or eight inches distance, the repellency is instantly destroyed, and the cork flies to the shot [discharge]. A blunt body must be brought within an inch, and draws a spark to produce the same effect. To prove that the electrical fire is drawn off by the point, if you take the blade of the bodkin out of the wooden handle, and fix it in a stick of sealing-wax, and then present it at the distance aforesaid, or if you bring it very near, no such effect follows: but slide one finger along the wax till you touch the blade, and the ball flies to the shot.

Few experiments are better known to people in general than the famous one in which Franklin, using a kite, showed that lightning is of electrical origin (1752). His report is in the form of directions to his friend Peter Collinson for performing the experiment. Collinson gave Franklin his initiation into this science by sending him a present of some apparatus, and Franklin was exceedingly grateful to him for all the fun he derived from his study of electricity.

[8] J. Priestley, *History of Electricity*, p. 10.
[9] This old spelling shows the derivation (Fr. *ciel*, sky) better than the modern form.

FIG. 279. Franklin's kite experiment.

Make a small cross of two light strips of cedar, the arms so long as to reach to the four corners of a large thin silk handerchief when extended; tie the corner of the handkerchief to the extremities of the cross, so you have the body of a kite; which being properly accommodated with a tail, loop and string, will rise in the air like those made of paper; but being of silk, is fitter to bear the wet and wind of a thunder-gust without tearing. To the top of the upright stick of the cross, is to be fixed a very sharp-pointed wire, rising a foot or more above the wood. To the end of the twine, next the hand, is to be tied a silk ribbon, and where the silk and twine join, a key must be fastened. This kite is to be raised when a thunder-gust appears to be coming on, and the person who holds the string must stand within a door or window, or under some cover, so that the silk ribbon may not be wet; and care must be taken that the twine does not touch the frame of the door or window. As soon as any of the thunder-clouds come over the kite, the pointed wire will draw the electric fire from them, and the kite, with all the twine, will be electrified, and the loose filaments of the twine will stand out every way, and be attracted by an approaching finger. And when the rain has wetted the kite and twine, so that it can conduct the electric fire freely, you will find it stream out plentifully from the key on the approach of your knuckle [!]. At this key, the [Leyden] phial may be charged; and from the electric fire thus obtained, spirits may be kindled, and all the other electric experiments be performed, which are usually done by the help of a rubbed glass globe or tube, and thereby the sameness of the electric matter with that of lightning completely demonstrated.

There is little in this account to intimate that Franklin knew perfectly well when he brought his hand near the kite-string that he stood face to face with death; but the thirst for knowledge has often braved that spectre.

If the fire of electricity and that of lightning be the same, as I have endeavoured to show, . . . may not the knowledge of the power of points be of use . . . in preventing houses, churches, ships &c. from the stroke of lightning, by directing us to fix on the highest parts of those edifices, upright rods of iron, made sharp as a needle, and gilt to prevent rusting, and from the foot of those rods, a wire down the outside of the building into the ground, or down round one of the shrouds of a ship, and down her side till it reaches the water? Would not those pointed rods probably draw the electrical fire silently out of a cloud before it came nigh enough to strike, and thereby secure us from the most sudden and terrible mischief?

Instead of Dufay's "two-fluid" theory of electricity, Franklin preferred a "one-fluid" theory:

A person standing on wax and rubbing the tube, and another person on wax drawing the fire, they will both of them (provided they do not . . . touch one another) appear to be electrified, to a person standing on the floor; that is, he will perceive a spark on approaching each of them with his knuckle. . . . If they touch one another after exciting the tube, . . . there will be a stronger spark between them. We suppose that . . . electrical fire is a common element, of which every one of the three persons . . . has his equal share . . . A, who stands on the wax and rubs the tube, collects the fire . . . from himself into the glass . . . B, who stands on wax likewise, passing his knuckle along near the tube, receives the fire which was collected by the glass from A . . . C, standing on the floor, . . . having only the middle quantity, . . . receives a spark . . . from B who has an over quantity; but gives one to A, who has an under quantity. . . . Hence have arisen some new terms among us; we say B . . . is electrized positively: A, negatively. Or rather, B is electrized plus: A, minus.

Franklin's greatness in practice needs no further token than his invention of the lightning-rod, the first application of electrical principles which was of economical importance. His greatness in theory, however, had a genius that lifts his name to a higher order of brilliance. No better accolade could be his than the fact that he introduced the terms *positive* and *negative* which came into universal use immediately and permanently. When we come to the modern epoch, we shall have further occasion to marvel at the signally prophetic insight of this man.

To suggest the violence of the shock produced by a group of Leyden jars combined, Franklin borrowed an artilleryman's name for a group of guns, a *battery*. In connecting the jars, he used two methods which foreshadow the modern terms "series arrangement" and "parallel" or "multiple:"

Upon this, we made what we called an electrical battery, consisting of eleven panes of large sash-glass, armed with thin leaden plates, pasted on each side, placed vertically and supported at two inches distance on silk cords, with thick hooks of leaden wire, one from each side, standing upright . . . and convenient communications of wire and chain, from the giving side of one pane, to the receiving side of the other [series] . . . . Another contrivance is to bring the giving sides, after charging, in contact with one long wire, and the receivers with another [multiple].

The instrument here employed by Franklin and his ingenious collaborator, Kinnersley, a glass plate between two metal sheets (an insulator between two conductors), is a simple form of a device which later came to have great importance, both practical and theoretical. It was christened *condenser* by Volta in 1774. The two sheets of tinfoil of a Leyden jar, with glass between them, constitute a condenser.

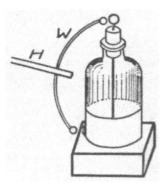

FIG. 280.
"Discharging tongs."

Franklin devised a simple form of *insulated conductor*, called *"discharging tongs,"* which proved very convenient and important in subsequent work, for instance in the researches of Galvani and Volta:

Place an electrized phial on wax [fig. 280]: take a wire [W] in form of a C, the ends at such a distance when bent, as that the upper may touch the wire of the bottle, when the lower touches the bottom: stick the outer part on a stick of sealing-wax [H], which will serve as a handle; then apply the lower end to the bottom of the bottle and gradually bring the upper end near the wire in the cork. The consequence is, spark follows spark till the equilibrium is restored . . . the crooked wire forming the communication.

Franklin demonstrated the production of *heat from electricity* and introduced the term *circuit* (1752):

Lightning melts metals. . . . We have also melted gold, silver, and copper in small quantities by the electric flash. . . . Take leaf-gold . . . cut off narrow strips. . . . Place one of these strips between two strips of glass. . . . Bind the pieces of the glass together; then place it so as to be part of an electric circuit (the ends of gold hanging out, being of use to join with the other parts of the circuit) and send the flash through it from a large jar. . . . The gold was melted.

This experiment is symbolic of a group of modern conveniences, electric toasters, curling-tongs, flat-irons, bed-warmers, etc., whose name is legion.

EPOCHS IN ELECTRICITY

Dividing the story of electricity into epochs is arbitrary but convenient. Various divisions could be made. It is not implied that on a certain date one epoch ended and another began. The epochs overlap and interlace; but each is distinguished by a certain broad aspect of the science. In the first epoch, which is symbolized by Thales' experiment, static charges were discovered, studied, and applied. It was the longest epoch, extending to 1729, and in it progress was least and slowest. The second epoch was characterized by the study of conduction. It was inaugurated by the work of Stephen Gray, but we found intimations of it in the work of von Guericke and as far back as Pliny and even Aristotle. Its duration was about two decades.

If it is true that, in Thales' experiment, the whole story of electricity was before us, then it should involve intimations of the second epoch. Let us search for these intimations by considering some questions which the new idea of conduction raised, new experiments it suggested, errors it corrected, and problems it solved which had previously defied solution.

Why did amber remain excited only temporarily or, in other words, why did it gradually lose its charge? We may suggest that the charge leaked away. Over what conductor? Over the amber and its support, or into the air or both? Because the process was relatively slow, we call amber and its support slow or poor conductors and air likewise. This raises also the question of the conductivity of air. Although Gilbert worked without the concept of conduction, he knew that the leakage of charge occurred more rapidly on humid days, and one of his experiments showed that flame increases the conductivity of air. A satisfactory answer to this question was not reached until about 1895. When it was reached, it became a harbinger of the modern epoch, of which the electron hypothesis is typical.

Why did touching the pith-ball in von Guericke's experiment discharge it? Von Guericke was far enough ahead of his time to reach a good answer, for it was his opinion that the charge flowed to the touching body.

Why did Gilbert classify metals as non-electrics? How would he test a brass rod? If he had been asked, he would probably have replied, "Exactly as I tested a glass rod, and why not?" With our knowledge of conduction, however, we can see his error and suggest the use of a glass handle for the brass rod. Then experiment shows, as Gray discovered, that a metallic body can readily be charged. The demonstration shows that metals are electrics. An error has been corrected.

From these considerations (and others) it can be seen that the idea of conduction was present in Thales' experiment, and that the germs of the second epoch were present in that early observation—whether or not anyone was astute enough to see them.

CHAPTER TWENTY

Influence or Induction Charges

ONE of Franklin's correspondents was a London schoolmaster, John Canton, who carried out many excellent experiments. We shall take Canton's work as marking the advent of the third epoch in electricity, characterized by the study and application of influence or induction charges. Such charges were glimpsed in the work of von Guericke and Hauksbee, and although the condenser is an application of influence charges, yet its inventors had only an intuitive knowledge of this topic. Once more, practice preceded theory.

In 1754, Canton read before the Royal Society a paper exposing in Dufay's terms "vitreous" and "resinous" such a fatal error that they had to be discarded:

Having rubbed a glass tube with a piece of thin sheet-lead and flower of emery mixt with water, till its transparency was entirely destroyed; after making it perfectly clean and dry, I excited it with new flannel, and found that it acted in all respects like excited sulphur or sealing-wax [resinous] . . . . If this rough or unpolished tube be excited by a piece of dry oiled silk, . . . it will act like a glass tube with its natural polish [vitreous]. Thus may the positive and negative powers of electricity be produced at pleasure [on the same substance] by altering the surface of the tube and the rubber. . . . For if the polish be taken off one half of a tube, the different powers may be excited with the same rubber at a single stroke.[1]

Since both kinds of charge can be obtained on glass, Dufay's term "vitreous" became meaningless. Furthermore, Canton showed that the kind of charge on a given electric depends on the substance of the rubber. Hence if charges are to be named by substances, both that of the electric and that of the rubber must be specified, e.g. the charge on glass rubbed with silk or on resin rubbed with fur. Dufay had overlooked the influence of the substance of the rubber.

In one of his letters to Franklin, Canton told of his discovery that the rubber which gives most intense electrification of glass is *chamois*

[1]*Phil. Trans. Roy. Soc. Lond.* (1754), XLVIII, ii, 780.

impregnated with tin amalgam. Tin, like all common metals (except iron) dissolves in mercury and the solution is called tin amalgam. The consistency of an amalgam is frequently putty-like, as many of us know from having a silver amalgam forced with sundry crunchings and squeaks into a cavity in a back tooth by our friend the dentist.

Spital Square, January 21, 1762,

Dear Sir,

Having formerly observed that the friction between Mercury and Glass in vacuo would not only produce the light of electricity, as in the luminous barometer, . . . but would also electrify the glass on the outside; I immersed a piece of dry Glass in a basin of Mercury, and found that by taking it out, the Mercury was electrified minus, and the Glass, electrified plus. . . . A small quantity of an amalgam of Mercury and Tin . . . being rubbed on the cushion of a globe or on the oiled-silk rubber of a tube, will excite the globe or tube to a great degree with very little friction.

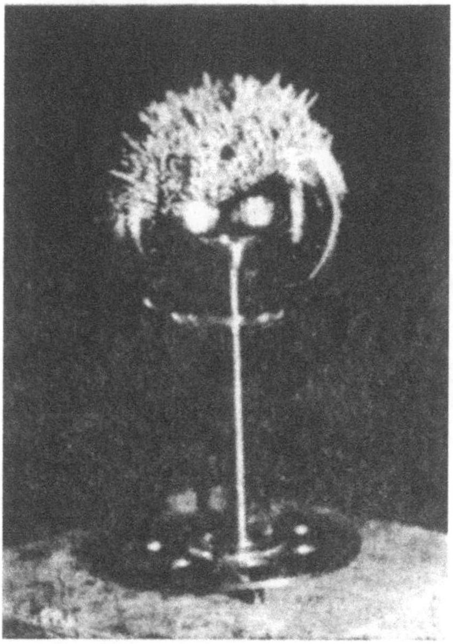

FIG. 281. Glass rubbed with tin-amalgam chamois.

This showed that charge can be produced merely by contact and separation without much or any friction.

Canton invented an electroscope which was the ancestor of a number of others (fig. 282):

The glass tube, when excited . . . will, in a few moments, electrify the air to such a degree, that (after the tube is carried away) a pair of balls, about the size of the smallest peas, turn'd out of cork, or the pith of the elder and hung to a wire by linen threads of six inches long, will repel each other to the distance of an inch and a half, when held at arm's-length in the middle of the room.[2]

The "glass tube" used by Canton as a source of charge was the one first described by Hauksbee (p. 401).

CANTON'S INFLUENCE CYLINDER

This new electroscope Canton used to good purpose in his greatest service to electricity—his study of influence charges (1753):

EXPERIMENT 1. From the cieling, . . . let two cork-balls, each about the bigness of a small pea, be suspended by linen threads of eight or nine inches in

[2]Ibid., p. 783.

length, so as to be in contact with each other. Bring the excited glass tube under the balls, and they will be separated by it when held at a distance of three or four feet: let it be brought nearer, and they will stand farther apart: intirely withdraw it, and they will immediately come together. . . . This experiment may be made with very small brass balls hung by silver wire.

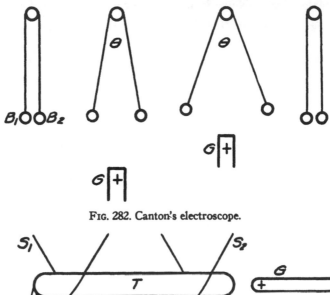

FIG. 282. Canton's electroscope.

FIG. 283. Influence charges.

EXPERIMENT 3.   Let a tin tube, of four or five feet in length, and about two inches in diameter, be insulated with silk; and from one end of it, let the cork-balls be suspended by linen threads. Electrify it by bringing the excited tube near the other end, so as that the balls may stand an inch and a half apart.

Thus the tin tube, an insulated conductor, was charged by entering the field of the charge on the glass but without touching it. The process is called influence (or induction) and a charge produced in this manner, an influence charge. Canton recorded a rather intricate series of observations of influence charges but did not reach a satisfactory explanation for all his facts. He suggested that if someone who was not a school-teacher and who could therefore have enough money to devote to experiments, would investigate this topic, he would find it a profitable research. Canton was right. A few years later, two scientists, Wilcke and Aepinus, verified his data and succeeded in deriving the laws of influence charges.

DECLINATION

Until 1724, the declination at any place on the earth was thought to be a constant; but in that year, George Graham, a London optician and instrument maker, found that it varies daily through about half a degree with the sun's motion. In Canton's report of his investigations on declination, Gilbert's term "variation" is used:

FIG. 284. John Canton.

The late celebrated Mr. George Graham made a great number of observations on the diurnal variation of the magnetic needle. . . . Since 1756, I have made near four thousand, with an excellent variation compass, of about nine inches diameter. The number of days on which these observations were taken is 603; and the diurnal variation was regular: that is, the variation [declination] . . . of the needle westward, was increasing from about eight or nine o'clock in the morning, till about one or two in the afternoon, when the needle became stationary for some time; after that, the variation . . . was decreasing and the needle came back to its former situation . . . by the next morning. Irregularities seldom happen more than once or twice in a month, and are always accompanied . . . with an aurora borealis.

NEUTRALIZATION

Robert Symmer described before the Royal Society in 1759 some amusing experiments with silk stockings and woollen underwear. In those days, men's silk stockings were not eclipsed by trousers, and there were men who felt that Nature had been so unkind to them in shaping their nether limbs as to warrant countermeasures. In one of his papers Symmer remarked casually that he "happened to wear woolen under the silk." He further related how, on pulling off a stocking and a drawer-leg together, and then the other stocking and the other drawer-leg, he discovered some important electrical facts. Demonstrations of these experiments were mercifully omitted from his lectures before the Society. Symmer, however, issued an invitation to the members to see the experiments in his apartments. One who accepted the invitation was Benjamin Franklin, for he was by no means a man to miss a chance of seeing something interesting.

The best way is to put the hand between the leg and the stockings and push them off together. Nothing more remains to be done, than to pull them asunder; for upon that, they both of them exhibit a degree of electricity, which is really . . . surprising. When I speak of the electricity in question, I mean such a power . . . that the stockings should appear . . . inflated; throw out an electrical wind to be felt by the bare legs; attract or repel another stocking visibly; and upon the touch, snap and emit or receive electrical fire. . . . I took a pair of white silk stockings, and having warmed them at the fire, put them both upon the same leg. After I had worn them about ten minutes, I took them off, and pulled them asunder, but discovered no signs of electricity. . . . I did the same with a pair of black silk, but to no other effect. I then proceeded to a decisive trial. I put a black and a white stocking upon my leg, and wore them likewise ten minutes. I waited with some impatience to see the success of my experiment, and in return, had the satisfaction of observing, upon their being pulled asunder, that each of them had acquired a stronger degree of electricity than I had before seen: they were inflated so much, that each of them showed the entire shape of the leg, and at a distance of a foot and a half, they rushed to meet each other. . . .

I had recourse to the little pocket electrometer of Mr. Canton's contrivance. By the terms positive and negative, I mean only to denote the opposition of the two different states. The particular allotment of the one or the other term appears to me arbitrary. . . . When the electrometer is placed on a non-electric [conductor, e.g. a metal] and the black stocking is presented to it, . . . the balls begin to be visibly attracted. . . . If instead of the black, the white stockings be presented at the same distances, it is found to have precisely the same effects. . . . From whence, it appears, that whatever difference there was between the electricity of the black and the white, under other circumstances, they each of them acquire an equal degree of electricity by being electrified together. . . . In the white, we find the positive and in the black, the negative electricity.

A white and a black stocking . . . when brought within the distance of three feet, . . . usually incline towards one another: within two and a half feet, . . . they catch hold of each other; and rush . . . together with surprising violence. As they approach, their inflation gradually subsides; and their attraction of foreign objects diminishes; when they meet, they flatten, and join . . . close together; and then the balls of the electrometer are not affected at the distance of a foot, nor even of a few inches. . . . But what appears more extraordinary, is, that when they are separated, and removed . . . from each other, their electricity does not appear to have been in the least impaired by the shock they had in meeting. They are again inflated, again attract and repel, and are as ready to rush together as before. . . . The electrical phial may be charged by the stockings, either positively or negatively, according as the wire . . . is presented to the white or the black; . . . but if the electricity of the white stocking be thrown into the phial, and upon that, the electicity of the black, . . . the phial will not be electrified at all.[3]

Both Franklin and Canton knew of the neutralization of opposite charges; but to Symmer goes the credit of a clear, vivid, and exhaustive demonstration. One of the advantages of Franklin's nomenclature is seen in the fact that the only way in which the addition of two quantities can yield zero is to have them opposite in sign and equal in magnitude. Symmer's neutralization experiment can therefore be written $+c +(-c) = 0$.

LAWS OF INFLUENCE CHARGES

Canton's work on influence charges, as has been said, led two physicists, Johann Wilcke, professor of physics at Stockholm, and Franz Aepinus, professor of physics at St. Petersburg, to investigate the subject further. Wilcke and Aepinus had been students together at Berlin. They achieved an epoch-making advance, namely, the deri-

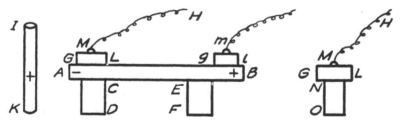

Fig. 285. Aepinus' experiment on influence charges, using his proof-plane.

vation of the laws of influence or induction charges. In 1759, Aepinus published a remarkable treatise on electricity, from which the following passage is taken. It contains incidentally, the invention of the *proof-plane*, a very useful form of insulated conductor (fig. 285):

Let the metal rod AB, about a foot long, be placed on glass supports CD and EF; and on its end A, place any small metal body GL, about an inch or half an inch long, furnished with a small hook M at its centre, to which is tied a silk thread HM, thoroughly dried. Then take the electric glass cylinder IK,

[3]*Phil. Trans. Roy. Soc. Lond.* (1759), LI, i, 357.

and when it is electrified by friction, bring it toward A, the end of the rod, to within a distance of about an inch and fix it in position there. Then by means of the silk thread HM, let the metal GL be lifted and placed on any glass block NO. If now the body GL be tested, it will be found to have an electric charge and of the negative kind. Then in a second experiment, let all the circumstances be as described, and let the glass tube IK again be brought near the end A, but let the body GL be placed on B, the opposite end of the rod, and if all the manipulations are repeated as in the preceding experiment, the body GL, on being placed on the block NO, will again be electrified but contrary to the previous case, it will now possess positive electricity. If a cylinder of sulphur be used in place of the glass tube, the outcome will be similar in every respect except that now on the contrary, GL when placed on the end A, will become positively charged but at B, negatively.[4]

Aepinus showed that a distal influence charge (the farther from the influencing charge) can be removed by touching or grounding, whereas a proximal charge cannot be thus removed as long as the inducing charge is near by. When Gray struck the iron plate on its "back Side" (p. 404), he was unwittingly applying these facts.

The knowledge of influence charges gained by Canton, Wilcke,[5] Aepinus, and their successors throughout the third epoch, enabled students of electricity, or "electricians" as they were then called, to understand facts which were previously incomprehensible or imperfectly understood—for example, the Florentine theorem and stage II in von Guericke's pith-ball experiment, which was really Thales' amber dust-test over again. The uncharged body, coming into the field of a charge, receives an opposite influence charge, and these charges attract each other, or, as it came to be stated, "induction precedes attraction." The new knowledge also enabled electricians to predict and seek new facts. Wilcke and Aepinus did valuable pioneer work in their study of the *condenser*, which is an application of influence charges. It was of crucial importance in the celebrated researches of Alessandro Volta, who gave it its name. In the modern world it is of broad utility.

Since the dust particle in Thales' amber experiment receives a positive charge by influence when coming into the field of the negative amber charge and these attract each other, it is plain to see that Thales' experiment foreshadowed the third epoch in the history of electricity.

Aepinus' book *Essay on Theory of Electricity and Magnetism*, was the first treatise on electricity to couch its ideas in the language of

[4]F. Aepinus, *Tentamen Theoriae Electricitatis et Magnetismi* (St. Petersburg, 1759), p. 127. The John Crerar Library, Chicago, has a copy.

[5]J. Wilcke, *Franklin's Letters*. This book cannot be sent out on the Interlibrary Exchange Plan. I wish to thank Dr. L. Carter, librarian of the Engineering Society, New York, for his courtesy in letting me read it "on the precincts"—and for a perfectly good excuse to visit New York.

algebra. This signified that the science was outgrowing the bibs and tuckers of its qualitative childhood and was developing toward an adult state, characterized by measurement, calculation, and mathematical expression.

BERGMAN'S SERIES

About a decade after Canton's experiments had exposed the error in Dufay's terms, "vitreous" and "resinous," T. Olof Bergman, professor of physics at Uppsala, Sweden, proposed in a letter to a Fellow of the Royal Society of London, a new classification of electrics which had several advantages. It enabled one to predict the kind of charge produced on an electric by rubbing it with any given electric; it helped the memory to recall which kind is produced in any particular case; and it gave a hint as to the intensity of charge produced.

Any body B, for example, a piece of rough glass, can become either negatively or positively charged according to whether it is rubbed with a piece of silk C, or with a small piece of wool A. Under given conditions, therefore, electrics may be arranged in a fixed series. Let $(+)$ A, B, C, D, E, ... $(-)$ be five bodies; any one of these if rubbed with any of those which precede it in the list, becomes negatively charged, but if with those which follow it, positively. Now, the less the distance in the list between two electrics, rubbed together, the weaker is the charge, *coeteris paribus*. Thus between A and E, it is stronger than when A and B are used. . . . Hence, with bodies of the same substance, it is impossible to produce any charge: on the other hand, the farther two electrics are apart in the series, the stronger is the charge produced when they are rubbed together.[6]

Such a method of classification is used in several branches of physics; for example, the metals can be arranged in a comparative or graded series with respect to their thermal or their electrical conductivities, each member of the series being a better conductor than any of its successors in the list. To give names to A, B, C, in Bergman's report, here is a Bergman series for a few common electrics:

(+) fur . . . wool . . . glass . . . white silk . . . black silk . . . resin . . . amber . . . vulcanite . . . celluloid (−)

CARBON A CONDUCTOR

That carbon is a conductor of electricity was discovered in 1766 by Rev. Joseph Priestley, LL.D., F.R.S., a Nonconformist minister of Warrington, Lancashire. The report of this discovery occurs in Priestley's *History of Electricity*, which he published in 1794.

On May the 4th, 1766, I tried charcoal . . . and found it to be . . . an excellent conductor of electricity. I placed a great number of pieces of charcoal, not

[6]*Phil. Trans. Roy. Soc. Lond.* (1764), LIV, xiii, 84.

less than twelve or twenty, of various sizes, in a circuit, and discharged a common jar through them, when to all appearance the discharge was as perfect as if so many pieces of metal had been placed in the same manner. . . . I took a piece of baked wood, which I had often used for the purpose of insulation, being an excellent non-conductor, and putting it into a long glass tube, I thrust it into the fire, and converted it into charcoal. . . . Upon trial, its electric property was quite gone, and it was become a very good conductor.

The influence of Desaguliers' false theorem can be seen in the wording of this article half a century after the theorem was enunciated, i.e. when Priestley used the term "electric" to mean non-conductor.

Priestley was a very able writer; his greatest contribution to electricity was his excellent *History of Electricity*, from which these excerpts are taken. In its introduction, he expresses an evaluation of the study of science, which to his mind includes the study of history of science.

Philosophy exhibits the powers of nature, discovered and directed by human art . . . and the idea of continual rise and improvement is conspicuous in the whole study, whether we be attentive to the part which nature, or that which men are acting in the great scene. . . . It is here that we see the human understanding to its greatest advantage, grasping at the noblest objects, and increasing its own powers, by acquiring to itself the powers of nature, and directing them to the accomplishment of its own views; whereby the security and happiness of mankind are daily improved. . . . If the exertion of human abilities, which cannot but form a delightful spectacle for the human imagination, give us pleasure, we enjoy it here in a higher degree than while we are contemplating the schemes of warriors, and the stratagems of their bloody art.

To whatever height we have arrived in natural science, our beginnings were very low, and our advances have been exceedingly gradual. And to look down from the eminence, and to see, and compare all those gradual advances in the ascent, cannot but give pleasure to those who are seated on the heights. And considering that we ourselves are, by no means, at the top of human science; that the mountain still ascends beyond our sight . . . a view of the manner in which the ascent has been made, cannot but animate us in our attempts to advance still higher, and suggest methods and experiments to assist us in our further progress.

Philosophical instruments are an endless fund of knowledge, such as the air-pump . . . pyrometer and electrical machine. . . . By the help of these machines, we are able to put an endless variety of things into an endless variety of situations, while nature herself is the agent that shows the results. . . . Human happiness depends chiefly upon having some object to pursue, and upon the vigour with which our faculties are excited in the pursuit. And certainly, we must be much more interested in pursuits wholly our own, than when we are merely following the track of others.

A Philosopher ought to be something greater, and better than another man. The contemplation of the works of God should give a sublimity to his virtue, should expand his benevolence, extinguish everything mean, base, and selfish in his nature, give a dignity to all his sentiments, and touch him to aspire to the moral perfection of the great Author of all things. The more we see of the wonderful structure of the world, and of the laws of nature, the more clearly do we comprehend their admirable uses, to make all the percipient creation happy: a sentiment which cannot but fill the heart with unbounded love, gratitude and joy.

And these are the words of a man who was persecuted for his religious and democratic views, calumniated, and driven into exile. He went to the United States where he found greater freedom.

THE HENLEY ELECTROMETER

A letter sent by Priestley to Franklin in 1770 records an improvement upon Canton's pith-ball electroscope made by William Henley, a London engineer. Three main purposes of any electroscope are (1) to detect a charge, (2) to determine its kind, and (3) to indicate that one charge is stronger than another. If the instrument performs the

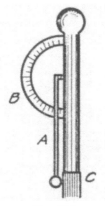

Fig. 286.
Henley's
electrometer.

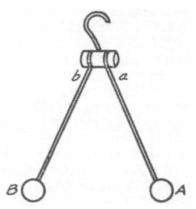

Fig. 287. Cavendish's electrometer.

third function quantitatively, it is called an electrometer, i.e. if it indicates *how many times* stronger one charge is than another. Priestley thought the new instrument merited the name electrometer. So did its inventor. The letter to Franklin runs thus:

I think myself happy in an opportunity of giving you a species of pleasure, which I know is peculiarly grateful to you as the father of modern electricity, by transmitting to you an account of some . . . improvements in your favourite science. The author of them is Mr. Henly, in the Borough. . . . In my history of electricity, . . . I have mentioned a good electrometer as one of the greatest desiderata among practical electricians, to measure both the precise degree of the electrification of any body, and also the exact quantity of a charge. . . . All these purposes are answered . . . by an electrometer of this gentleman's contrivance [fig. 286]. The whole instrument is made of ivory or wood. A is an exceedingly light rod, with a cork ball at the extremity, made to turn upon the centre of a semicircle B. . . . C is the stem that supports it. . . . The moment that this little apparatus is electrified, the rod A is repelled by the stem C, and consequently begins to move along the graduated edge of the semicircle, so as to mark . . . the degree in which the prime conductor is electrified. . . . I doubt not that you and all other electricians will join with me in returning our hearty thanks to Mr. Henly for this excellent and useful instrument.[7]

ZERO POTENTIAL

The Honourable Henry Cavendish (1731-1810) was the elder son of Lord Charles Cavendish and nephew of the third Duke of Devonshire. His aim in life was to study science and avoid society. He was signally successful in both aims. He lived in his laboratory and his library, except for a weekly splurge on Thursday evenings when he dined with fellow members of the Royal Society. He seldom spoke. He ordered his dinner by a note left on the hall table, and it was the housekeeper's duty to remove the note in his absence, prepare and serve the dinner and get out of sight and keep out of sight on pain of dismissal. Most of the stories of Cavendish's misogyny and other idiosyncrasies simply indicate that he was morbidly shy. He never married. By bequests from uncles and aunts he became very wealthy; not that it made much difference to him. His banker, becoming alarmed at the size of his client's bank account (some quarter of a million dollars), urgently requested instructions. Cavendish bade him stifle his yammerings and ululations or he would change bankers. "A pretty howdyedo," said Henry, "if a man can't have a few pounds on deposit without being pestered." On appointed days, he was present in his ancestral library in Dean Street to lend books to any who were properly attested, and when he used one of his own books, he entered his name in the list just as he would for any other borrower, and cancelled the entry when he returned the book to its place.

Although most of Cavendish's electrical discoveries had little influence on the advancement of the science because they were not published at the time, yet a few quotations from his papers will be given here. Clerk Maxwell edited them in 1879 and they showed that Cavendish was easily a half century ahead of his contemporaries. Our first quotation from Cavendish's papers discloses that brilliant recluse's remarkable genius for tacking together a few odds and ends and, from the rather rough-looking apparatus thus fashioned, obtaining results that were far in advance of his times (fig. 287):

I made use of a more exact kind of electrometer, consisting of two wheaten straws Aa and Bb, eleven inches long, with cork balls A and B at the bottom, each one-third of an inch in diameter, and supported at a and b by fine steel

[7]*Phil. Trans. Roy. Soc. Lond.* (1772), LXII, xxvi, 359.

pins bearing on notches in the brass plate. . . . A piece of paste-board with two black lines upon it, was placed six inches behind the electrometer . . . in order to judge of the distance to which the balls were separated. . . . The electrical machine was turned [by Richard, his valet] till the balls appeared even with those lines. By these means, I could judge of the strength of the electricity to a considerable degree of exactness. In order to make the straws conduct the better, they were gilt over.[8]

Cavendish seldom invented a new machine or coined a new term, but frequently he improved an instrument or defined an old term and used it with new precision and significance. Although the terms "potential" and "electrical pressure," or "tension," were first introduced half a century later by Volta and Poisson, nevertheless, there is no doubt that Cavendish reached these concepts long before they became current. Potential in electricity corresponds to temperature in thermics and to level or pressure in hydraulics. Cavendish's expression for this fundamental concept was "degree of electrification."

If several insulated conductors are connected by conductors, and one of them . . . is positively electrified (or negatively), all the others must be electrified in the same degree [to the same potential] . . . . Let any number of excellent conductors be connected by conductors. It is plain that the electric fluid must be equally compressed in all these bodies, for if it was not, electricity would move from those in which it was more compressed to those in which it was less compressed till the compression became equal in all.

It is impossible for any grounded conductor to be either positively or negatively electrified: for the earth, taking the whole together, contains just fluid enough to saturate it [his concept of a neutral body] and consists in general of conductors: and consequently, though it is possible for small parts of the surface of the earth to be charged positively or negatively, by the approach of electrified clouds or other causes; yet the bulk of the earth, and especially the interior parts must be neutral. Therefore, assume any part of the earth which is neutral, . . . any conductor which is grounded, . . . is not at all electrified.

To compare two condensers or two bodies as to their ability or rate in storing electricity, Volta introduced the term "electrical *capacity*," corresponding to thermal capacity. The electrical capacity of a body is measured by the charge necessary to change its potential one unit. Cavendish's records show that he employed this idea in his work in 1772. He also anticipated Coulomb's discovery that the force of attraction or repulsion between two charges varies inversely as the square of their distance, a relation which is analogous to the law of gravitational attraction.

LAW OF INVERSE SQUARES

Cavendish's argument ran as follows: If two equal like charges, at a distance of 1 cm., in air, repel each other with a force of 1 dyne, each

[8] *The Electrical Researches of the Honourable Henry Cavendish*, ed. J. Clerk Maxwell.

is called an *electrostatic unit of charge* (abbrev., e.s.u.). Therefore at the same distance the force exerted on each other by two charges q' and q'',

Eq. 1. $$f \propto q' q''.$$ (distance constant)

If the distance, d, increases, the force diminishes. The relation is therefore inverse, but the exact form of the function is still to be found. If f varies inversely as a power of d, it may be represented thus

$$f \propto 1/d^n$$ where n is to be determined.

Cavendish's proof that $n = 2$ was as follows:

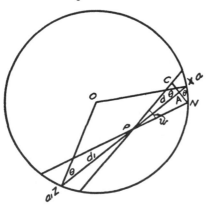

FIG. 288. Law of inverse squares.

Let XPZ be the axis of a very slender double slant cone with vertical angle w, at P, any point inside a hollow spherical charged conductor (fig. 288). Let a and a_1 be the areas it cuts from the surface of the sphere. The area of the base of the right cone PCN varies directly as the solid angle w and directly as the square of its height d. $A = cd^2w$, where c is a variational constant. Let the base of this cone, CN, meet the surface of the sphere at angle θ (XNC), $\therefore a = cd^2w/\cos\theta$. Similarly

$$a^1 = cd_1^2w/\cos\theta, \because \angle OZX = \angle OXZ = \theta.$$

Let the uniform distribution of charge on the sphere be σ units per sq. cm. $\therefore$ charge on $a = q = cd^2w\sigma/\cos\theta = kd^2$, where $k = cw\sigma/\cos\theta$, and charge on $a_1 = q_1 = cd_1^2w\sigma/\cos\theta = kd_1^2$

$\therefore$ the action of q on unit charge U at P, $I, = (1)kd^2/(d)^n$ and the action of q_1, $I_1 = (1)kd_1^2/(d_1)^n$.

These two forces act in opposite sense. If $n = 2$, $I = I_1$. Then U will not be moved. Since the whole space inside the sphere is the sum of a series of such cones, the charge on the sphere will not exert a force on a charge placed in the interior, and this is what Cavendish by experiment found to hold true. Therefore

Eq. 2. $$f \propto \frac{q'q''}{d^2}$$

or, in words, the force varies directly as the product of the charges and inversely as the square of their distance. These laws were first demonstrated in the literature by Coulomb(1785)and are known as Coulomb's laws of charges.

BE YOUR OWN INDICATOR

The term *conductivity* when used quantitatively denotes the degree to which a conductor possesses the ability to conduct electricity. Cavendish compared the conductivities of various conductors, for instance, saturated and unsaturated brine. The two conductors were placed in tubes like T in fig. 289. The discharge from a Leyden jar, J, was sent through each in turn.

At this point in the experiment, Cavendish needed an indicator to show when the flow of electricity in the two circuits was the same.

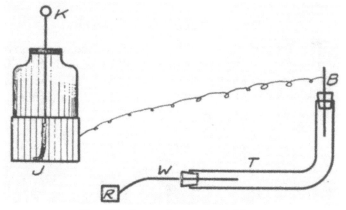

FIG. 289. Comparing conductivities.

When Ohm came to the same problem in 1827, he had at his disposal a suitable indicator, invented by Schweigger in 1820, which was named by Ampère a "galvanoscope" or "galvanometer" in honour of Galvani. But the Honourable Henry had no such instrument. With notable intrepidity he deliberately made himself part of the circuit, and not only took the shocks but also noted carefully in each case the severity of the thrill he suffered, whereas Musschenbroek refused to take a second taste of such "Tonner und Blitzen" even for "the whole kingdom of France!" Then Cavendish compared shock I and shock II. If they were unequal, he changed the effective length of the liquid by moving the wire W. Now two more shots! And so on until he received the same jolt from both tubes, as may be seen from his tabulated results:

Exp. 1.   In tube 14,—sat'd sea salt; tube 15,—salt in 69 of water.
Distance of wires

tube 15	tube 14	Shock,—
6.5 inch.	40.7 inch.	very sensibly less in short tube than in long one.
5.8 "	"	sensibly less in short tube than in long one.
3.5 "	"	sensibly greater
4.2 "	"	scarce sensibly greater
5.3 "	"	just sensibly less.

Straw electrom.–4. Resistance of 4.7 inches in tube 15 supposed equal to 40.7 in 14. Therefore, sat'd soln. conducts 8.6 times better than salt in 69 of water Duplicate—8.6, electrom. 1¼.

The use of the term *resistance* in this passage was a remarkable anticipation of Ohm. In these investigations, Cavendish became aware of the relation among the three quantities, electric pressure, strength of flow, and resistance of conductor, which is now known as *Ohm's law*. What a loss it was to the advancement of science when Cavendish stuffed his notes into an old barrel and let the world go hang!

THE CAVALLO ELECTROSCOPE

A further modification of Canton's electroscope was invented in 1777 by Tiberius Cavallo of London, and described in his *Treatise on Electricity* which was published nine years later. His discussion shows how to charge the instrument by contact and by influence:

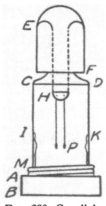

Fig. [290] is a representation of my new electrometer . . . which, after various trials, I brought to the present state . . . as long ago as the year 1777 . . . The principal part . . . is a glass tube *CDM*. . . . The conical corks *P*, . . . which, by their repulsion, shew the Electricity ,. . . . are as small as they can possibly be made and they are suspended by exceedingly fine silver wires *HP*. . . . It is electrified by touching the brass cap *EF*. To communicate any Electricity to this electrometer, by means of an excited electric, e.g. a piece of sealing-wax . . . the best method . . . is to bring . . . the wax so near the cap, that one or both of the corks may touch the side of the bottle . . . after which, they will soon collapse . . .; if now the wax is removed, they will again diverge and remain electrified positively. In this operation, the wax . . . acts by means of its atmosphere.

FIG. 290. Cavallo's electroscope.

Since atmosphere is an old word for "field," Cavallo is here referring to the procedure of charging the electroscope by influence or induction. Cavallo was a friend and correspondent of Volta. As a young man he went to England from his native Italy to continue his education, and he liked life in England so well that he remained for the rest of his days.

As a science advances, the number of researchers and discoveries continually growing, it becomes increasingly difficult to make contributions of major importance. In the building of a cathedral, not all the work will be on corner-stones, facades and pinnacles; the bulk of the work consists of the laying of thousands of inconspicuous bricks, or stones, well and truly placed. Although Cavallo's contribution consists of a miscellany of minor items, yet it was valuable and well worth our consideration.

Economy is often gained by using the ground as part of a long circuit as in telegraphy. Sir William Watson of London used the water of the Thames in this way in 1747. Cavallo's text describes the method (fig. 291):

> The object for which those experiments were performed was to fire gunpowder . . . from a great distance by means of electricity. At first, I made a circuit with a very long brass wire, the two ends of which returned to the same place, whilst the middle of the wire stood at a great distance. In this middle, an

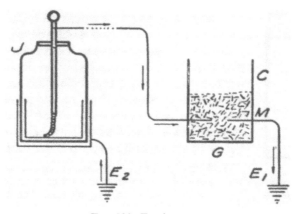

FIG. 291. Earth return.

interruption was made in which a cartridge, C, of gunpowder mixed with steel filings (M) was placed. Then, . . . by touching one wire to the knob of a charged Leyden phial, whilst the other was connected with the outside coating, the cartridge was fired. . . . A brass wire . . . was laid on the ground, and its extremity was inserted in the cartridge. . . . Another piece of wire was likewise inserted with one end into the cartridge, and its other end thrust into the ground E . . . applying the knob of the charged jar J, to that extremity of the long wire which was remote from the cartridge . . . the outside of the jar communicated with the ground. . . . That the charge of the jar passed through the wire and through the ground was evidently proved by the powder being sometimes fired.

Cavallo's treatise describes his invention of the original *spark-plug* (fig. 292). In those days the term "air" was often used to refer to any gas, and "inflammable air" meant hydrogen. The term "gas" was coined by the Belgian chemist van Helmont about 1640; he may have derived it from the German word *Geist*, which is related to our word ghost. Another proposed derivation is from the Greek *chaos*. It was Cavendish who discovered hydrogen in 1766 and named it inflammable air.

The inflammable air must be contained in a two or three ounce phial. . . . By striking . . . with a pointed thick wire, a hole will be made in the bottom of the phial. In this hole, a piece of thin wire must be cemented. . . . A very sound cork must be fitted to the mouth of the phial, and another piece of wire must be passed through it . . . to come . . . to the distance of about the fortieth part of an inch from the extremity of the wire that proceeds from the bottom. . . . Remove the cork . . . and place the phial inverted for four or five seconds over the mouth of a large phial . . . full of inflammable air; then slip it off, and cork it up as fast as you can. . . . If the least electric spark be passed from wire to wire, . . . the air will instantly explode.

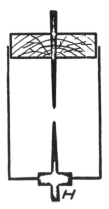

Fig. 292. Cavallo's spark-plug.

Pure hydrogen in the tube, of course, would not be explosive, but with rather primitive methods of handling gases, some air would be included and the resulting gaseous solution would be explosive.

Insulated copper wire is now a regular commodity, of course. Many of us probably use it without pausing to realize what a convenience it is, or how our work would be slowed if we had to invent it or even make some for our own use. In 1786 it was a different story; Cavallo describes his invention of insulated wire and the technique seems to us even laughable:

A piece of copper . . . or brass wire . . . being stretched from one side of a room to the other, heat it successively from one end to the other by means of the flame of a candle, or of a red-hot piece of iron, and as you proceed with the candle, rub a lump of pitch over the heated part of the wire. When the wire has been thus covered with pitch, a slip of linen rag must be put round it, which can be easily made to adhere to the pitch, and over this rag another coat of pitch must be laid with a brush, the pitch being melted in a pipkin or other convenient vessel. This second layer of pitch must be covered with a slip of woollen cloth, which must be fastened by means of needle and thread. Lastly, the cloth must be covered with a thick coat of oil paint, and when the paint is dry, the covered wire may be used for the experiment. In this manner, many pieces of wire, each of about . . . twenty feet in length, may be prepared, which may afterwards be joined together, so as to form one continued metallic communication; but care must be taken to secure the places where the pieces are joined, which is most readily done by wrapping a piece of oil-silk over the painted cloth, round the two contiguous extremities, and binding it with thread.

COMMUNICATION

The idea of electrical long-distance communication was envisaged by a number of people long before Morse and Bell. In 1637, Galileo humorously discussed it:

SAGREDUS.   You remind me of a man who offered to sell me a secret for enabling one to speak to a person two or three thousand miles away by means of the attraction of a certain magnetic needle. I said to him that I would be willing to purchase it but that I should like to witness a trial of it and that it would please me to test it, I being in one room and he in another. He told me that at such a short distance, the action could not be witnessed to advantage; so I sent him away, saying that I could not take time just then to go to Egypt or Muscovy to see his experiment but that, if he would go there himself, I would stay here in Venice and attend to the rest.[9]

Cavallo discovered one answer to the problem:

If this experiment be tried without any inflammable air in the phial, the only effect will be that a spark will be seen between the two wires. . . . By sending a number of sparks at different intervals of time, according to a settled plan [code], any sort of intelligence might be conveyed instantaneously from the place in which the operator stands to the other place in which the phial is situated. . . . I can only say, that I never tried the experiment with a wire longer than about 250 feet; but from the results of those experiments, . . . I am led to believe that the above mentioned sort of communication might be extended to two or three miles [!] and probably to a much greater distance.

Thus Cavallo preceded Morse by almost a century in the electrical transmission of messages.

THE BENNET GOLD-LEAF EXPERIMENT

The gold-leaf electroscope, invented in 1786 by Rev. Abraham Bennet, was an advance on Cavallo's instrument. In a letter to Priestley, Bennet described the new indicator (fig. 293) and also reported some discoveries which he made by its help and which gained him election to the Royal Society.

Wirksworth, September 15, 1786.

Rev. Sir,

I send you a description of my electrometer which, having the honour of your approbation, may be communicated to the Royal Society. . . . The following experiments will shew the sensibility of this instrument. If a metal cup be placed upon the cap *C*, with a redhot coal in it, a spoonful of water, thrown in, electrifies the cup negatively; and if a bent wire, *W*, be placed in the cup [of another electroscope] with a piece of paper, *P*, fastened to it to increase its surface, the positive electricity of the ascending vapour may be tried by introducing the paper into it. Perhaps the electrification of fogs and rain is well illustrated by pouring water through an

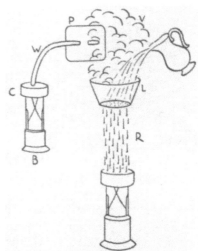

Fig. 293. Bennet's gold-leaf electroscope.

9G. Galilei, *Dialoghi delle duo nuove scienze*, Giornata prima.

insulated cullender, *L*, containing hot coals, where the ascending vapour *V* is positive and the falling drops *R* negative.[10]

COULOMB'S TORSION BALANCE

Charles Coulomb of Paris (1736-1806), who discovered the laws of friction and of torsional elasticity, is still better known for his researches in electricity. These are held in such high esteem by scientists that an important unit quantity of electricity or unit charge is called the *coulomb*. The excerpts quoted here are from a paper that Coulomb read before the Royal Academy of Paris in 1785. He describes the sensitive balance which bears his name. He invented it and used it in measuring the forces of repulsion or attraction between two electric charges and/or between two magnet poles:

In a memoir, presented to the Academy in 1784, I determined experimentally the laws of the force of torsion in a metal wire and I found this force was proportional to the angle of torsion. . . .
To-day, I present for the inspection of the Academy, an electrical balance, constructed on the same principles: it measures with extreme accuracy the force of the electrical charges on a body, however weak may be the degree of electrification.

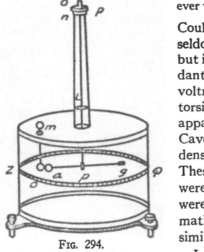

FIG. 294.
Coulomb's torsion balance.

Coulomb's torsion balance (fig. 294) is seldom used now for its original purpose, but it has a number of important descendants, for instance, the ammeter and the voltmeter. The similarity between the torsion balance used by Coulomb and the apparatus used by his contemporary, Cavendish, to determine the mass and density of the earth is quite evident. These two men though not acquainted were kindred spirits in several ways. They were both master experimenters and good mathematicians, and they investigated similar topics.

In order to transfer charges from one body to another, Coulomb used a modified form of Aepinus' proof-plane, namely, the rigid form which has remained in common use ever since Coulomb. "Proof" here means "test" as in the adage, "the proof of the pudding. . . ."

EXPERIMENT. We electrify a small conductor which is nothing more than a large-headed house-pin that is insulated by inserting its point in the end of a stick of Spanish sealing-wax. We insert this pin in the aperture *m* and make it touch the ball in contact with the sphere *a*. On withdrawing the pin, the two spheres are charged with the same kind of electricity and they mutually repel each other to a distance that we measure by looking at the suspension wire, the centre of the ball *a*, and the corresponding division of the circle *ZOQ*. Now, turning the index of the micrometer in the sense *pno*, we twist the suspension wire *LP*, and produce a force proportional to the angle of torsion, which tends to force the ball *a* toward the [other] ball . . . .[11]

LAW OF INVERSE SQUARES

On the day that Coulomb read his paper, the members of the Academy had the privilege of witnessing the first public demonstration of the famous law of inverse squares:

EXPERIMENT 1. Having charged the two spheres with the head of the pin [proof-plane], the index of the micrometer reading 0°, the ball *a* on the needle *ag* moves away from the [other] through the angle 36°. Having twisted the suspension wire by means of the micrometer head *O* through 126°, the two spheres approached each other and stopped at a distance of 18° from each other.

FIG. 295. Coulomb's proof-plane.

EXPERIMENT 3 [similarly]. . . . We find in our first experiment, in which the micrometer index stood at 0°, that the two spheres separate through 36° producing at the same time a force of tension of 36°. In the second case, the distance of the spheres was 18°. . . . Hence, at a distance of 18°, the repelling force was (126 + 18) or 144°; thus at half of the first distance, the repulsion of the two spheres is quadrupled. . . . It follows then from these three experiments that the repelling action which the two spheres charged with the same kind of electricity, exert upon each other, is *inversely proportional to the square of their distance*.[12]

An algebraic expression of Coulomb's laws of charges is

$$f \propto \frac{qq''}{r^2} \quad \text{or} \quad f = \frac{q'q''}{kr^2} \qquad \text{Eq. 2.}$$

where *k* or $(1/k)$ is a variational or proportionality constant.

Coulomb's experiments with magnet poles yielded a similar result

$$\text{Eq. 3.} \qquad f \propto \frac{s's''}{r^2} \quad \text{or} \quad f = \frac{s's''}{\mu r^2}$$

where *s'* and *s''* are the strengths of two poles, *r* their distance and μ a proportionality constant. In both *Eq.* 1 and 2 it is assumed that *r* is

[11]*Mémoires de l'Académie Royale des Sciences: Collection relative à la Physique*, I, 107.
[12]Ibid, p. 111.

large in comparison with the diameters of the charges or poles concerned. The resemblance between *Eq.* 2 and 3 strengthened the suspicion that in some way magnetism and electricity were related. The further resemblance between these equations and the algebraic expression of Newton's law of gravitation increased the excitement with which Coulomb's results were received.

Aepinus' use of algebra in his discussion of electricity was indicative of important advancement. Now the labours of Coulomb were epochal in bringing this subject to the stage of a physical science characterized by quantitative law.

On the basis of Coulomb's laws of charges we are in a position to choose a unit quantity of electricity or unit charge (p. 428). The electrostatic unit of charge (e.s.u.) is a quantity of electricity. It is that charge which can repel an equal like charge at a distance of 1 cm. in air (or vacuum) with a force of 1 dyne. For practical purposes the e.s.u. charge is inconveniently small; so a larger practical unit was chosen and named in honour of Coulomb.

1 coulomb (abbrev., cl.) $= 3 \times 10^9$ e.s.u. (charge).

CHAPTER TWENTY-ONE

Electric Current

AN important treatise, entitled *The Role of Electricity in Muscular Motion*, was published in 1791 by Luigi Galvani (1737-98), professor of biology at the University of Bologna. Its main experiment, which is named after him, paved the way for the epoch-making researches of his great countryman, Alessandro Volta:

> I dissected a frog, prepared it as in fig. [296], and placed it with all other requisite accessories on a table on which there was an electric machine . . . separated from it by a not inconsiderable distance. When, now, one of my assistants chanced to touch one inner crural nerve (c) of the frog very lightly with the point of a scalpel, all the muscles of the joints were immediately contracted as if seized with violent tonic spasms. The other person, however, who assisted us in our electrical experiments, thought he had noticed that this occurrence took place when a spark occurred at the spark-gap of the machine. Astonished at this novel phenomenon, he brought it vividly to my notice although I was quite absorbed, thinking about something altogether different. Thereupon, I was seized with an overpowering zeal and yearning to investigate the problem and bring to light whatever was hidden in it. I touched, therefore, with the same scapel point one or other crural nerve while one of those present produced the spark. The phenomenon occurred persistently in the same manner.[1]

TWO DIFFERENT METALS

Galvani, being a humane man, gave his frog specimens instantaneous death by the method called *pithing*: a sharp brass hook is suddenly forced into the medulla oblongata or spinal cord. This hook proved to be specially significant in some of Galvani's experiments.

> But when I took the frog into a closed room, laid it on an iron plate and had begun to press against the latter, the hook which was fixed in its spinal cord, lo and behold, the same contractions and the same kicks! . . . Finally, we thought of using a non-conducting plate of glass. . . . Then, no muscular

[1]L. Galvani, *De viribus electricis in motu musculari*. A copy of the treatise is (or was) in the University Library, Göttingen, Germany. I am grateful to Com. Leon Leppard, Ph.D., for obtaining this material for me when he was studying at Göttingen on an 1851 Exhibition Scholarship.

spasms were to be seen. Naturally, such a result excited in us no small wonder and led us to the opinion that electricity originates in the animal itself. Both instances supported us in assuming a very subtle nerve-fluid which during the action flows from the nerves to the muscles, resembling the electric stream from

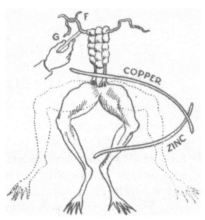

FIG. 296. Galvani's experiment.

a Leyden flask. . . . But in order to make this clearer, I obtained good results by laying the frog on an insulating plate of glass . . . and applying to it first a conducting and then a non-conducting bow, one of whose ends touched the hook, fixed in the spinal cord, and the other, the leg muscle or that of the foot.

Most important of all, we succeeded in observing . . . the noteworthy fact that having the two wires of different metals is of prime importance in producing the stimulation. If the whole bow was of iron, the spasms generally did not occur or if so, they were weak. If, however, one wire is iron and the other of brass, or still better, of silver, . . . immediately, a far greater and more persistent convulsion occurs. If zinc foil is applied to one point and brass foil to the other, the contractions become still far greater.

The "bow" used by Galvani was a modification of Franklin's discharging tongs. It is interesting to note that in some experiments in which the frog leg responded to distant lightning, Galvani may have been experimenting with what we call radio waves; but, of course, such ideas did not arise until a century later.

THE ELECTROPHORUS

Until the end of the eighteenth century, the study of electricity brought little improvement in the living conditions of people in general. In other words, there were no electrical engineers. All the flows of electricity obtained had been momentary or intermittent, and so had been most electrical effects. But if streets are to be lighted electrically or street-cars, telephones, and factories operated, a continual steady flow of electricity must be available. It was the work of Alessandro Volta of Como and Pavia (1745-1827) which achieved a new source of electricity that yields continual flow, and which thereby brought the Electrical Age within the range of possibility and almost within view. Volta was like Moses in being vouchsafed a sight of the promised land from his mountain height without the privilege of entering it. Volta was such a prodigious worker that the dozen passages from his writings quoted here can give only a faint idea of the magnitude of his labours. All his writings are in the form of letters to his many scientific friends. The first excerpt describes an application of influence charges which he invented when he was a teacher of physics in the Gymnasium at Como, his home town. It is symbolic of his whole life's work. Being an application of influence charges, it is representative of the third epoch in electricity and it is also the prototype of the modern electric machine.

Fig. 297. Alessandro Volta.

Como, June 10, 1775.

To Signor Dr. Giuseppe Priestley, F.R.S.,

Illustrious Sir,

I do not know how far I dare promise that my experiments will please you . . . even though you have earnestly requested the account. With my nose to the grindstone, pursuing my own researches, a new development may occur without my knowing of it, just as happened in the case of your experiment with charred wood, of which your excellent *History of Electricity* informed me, whereas I should have known of it earlier, but I was busily occupied with other work. . . .

I hereby draw your attention to a body that, after being electrified by a single brief rubbing, not only does not lose its electricity but retains obstinately the indications of its active force in spite of being touched repeatedly any number of times. . . . With your sanction, I should like, then, to apply a name to this small apparatus, and that name would be "Elettroforo perpetuo." In fig. [298], *AA* is the plate . . . in which is contained a cake of sealing-wax. . . . *CC* is a Shield of gilded wood . . . *E* is an insulating handle, i.e. a little rod of glass, coated with sealing-wax . . . by which the Shield can be lifted up.[2]

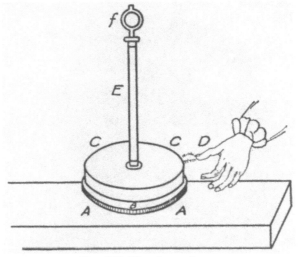

FIG. 298. The electrophorus.

This instrument is now called the *electrophorus*. Of the two charges which the resin plate *A* induces on the shield *C*, the distal is removed by touching or grounding. When *C* is withdrawn from *A*'s field, the proximal influence charge on *C* can be drawn off as a spark. This does not reduce the initial charge on *A*, so the operation can be repeated any number of times("perpetuo"). It may seem at the first glance that we are here obtaining electrical energy for nothing: but the energy of the sparks is being furnished as mechanical energy by the arm that operates *C* and some of this mechanical energy is being transformed into electrical energy. Similarly when we turn an electric machine the mechanical energy we put into the machine is transfomed into electrical.

VOLTA'S CONDENSING ELECTROSCOPE

Seven years later, Volta sent two letters to the Royal Society of London, describing, among other results, an instrument that he called

[2]*Collezione dell' Opere del Cavaliere Conte Alessandro Volta* (Piatti, 1816), I, i, 107. This book may be examined in the John Crerar Library, Chicago.

a "condensator." The translator for the *Transactions* shortened Volta's term to "condenser." The instrument which Volta describes in his letter is known as his condensing electroscope. Without it, he could not have made his discoveries which inaugurated the fourth epoch in electricity.

Read before the Royal Society, March 14, 1782.

When the sky is free from thunder-clouds, an ordinary conductor erected in the best manner, . . . seldom gives any indications of electricity. . . . But by means of the apparatus I am going to describe, it is found that the said conductors are never entirely devoid of electricity, and consequently, it must be concluded that the air which surrounds them is also at all times electrified. . . . The whole method may be reduced to the following few observations. . . . An electrophorus must be prepared whose resinous coat is very thin and not at all electrified. . . . A wire must be brought from the atmospherical conductor to . . . the metal plate of the electrophorus. . . . Lastly, the wire must be removed: the metal plate is then separated from the resinous one, by lifting it up by its insulating handle, after which it is found to electrify an electrometer or even to give sparks. . . . This method not only detects the existence of electricity but also enables one to ascertain whether it is positive or negative and this even when the conductor itself is not capable of attracting the finest thread. The electrophorus in this case might perhaps better deserve the name of. . . . micro-electroscope, but I prefer to call it a condensator of electricity so as to use a term which expresses both the reason and the cause of the phenomenon under examination. . . . Before the invention of my condensator [or condenser] and of Mr. Cavallo's sensitive electrometer, we were very far from being able to detect such weak charges; whereas at present, we can observe a quantity of electricity incomparably smaller than the smallest previously observable.[3]

CAPACITY

The same letter contains the term "capacity," which Volta used for the first time in a letter to his Swiss friend Horace Saussure, professor of philosophy at the Academy of Geneva. The letter was written during Volta's summer holidays of 1778.

I have thought of substituting for the resinous plate a plane which should not be a perfect electric [not an excellent insulator] as for instance, a clean dry marble slab, or a piece of wood, . . . covered with a coat of varnish. . . . The various experiments that can be made with my condensing apparatus, throw considerable light on the theory of electric atmospheres [fields]. The whole matter . . . can be reduced to this, viz., that the metal plate has a much greater capacity for holding electricity . . . when it lies on a suitable plate, than when it is quite isolated, as when it is suspended in mid-air by silk strings or by an insulating handle. . . . Wherever the capacity for holding electricity is greater, there, the intensity is proportionately less, that is, a greater quantity of electricity is required . . . in order to raise its intensity [potential] to a given degree; so that the capacity is inversely as the intensity.

[3] *Trans. Roy. Soc. Lond.* (1782), LXXII, 237.

The idea which Volta expressed by "intensity" and Cavendish by "degree of electrification" was given the name *potential* by Lagrange and Poisson about 1820. In 1861, a British committee named the unit of potential the "volt"—in honour of Volta, of course. In equations, a symbol often used for potential is V, which is Volta's initial and suggests volts. The relation whichVolta stated in his letter to Saussure may be expressed thus:

Eq. 1. $$c = \frac{q}{V}$$

where c is capacity (or capacitance) of a body, q, the charge or quantity of electricity on it and V, its potential or the difference of potential produced by the charge. Thus as Volta said, the capacity is inversely proportional to the potential difference produced.

Volta's phrase, "a perfect electric," is the last vestige we shall meet of Desaguliers' meretricious statement about electrics being non-conductors and conductors non-electrics.

VOLTA ON GALVANISM

After the publication of Galvani's treatise, a great many of Volta's researches were devoted to investigating the rich fields it opened. One of Volta's references to the work of Galvani occurs in a letter to his friend Baronio which was read at the commencement exercises of the University of Pavia, May 5, 1792:

An essay appeared a few months ago on the action of electricity in muscular motion, written by Dr. Galvani, a member of the Institute at Bologna and professor in the University of that place, who has already won a wide reputation by his other anatomical and physiological discoveries. It contains one of those splendid major discoveries which, in the annals of physical and medical sciences, serve to usher in new epochs; not only because it is new and wonderful but also because it opens a broad field of experiments that are especially and outstandingly capable of application. . . . A discovery of such calibre must excite admiration everywhere it becomes known, but more especially amongst us Italians since it is the contribution of one of our fellow-countrymen. Immediately, many investigators vied with each other in repeating Galvani's experiments. . . . I did the same thing myself during my Easter holidays. I must confess that it was with scepticism and faint hopes that I proceeded to the first ones, so incredible did the accounts seem to me. . . . I am not ashamed to beg the discoverer's forgiveness for my unbelief and obstinacy. . . . Now I am converted . . . indeed I have performed the miracles myself and have swung over from incredulity to enthusiastic belief.

Volta's results led him to doubt that Galvani's assumption of animal electricity was necessary or even tenable. There is a passage in his letters which marks the place where Volta's work took this new turn:

The foot of a large frog is separated from the torso and the sciatic nerve exposed . . . and cleared of all flesh. Then its upper end is bound with tinfoil. . . . The same technique is performed . . . lower down so that the two coatings are only a few lines apart, and below the second one, a part of . . . the nerve remains exposed. Now if an extremely weak discharge, hardly capable of giving a spark, is conducted to these coatings, . . . so that the part of the nerve between them is traversed by it, all the muscles stir into motion and the whole foot twitches. . . . It is therefore, not necessary that the electricity reach as far as the muscles and traverse them. All that is required is to stimulate the nerves that control the motions of the voluntary muscles concerned.

Sometimes the professor was his own guinea-pig. In one of his letters to Cavallo, Volta described an experiment in which he used his own body as a test-object or indicator:

Similarly, it occurred to me that in the case of the tongue, we have a naked muscle. . . . Here, I said, all the conditions are fulfilled. With this in mind, I performed the following experiment upon my own tongue. I covered the point of the tongue . . . with a strip of tin. . . . With the bowl of a silver spoon, I touched the tongue further back; then, I inclined the handle of the spoon to touch the tin. I expected . . . a twitching of the tongue, so I did the experiment in front of a mirror. . . . The expected sensation, however, I did not perceive at all; but instead, a rather strong acid taste at the tip of my tongue. This surprised me at first; but after a little thought, I realized that the nerves which end at the tip of the tongue, are for receiving taste sensations and have nothing to do with moving this very nimble muscle. Hence it was very natural that the electrical stimulus produced a taste and nothing else.

PERSISTENT FLOW

In discussing the tongue experiment, Volta showed that he had a clear view of the chief feature of the new epoch which his work was bringing into the world and of the coming of the Electrical Age:

In order to experience this taste, one must place a clean smooth strip of tinfoil against the tip of the tongue and press firmly; . . . on some other part of the tongue, a gold or silver coin is placed; then these two applied bodies are brought into contact with each other. . . . It is also worthy of note that this taste lasts as long as the tin and silver are in contact with each other. . . . This shows that the flow of electricity from one place to another, is continuing without interruption.

The name given to a steady flow of electricity is the word "current," and from this date Volta used it more and more. His great service to the world was his leadership into the fourth epoch of electricity, which was characterized by the study and application of electric currents. This new department of electricity is referred to as electrokinetics (or electrodynamics). In a course in electricity, the change from the study of static electricity to that of electric currents is so radical, as indicated by the apparatus used, that one might think the course had switched to a different subject.

By his next observation, Volta turned the first sod for a new science called electro-chemistry; but he was in no position to realize the full significance of his discovery.

Not less remarkable is the fact that if the experiment is reversed so that the silver touches the point of the tongue and the tin . . . its middle, a very different taste is perceived at the tip of the tongue, which is no longer sour, but more alkaline, sharp and approaching bitter.

This experiment and others rising from it afford a method of determining the direction or sense of an electric current.

THE VOLTAIC PILE AND CELL

Of all Volta's writings, the most momentous were those in which he recorded how he made his Pile, an instrument which yielded a continuous flow of electricity. The announcement was made in a letter to the President of the Royal Society of London, of which Volta was a Fellow:

Como, March 20, 1800.

SIR JOSEPH BANKS, P.R.S.,
LONDON,

Illustrious Sir,

After a long silence, I have the pleasure of communicating to you . . . some results . . . that I have obtained in pursuing my experiments on electricity,—excited simply by the mutual contact of different kinds of metal. The chief of these results . . . is the construction of an apparatus which resembles in its effects . . . a feebly charged . . . battery of Leyden jars. It works continuously, however, that is, after each discharge, its charge re-establishes itself. The apparatus is merely . . . an assemblage of a number of good conductors of different sorts. . . . I obtained some little round discs . . . of silver . . . of a thumb's breadth and an equal number . . . of zinc. . . . I prepare . . . a number of cardboard discs . . . well soaked in brine. I place them horizontally upon a table . . . first one of silver, and on this I fit a second one of zinc; on the second, I lay one of the wet discs; then another silver plate. . . . I continue in the same manner . . . to build up a tower . . . as high as can stand without toppling [fig. 299]. Now if it will stand up to reach about twenty of these stories of metal couples, it will already be capable not only of making the electrometer signals of Cavallo, aided

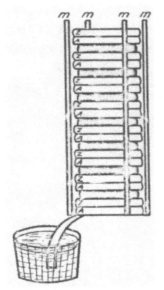

FIG. 299. Voltaic pile.

FIG. 300. Huge storage battery.

by the condenser, . . . but even to the point of giving a spark. Every one of these sparks resembles perfectly the slight effect given by a weakly charged Leyden jar or a battery of jars even more weakly charged, or, finally, an extremely exhausted torpedo which imitates even better the effects of my apparatus on account of its ability to give shocks one after the other. This apparatus, similar to the electric organ of the torpedo . . . I should like to call the "artificial electric organ."[4]

In spite of Volta's suggestion, the instrument has always been called Volta's Pile or the *voltaic pile*. It caused a real furore among the electricians of Europe and America, but its career was briefer than that of Volta's most important modification of the pile, namely, the Voltaic Cell. A group of such cells properly connected is called a voltaic battery. The name is obviously taken from the terminology of the Leyden jar but, according to Volta's description of the instrument, the term has lost all its original meaning of violence. The voltaic cell was the source of the electric currents used in the early decades of the nineteenth century. It still serves that purpose, though to a minor degree, principally in the forms called the Dry Cell and the storage cell.

[4]Quoted in Gerland and Traumüller, *Geschichte der Physik.*, p. 361.

The letter to Sir Joseph Banks contained a description of a battery of voltaic cells (fig. 301): Volta called each of his cells a cup.

> In the accompanying diagram, I represent this new apparatus which I call a "crown of cups." A row of several cups is arranged; little drinking glasses or goblets are most suitable. Each is half filled with brine . . . and they are joined to form a sort of chain by means of metallic bows, in each of which . . . the extremity *A* dips into one of the goblets and is of copper . . . and the other end *Z* is of zinc. . . . The two metals composing each bow are soldered together.[5]

In a memoir addressed to the French Royal Academy, Volta showed that any two different solid conductors might be used as the plates in a voltaic cell. His discussion contained the source of the term *electric pressure*. He used the word "tension," which was probably a source of

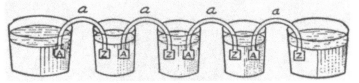

FIG. 301. Battery of voltaic cells.

the rather questionable term "electromotive force," and led also, though less directly, to the concept of potential difference (abbrev., p.d.). The same memoir also stated the general conclusion called Volta's law.[6]

> Several other metals differ less in the power to produce the electric fluid, yielding a tension which is less according as they are nearer in the following series . . .
> silver, copper, iron, tin, lead, zinc . . .
> a scale in which the first drives the electric fluid to the second, the second to the third, etc. There are, however, other substances that appear to force the electric fluid to the metals, especially zinc, with greater pressure than silver; these substances are plumbago, . . . and especially, black manganese.[7]

As we have seen, the esteem in which Volta's name is held among scientists is witnessed by the fact that the unit of potential difference is named in his honour the *volt*. It is his "monument more lasting than brass."

INSIDE INFORMATION

There seemed to be hardly any limit to which Volta's curiosity would not impel him in using his body as a test object:

> Concerning the sense of hearing. . . . I have finally succeeded in stimulating

[5] *Phil. Trans. Roy. Soc. Lond.* (1800), XC, 410.
[6] Quoted in Silvanus Thompson, *Notes on De Magnete*, p. 86.
[7] *Annales de Chimie et de Physique* (1801), XL, 243.

this sense with my new apparatus, composed of ... 40 couples. ... I introduced ... into my ears ... two metal rods ... with rounded ends and joined them to the terminals of the apparatus. At the moment when the circuit was completed, I received a shock in the head: and a few moments afterwards, the circuit remaining closed, I began to hear a noise, which I cannot well describe. It was a sort of crackling or bumping ... as if some paste or viscous material [perhaps gray matter?] were boiling. ... This disagreeable sensation which I feared might be dangerous [!] ... has deterred me so that I have repeated the experiment only six or seven times.[8]

In 1801, Napoleon invited Volta to address the Institut de Paris on his electrical researches. The French struck a gold medal in his honour and Napoleon accorded him signal distinctions.

How different was Volta's fate from that of his brilliant friend Galvani, who was really the discoverer of current electricity, or "galvanism" as it was first called. After the death of Galvani's beautiful and accomplished wife, his star declined and he cared little what happened; yet because of his elysian marriage he would not have changed destinies even with Volta. Galvani considered Napoleon a marauder on Italian territory, and for his objections was dismissed from his professorship. After a public outcry, he was reinstated but he died before resuming the chair.

When Volta resigned because of waning strength, Napoleon refused to accept his resignation, saying that the university should not suffer such a loss of prestige. He retained Volta in his position at an increased stipend and with his duties reduced to the delivery of one lecture per annum!

ELECTROLYSIS OF WATER

Sir Joseph Banks showed Volta's letter to William Nicholson,[9] an engineer and publisher whose hobby was science. In collaboration with a chemist, Sir Anthony Carlisle, he assembled a large voltaic pile and, by means of it, discovered the electrolysis of water (fig. 302).[10] The priority for this observation goes to three others, Ash, Fabbroni, and Crève, but the work of Nicholson and Carlisle identified the gases produced. Some idea of the fever of excitement caused by Volta's invention of the pile may be gathered from the fact that some of Nicholson's results were published actually before Volta's letter appeared in the *Transactions* of the Society. It would be difficult to find a parallel case.

The contacts being made sure by placing a drop of water upon the plate, Mr. Carlisle observed a disengagement of gas around the touching wire. This

[8]*Phil. Trans. Roy. Soc. Lond.* (1800), xc, 427.
[9]Magie, *Source Book of Physics*, 432.
[10]Gerland and Traumüller, *Geschichte der Physik.*, p. 365.

gas, though very minute, . . . seemed to me to have the smell afforded by hydrogen when the wire of communication was steel. . . . On the second day of May, we, therefore, inserted a brass wire through each of two corks, inserted in a glass tube. . . . [fig. 303]. The tube was filled with new river water, and the distance between the points was one inch and three quarters. This compound discharger was applied so that the . . . external ends of its wires were in contact with the two extreme plates of a pile of thirty-six half-crowns [silver] with the correspondent pieces of zinc and pasteboard. A fine stream of minute bubbles immediately began to flow from the point of the . . . lower wire . . . which communicated with the silver and the opposite point of the upper wire became

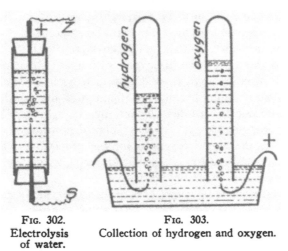

FIG. 302.   FIG. 303.
Electrolysis   Collection of hydrogen and oxygen.
of water.

tarnished. . . . [copper oxide?] On reversing the tube, the gas came from the other point. . . . The product of gas . . . was mixed with an equal quantity of common air and exploded by the application of a lighted wax thread. We had been led by our reasoning on the first appearance of hydrogen to expect the decomposition of water: but it was with no little surprise that we found the hydrogen extricated at the contact with one wire, while the oxygen fixed itself in combination with the other wire at the distance of almost two inches. . . .

This new fact still remains to be explained and seems to point at some general law of the agency of electricity in chemical operations.

The simple decomposition of water by platina [platinum] wires without oxidation offered a means of obtaining the gases separate from each other. A cloud of gas arose from each wire but most from the . . . minus side. It was found that the quantities of water displaced [in the collection tubes] by the gases were, respectively, 72 grains by the gas from the plus side [oxygen] and 142 grains by the gas from the minus side [hydrogen].[11]

This experiment marked the beginning of the science of electro-chemistry.

[11]Quoted in Magie, *Source Book of Physics*, p. 436.

OERSTED'S EXPERIMENT

We have noted the growing conviction among electricians that there must be some relation between magnetism and electricity. A professor of physics at Copenhagen, Johann C. Oersted (1777-1851), published, in 1819, a treatise that brought to light the long-suspected relation between the two sciences. From then the two subjects marched together and the combination was more powerful than the sum of its two components.

Knowing that Danish would be inconvenient to many scientists, Oersted wrote his report in Latin. The only subsequent Latin paper in

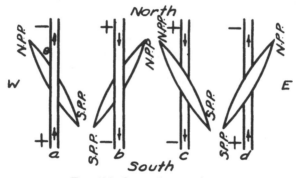

FIG. 304. Oersted's experiment.

this science was Gauss's paper of 1832 on magnetism. Although Oersted was a good linguist yet his Latin was hardly Ciceronian, for with the purpose he had in hand, it was of little moment whether a certain verb took *ut* and the subjunctive or accusative and infinitive.

The first experiments on the topic of this paper, I made last winter while giving a series of lectures on electricity, galvanism, and magnetism in the university. By those experiments it seemed demonstrated that the magnetic needle is deflected by the galvanic apparatus [voltaic pile or battery], provided the galvanic circle [external circuit] is complete. Let the opposite ends of a galvanic battery be joined by a metallic wire, which, for brevity's sake, we shall call the "uniting conductor" [external circuit]. To the effect produced in this conductor and in the surrounding space, we shall apply the name, electric storm or cyclone ("electricus conflictus").

Let a straight part of this wire be placed horizontally above the magnetic needle, properly suspended and parallel to it. . . . The needle will be deflected and the end of it which is under the part of the uniting wire conductor that is next to the negative side of the battery will turn westward. . . . If the wire . . . be placed under the needle, all the effects are the same as when it is above . . . except that the directions [or senses] are reversed; for the pole of the needle next to the negative end of the battery declines to the east. . . .

If the distance of the uniting wire is increased, the angle of deflection . . . diminishes. . . . The declination likewise varies with the power of the battery.

... The nature of the metal of the uniting wire does not altei the effect in kind but merely in degree.

The uniting wire may shift eastward or westward parallel to itself without any change of the effect other than with respect to its quantity. Hence, the effect cannot be ascribed to attraction; for the same pole of the magnetic needle that approaches the uniting wire while placed on its east side, would have to recede from it when on the west side, if these deflections depended on attractions and repulsions. ...

The effect of the uniting wire passes to the needle through glass, metals, wood, water, resin, stoneware, etc., for it is not prevented by interposing bodies of these substances.

It is sufficiently evident from the preceding facts that the electric cyclone is not confined to the conductor, but extends pretty widely in the circumjacent space. From the preceding facts, we may also conclude that this cyclone moves in circles. ... The electric storm acts only on the magnetic particles of matter. All non-magnetic bodies appear transparent to the electric storm. ... If the joining wire is placed perpendicular to the pole of the needle, and the upper end of the wire receives electricity from the positive end of the galvanic apparatus, the pole is moved towards the west:[12]

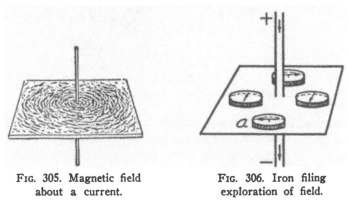

FIG. 305. Magnetic field   FIG. 306. Iron filing
  about a current.   exploration of field.

Figures 305 and 306 indicate where Oersted obtained his term electric cyclone. Because the needle remains deflected while the circuit is closed or complete, Volta, Ampère and others called such a flow a *current*.

Oersted's discovery gave a powerful stimulus to the study of electricity. It opened the door to the study of electric currents (electrokinetics) and marked the beginning of the study of electro-magnetic effects. It would not be illogical to consider 1819 as the beginning of a new epoch. The whole aspect of the science changed from that date. Amber rods, electroscopes, Leyden jars, and electric machines faded into the background and voltaic batteries, galvanometers, electromagnets, generators, telephones, and vacuum tubes came to the fore.

[12]H. C. Oersted, *Experimenta circa effectum conflictus electricae* (1820). The book is in the University Library, Göttingen, Germany, from which this material was obtained through the courtesy of Dr. Leon Leppard.

Most of the scientists we have met heretofore have been proficient in technique as well as in thinking and questioning, for example, Newton, Galileo, Cavendish, Coulomb, Rowland, and Michelson, but a few have been deficient in manual dexterity, for instance, Kelvin —and now Oersted whose clumsiness in technique necessitated the help of an assistant to perform the manipulative parts of his experiments while Oersted directed. Yet he rendered invaluable service to science and his name is well known in electricity while the name of his deft and practical assistant is generally forgotten. The unit of magnetic field strength is now called the *oersted*; previously it was called a gauss.[18]

As an indication of the potency of the influence of Oersted's paper, we can note that it was only a few months after its publication that a German professor of physics, J. S. C. Schweigger of Halle (1779-1857), modified Oersted's experiment so as to produce a new indicator (1820). To increase its sensitivity—a desideratum after which physicists are perpetually striving—Schweigger turned the wire back so as to let the same current pass the needle a number of times. This was the advent of the *coil* or helix in electricity and it is now long since in universal use. The new instrument served three purposes (1) to detect a current, (2) to indicate its sense, and (3) to compare qualitatively the strengths or intensities of two currents (galvanoscope). When later developed to compare two intensities quantitatively, the name was changed to galvanometer.

AMPÈRE'S SWIMMER RULE

Oersted's rule for predicting the sense of the needle's deflection, though correct, was not satisfactory from the standpoint of simplicity, clarity, and ease of remembering and applying. A better rule was quickly formulated by the brilliant scientist M. Ampère of Paris (1775-1836). Here we have another instance of the process of simplification or advancement from cumbersome to elegant. Ampère's rule is given in a monumental treatise on electro-magnetism which he published in 1823. This treatise is a masterpiece. It is a model of beauty and clarity of style; Maxwell described it as "perfect in form and unassailable in accuracy." It is probable that such an appraisal can be truer of a treatise in French than of one in any other language.

The ordinary electroscope indicates when there is tension and the intensity of that tension; we used to lack an instrument to detect the presence of an electric current . . . in a conductor and to indicate its energy and direction. To-day that instrument exists; it is sufficient to place the pile or any part of the conductor approximately in the magnetic meridian horizontally and to put . . above or below any part of the conductor, an apparatus similar to a

[18]See Robert B. Lindsay, *General Physics for Students of Science* (Wiley, 1940), p. 353.

compass, differing only in the use we make of it. As long as there is any break
in the circuit, the compass needle rests in its ordinary position; but it is deflected
from that position as soon as the current is established and with an angle proportional to its energy. Also it indicates the sense according to the following
general rule, that if one imagine himself placed in the direction of the current

FIG. 307. André M. Ampère.

so that it is directed from the feet to the head of the observer, and that the
latter has his face turned toward the needle, then, it is constantly toward his
LEFT that the action of the current deflects from its normal position that end
of the compass which points northward and which I shall always call the austral
pole of the magnetized needle, for it is analogous to the austral pole of the earth.
This I shall express more briefly by saying that the austral pole of the needle
is deflected to the left of the current that moves the needle. I think that to

distinguish this instrument from the ordinary electroscope, we should give it the name galvanoscope.[14]

To honour Galvani thus was excellent but it does seem that the inventor, Schweigger, should have received more credit than he did. Ampère was a prodigious worker, for, like Galileo and many other brave human souls, he sought to drown sorrow in hard, unremitting labour. His life was tragic. Unmerciful disaster struck him down once and once again beyond recovery. In the Revolution, young Ampère, whose father was his teacher and chum, was forced to witness the dreadful death of that beloved and blameless parent at the guillotine. The youth was mentally paralysed. He would wander speechlessly about the house or sit in the garden and trickle some sand through his fingers. From this apathy Ampère's happy marriage spelled recovery, but a few years later, he was stricken by the untimely death of his young wife. From this blow he never recovered and the feverish pace at which he worked may well have been a defence mechanism. The whole treatise from which we have read was composed within two or three years of Oersted's paper, which was its starting-point. Its "swimmer-rule" is often expressed as a Right-hand-Rule: imagine the right hand in the current, pointing with the current and with the palm toward the needle, then the thumb indicates the sense in which the north-pointing pole of the needle is deflected.

That Ampère's prestige is great is attested by the fact that the common unit of electrical current strength or intensity is called in his honour the *ampere*. If one coulomb of electricity flows past a point in a circuit in one second the strength of the current is one ampere.

The same treatise deals with Ampère's discovery of the laws by which two currents act upon each other.

But the differences that I have just recalled are not the only ones that distinguish these two states of electricity. I have discovered some still more remarkable ones by arranging in parallel directions two rectilinear parts or two conducting wires joining the terminals of two voltaic piles. One of them was fixed and the other, suspended from points and rendered very mobile by a counterpoise, was able to approach the first one or recede from it, keeping its parallelism. I observed then that when an electric current was sent through both of these at the same time, they [parallel currents] mutually attracted each other when the two currents were in the same sense, and that they repelled each other when the currents flowed in opposite sense.

THE ELECTRO-MAGNET

One direct outcome of Oersted's work was the invention of the electro-magnet. Schweigger's invention of the galvanoscope was simply

[14]*Annales de Chimie et de Physique* (1820), xv, 66-9.

a matter of modifying Oersted's experiment to convenient laboratory form and improving the sensitivity. That invention introduced the *helix* or coil of wire into the technique of electricians. Suppose we place an iron rod in a helix of insulated copper wire through which current from a voltaic battery flows. The rod or "core," being a magnetic body may be considered as composed of minute bipolar magnets. Some of these at least would probably turn like compass-needles in accordance with Ampère's swimmer or right-hand rule. The resulting change of arrangement so that a considerable portion of the rod's particles would have like poles pointing toward the same end of the rod, is the particle picture for the process of magnetization. Experiment corroborates this prediction, and such a configuration is called an electro-magnet (fig.

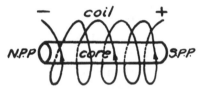

FIG. 308. Electro-magnet.

FIG. 309. Arago's experiment with electro-magnet.

308). Without this fundamental instrument, there could be no Electrical Age, for the generator, which is the fundamental source of current, is largely an application of the electro-magnet. And so are other important instruments—the motor, transformer, electric bell, telegraph, telephone, etc. Fig. 310 shows an electro-magnet at work, lifting scrap iron. The credit for the invention of the electro-magnet is shared by four or five men, Arago, Ampère, Davy, and Henry, and a good share of the credit for its practical improvement goes to William Sturgeon of London, who gave the instrument its name in 1825. An interesting account of the invention (fig. 309) occurs in a paper written by François J. Arago, Astronomer Royal at Paris and Secretary to the Royal Academy of Sciences:

The brilliant discovery that M. Oersted has just made has to do, as we have seen, with the action that the voltaic current exerts upon a steel needle previously magnetized. While repeating the Danish physicist's experiments, I found that the same current produces strong magnetism in wires of iron or steel originally quite devoid of it. . . .[?] I succeeded . . . in magnetizing . . . a sewing needle. . . . In order to support it more conveniently, I wound a few turns of the connecting wire around its ends. . . .

M. Ampère, to whom I showed these experiments, had just made the important discovery that two . . . parallel currents . . . attract each other when in the same sense . . . and repel when in opposite sense. . . . He deduced . . . by analogy, the consequence that the attraction and repulsion of magnets depend on electric currents circulating about the molecules of iron and steel in a direction perpendicular to the axis of the magnet. . . . These theoretical views suggested to him at once the idea that one might obtain a stronger magnetization by substituting for the straight connecting wire I had used a wire wound into a helix, in the middle of which a steel needle is placed. . . . A copper wire, bent into a helix, was terminated by two straight portions that could be connected to the opposite poles of a powerful voltaic pile. . . . A steel needle, wrapped in paper, was introduced into the helix. . . . After a few moment's sojourn in the helix, the steel needle had received a rather strong "dose" of magnetism; moreover, the location of the north and south poles was found to conform perfectly with the results that M. Ampère had predicted.[15]

Significant words these!

FIG. 310. Electro-magnet lifting scrap-iron.

OHM'S LAW

Georg Simon Ohm, though miserably hampered by poverty (he was a teacher), struggled on with a series of researches on the conduction of electric current, and ultimately discovered the famous relation known as Ohm's law, which transformed chaos into cosmos in the science of electric currents. The appreciation of his labours is attested by the fact that the unit of electrical resistance is named the *ohm*. His chief source of inspiration was Fourier's treatise on the conduction of heat[16] published in 1822. Ohm's father was a locksmith and a mathematician. He gave his son a training in mathematics which stood him in good stead throughout his career. Without this training, for example, he could not have read Fourier's book. Some of Ohm's writings are rather difficult reading for beginners but a few excerpts will be given here. They are well worth any struggle. His 1825 paper arranged a number of conductors in order of their conductivities. It was addressed to Schweigger, the inventor of the galvanometer, without which Ohm could not have done his work unless he had been willing to resort to Cavendish's heroic method.

I am very glad to be able to confirm by a series of observations what you

[15]Ibid., p. 94.
[16]Joseph Fourier, *La Théorie Analytique de la Chaleur.*

published in section 5 of your *Chemical and Physical Yearbook* for 1825, page 121, namely, that the magnetic action of the galvanic current may offer an excellent means of determining the conductivities of the metals quantitatively. One can hardly believe the exactness with which the evidence turns out by this method. The series of metals from the best to the worst conductor is as follows, copper, gold, silver zinc, brass, iron, platinum, tin, lead, so that indeed copper conducts about 10½ times better than lead.[17]

FIG. 311. G. S. Ohm.

Later, Ohm found an error in his measurement for silver. The wire he used first, he discovered, had a coat of varnish obtained from the gloves of the man who drew the wire. Upon repeating the experiment with a clean wire, he found silver to be at the head of the list as the best of all conductors of electricity. Then he published a correction of his error. He may have grown suspicious of his first result with silver, from his knowledge of thermal conductivity, for there is a parallelism, as may be seen from the following thermal conductivities: silver 1.006

[17]G. S. Ohm, *Collected Works* (ed. Lommel), p. 9. Also quoted in Schweigger's *Journal für Chem. und Physik.* (1825), XLIV, 245.

copper 0.918, gold 0.70, zinc 0.265, brass 0.26, platinum 0.166, iron 0.16, tin 0.155, lead 0.083.

I had prepared 8 different conductors . . . which were respectively 2, 4, 6, 10, 18, 34, 66, and 130 inches long, 7/8 lines thick [1 inch = 12 lines] and cut from the same piece of so-called plated copper wire. . . . These wires were inserted in the circuit one after the other. . . . I thus obtained the following results: Jan. 15th.

length	2 in.	4 in.	6 in.	10 in.	18 in.	34 in.	66 in.	130 in.
deflection	305¼°	281½°	259°	224°	178½°	124¾°	79°	44½°

These numbers can be represented very satisfactorily by the equation $X = \dfrac{a}{b + x}$ in which X is the strength of the magnetic action when the conductor is used whose length is x, and a and b are constants which represent magnitudes depending on the exciting force and the resistance of the rest of the circuit. If, for example, we set b equal to 20¼ and a equal to 6800, we obtain by calculation the following results:

x =	2 in.	4.0	6.0	10.0	18.0	34.0	66.0	130 in.
X =	305½°	280½°	259°	224⅜°	177¾°	125¼°	79°	45°

If we compare these numbers, found by calculation, with the former set, found by experiment, it is seen that the differences are very small and are of the order of magnitude that one might expect in researches of this kind.[18]

In Ohm's equation, X is proportional to the strength of the current or intensity; the usual symbol for it is I or, in some older texts, C. The quantity a is proportional to the potential difference between the ends of the wire or the electrical pressure or electro-motive force (symbol E or V); x is the length of the wire and proportional to its property which Ohm called its *resistance*, and b is the resistance of the rest of the circuit. Hence $b + x$ represents the resistance of the whole circuit (symbol R). The equation may therefore be written

Eq. 2. $$I = \frac{E}{R}$$

Ohm's 1827 paper, *The Electric Circuit Mathematically Treated*,[19] is the most famous of his writings. It brought him vilification at home as an irresponsible visionary, but the Copley medal of the Royal Society of London and other honours abroad. Then came recognition in his own country, an associate professorship and later a professorship at Munich. The latter distinction had always been his ambition but he died a few years after reaching that elevation. It is seldom that a scientist's paper complains about his circumstances, but in Ohm's preface we see his bitterness at the thought of what he could have done if he had had enough money to buy the books and apparatus he needed:

[18]G. S. Ohm, *"Bestimmung des Gesetzes nach welchem Metalle die Contaktelektrizität leiten"* (Schweigger's *Journal*, 1826), XLVI, 137.
[19]*Die galvanische Kette mathematisch bearbeitet* (Gesammelte Abhandlungen) p. 61.

I herewith present to the public a theory of galvanic electricity. . . . The circumstances under which I have lived until now have not been conducive to fanning up my enthusiasm after the threatened cooling off, caused by the daily grind, nor to allowing me to become familiar, as is certainly necessary, with the whole range of the literature relating to cognate researches. Consequently, I have chosen . . . an 'Arbeit' [problem] in which I need least fear that I am intruding upon another man's research. . . . May the gentle reader receive my performance with the same love for the subject as that with which it is sent out into the world.

The endeavour of this memoir consists in the attempt to derive in a rigorous manner an all-embracing relation from a few principles, obtained mostly from experiment, coordinating the essential ideas about those electrical phenomena which are . . . designated by the term galvanic. Its aim will have been achieved if the multiplicity of the facts is reduced in this manner to a simplicity of concept.

Thus again we hear a scientist declaring his purpose to advance from the Many to the Few or to the One. In this project Ohm could not have wished better success, for his law has been tested closely, often with an accuracy of 1 in 100,000, and it has successfully withstood all tests.

The strength of the current in any homogeneous part of the circuit is determined by the quotient which is formed from the difference of electrical pressures (*Spannung*, tension) at the two ends of this part, divided by its resistance. . . . The second of the former equations, by the proposed change, assumes the form $I = \dfrac{E}{R}$, which is true in general and by its own form reveals the uniformity of the current strength throughout the circuit. In words, it is expressed thus:

The strength of the current (I) in a galvanic circuit is directly proportional to the sum of all the pressures (E) and inversely as the total resistance of the circuit (R), bearing in mind that by resistance is understood the sum of all the quotients formed by dividing the actual lengths of homogeneous parts by the products of the corresponding conductivities and their cross-sections.

Ohm had previously shown that the intensity of a current at any instant throughout a series circuit is uniform or constant (Ohm's theorem).

The same paper gives an excellent discussion of the fundamental problem of *divided circuits*. Here is one of its most crucial statements:

From which it follows immediately that the current strengths . . . in the branches, are to one another inversely as their resistances.

It is important to note that it is only when electricity is in motion that it produces magnetic effects.

CHAPTER TWENTY-TWO

Induced Currents

THE germs of the first four epochs in electricity, typified by the ideas of charge, conduction, influence, and current, we have found in the words and work of Thales, Gray, Canton, and Volta. In the present chapter occur the pregnant words which brought the fifth epoch of electricity into the world and the Electrical Age with all its blessings and abuses. This advancement was achieved largely by the application of Michael Faraday's discoveries in the world's affairs. Most of the currents used in the present age for household, industrial, and municipal purposes are induced currents, obtained from generators.

FARADAY'S DISCOVERIES

Although the voltaic battery yields continuous flow of electricity, yet it could not bring the electrical age because there was not enough zinc in the world to light all the streets and buildings and to run the factories. What was needed was some supply of electricity of much greater magnitude. The chief source of energy available, of sufficient magnitude, is mechanical energy (kinetic and potential), for example, the energy of waterfalls, winds, steam engines, and so forth.

Faraday conjectured that a magnet should have in some way an electrical effect. He was searching for a new relation between magnetism and electricity. In Oersted's experiment the energy of electricity is converted into the kinetic energy of the needle, which is a magnet. Faraday knew that many energy transformations can be reversed. Thus in a steam engine, heat is transmuted into kinetic energy and contrariwise, when a boy slides down a bannister, kinetic energy is transformed into heat. Faraday asked himself how he could reverse the conditions of Oersted's experiment so as to obtain electrical energy from the kinetic energy of a magnet. By his experiments, he asked Nature this question. It took him seven years of patient and discouraging work to find the answer. When in 1831 he did succeed, realizing the tremendous significance of his discovery, he was quite overcome with joy. It is said that he gambolled about the laboratory like a youngster and even jumped right over the table. Such an antic on the part of a staid English gentleman was certainly a "cutting loose"—though still milder than Archimedes' mode of expressing joy. In reporting his discoveries to the august Royal Society, Faraday was somewhat less boisterous:

It appeared very extraordinary that as every electric current was accompanied by a corresponding intensity of magnetic action, . . . good conductors of electricity when placed within the sphere of this action, should not have any current induced through them. . . . These considerations with their consequences, the hope of obtaining electricity from ordinary magnetism, have stimulated me at various times to investigate experimentally the inductive effect of electric currents. I lately arrived at positive results. . . .

(6) About twenty-six feet of copper wire, one-twentieth of an inch in diameter, were wound round a cylinder of wood as a helix, the different spires of which were prevented from touching by a thin interposed twine [!]. . . . This helix was covered with calico, and then a second wire applied in the same manner. In this way, twelve helices were superposed, each containing an average length of wire of twenty-seven feet, and all in the same direction. The first, third, fifth, seventh, ninth, and eleventh of these helices were connected at their extremities, end to end to form one helix; the others were connected in a similar manner: and thus, two principal helices were produced, closely interposed, having the same direction, not touching anywhere, and each containing one hundred and fifty-five feet of wire.

A combination of helices like that already described in (6) was constructed upon a hollow cylinder of pasteboard; there were eight lengths of copper wire, containing altogether 220 feet; four of these helices were connected end to end and then with the galvanometer. A cylindrical magnet three quarters of an inch in diameter and eight inches and a half in length was used. One end of this magnet was introduced into the axis of the helix . . . and then, the galvanometer needle being stationary, the magnet was suddenly thrust in; immediately, the needle was deflected. . . . Being left in, the needle resumed its first position, and then the magnet being withdrawn, the needle was deflected in the opposite direction.

A welded ring was made of soft round bar-iron, the metal being seven-eighths of an inch in thickness, and the ring, six inches in external diameter. Three helices were put around one part of this ring, each containing about twenty-four feet of copper wire one twentieth of an inch thick; they were insulated from the iron and from each other, and superposed in the manner before described (6), occupying about nine inches in length upon the ring. They could be used separately or arranged together; the group may be distinguished by the mark A [fig. 312]. On the other part of the ring, about sixty feet of similar copper wire, in two pieces, were applied in the same manner, forming a helix B, which had the same common direction with the helices of A . . . but being separated from it at each extremity by about half an inch of the uncovered iron. . . . The helix B was connected by copper wires with a galvanometer three feet from the ring. The wires of A were connected end to end so as to form one long helix, the extremities of which were connected with a battery of ten pairs of plates four inches square. The galvanometer was immediately affected, . . . but though the contact was continued, the effect was not permanent, for the needle soon came to rest in its natural position. Upon breaking the contact with the battery, the needle was again powerfully deflected, but in the contrary direction to that induced in the first instance.

The various experiments of this section prove, I think, most completely the production of electricity from ordinary magnetism.[1]

How many Fellows of the Royal Society, as they listened to Faraday on that unique occasion, realized the far-reaching changes his paper would make in human affairs? Faraday himself saw farthest; and it was an abiding joy to him to know that he had rendered the world an immeasurable service.

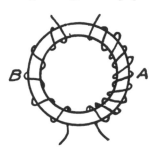

FIG. 312. Iron core with two coils.

Faraday's researches on electro-magnetism, though tremendous, were only a fraction of his contribution to electricity. He discovered the *laws of electrolysis* frequently referred to as Faraday's laws of electrolysis. With the help of his friend, Dr. William Whewell, he coined and introduced a number of useful terms:

In place of the term pole, I propose using that of *Electrode*, and I mean thereby that . . . body, which bounds the extent of the decomposing matter in the direction of the electric current.

The *anode* is . . . that surface at which the electric current, . . . enters: it is the *negative* extremity of the decomposing body; . . .

The *cathode* is that surface at which the current leaves the decomposing body, and it is its *positive* extremity;

Many bodies [substances] are decomposed directly by the electric current . . . ; these I propose to call *electrolytes*.

Substances are frequently spoken of as being *electro-negative*, or *electro-positive*. . . . But these terms are much too significant for the use to which I should have to put them; for though the meanings are perhaps right, they are only hypothetical, and may be wrong: and then, through a very imperceptible, but still very dangerous, because continual, influence, they do great injury to science, by contracting and limiting the habitual views of those engaged in pursuing it. I propose to distinguish such bodies by calling those *anions* which go to the anode . . . and those passing to the *cathode*, *cations*; and when I have occasion to speak of these together, I shall call them *ions*.[2]

Since he was both physicist and chemist, Faraday was especially well prepared for the investigation of electro-chemical problems. He and Gay-Lussac were complementary cases; for when Faraday was younger

[1] *Trans. Roy. Soc. Lond.* (132), cxxii, 125.
[2] *Experimental Researches in Electricity.* See p. 112 of Everyman's Library edition.

he was a chemist but as he matured he advanced to the status of a physicist.

Two laws of electrolysis that were discovered by Faraday, he reported as follows:

Fig. 313. Michael Faraday.

When water is subjected to the influence of the electric current, a quantity of it is decomposed exactly proportionate to the quantity of electricity which has passed.

In these experiments several substances were placed in succession [series] and decomposed simultaneously by the electric current:... the tin, lead, chlorine, oxygen, and hydrogen evolved being *definite in quantity* and electrochemical equivalents to each other.

A third law is what one might expect. The weight of ion liberated by a steady current is proportional to the *time*. These three laws can be expressed algebraically $w \propto ite$ where w is the weight of ion liberated, i, the current intensity, t, the time, and e, the equivalent of the ion; or

$w = kite$, where k is a variational constant. Also

$$i = \frac{w}{ket}.$$

In an experiment, w and t can be measured and k and e can be taken from tables. Hence we can calculate i. An electrolytic cell used for measuring current intensity is called a *voltameter*. Faraday gave it its name:

The instrument [electrolytic cell] offers the only *actual measurer* of voltaic electricity which we at present possess. . . . It takes note with accuracy of the quantity of electricity which has passed through it. I have therefore named it a VOLTA-ELECTROMETER.

The name was subsequently shortened.

In connection with Faraday's mention of the flow of a quantity of electricity, let us compare the flow of electricity with that of a water stream. This is only an analogy, of course, and proves nothing; but it can help not only the imagination and the memory but also our predicting faculty. In other words, this water analogy is another mechanical model, drawn, this time, from hydraulics or hydrodynamics. If a water current of intensity $I = 5$ gal. per min. flows through a channel for a time $t = 3$ min., the quantity of water passing any point in the channel $Q = 5 \times 3 = It$. Similarly if an electric current of intensity I amp. flows for t sec., the quantity of electricity passing any point in the circuit $Q = It$. Thus if the intensity is 6 amp. and the time 2 min. (120 sec.) then the quantity of electricity flowing is $6 \times 2 \times 60 = 720$ coulombs. An ampere liberates 0.001118 g. silver per sec.

In his discussion of electrolysis, Faraday seemed sometimes to be edging toward the modern electron theory. Noting the tremendous magnitude of the forces that combine atoms, he conjectured that if molecules could be somehow disrupted, scientists would tap a source of energy which would be far beyond the magnitude of those available in his day.

On the absolute quantity of Electricity associated with the particles or atoms of Matter

It is impossible, perhaps, to speak on this point without committing oneself beyond what present facts will sustain. . . . Although we know nothing of what an atom is, . . . and though we are in equal, if not greater, ignorance of electricity, . . . yet there is an immensity of facts which justify us in believing that the atoms of matter are in some way endowed or associated with electrical powers, to which they owe their most striking qualities, and amongst them their mutual chemical affinity.

Now it is wonderful to observe how small a quantity of a compound body [substance] is decomposed by a certain portion of electricity. . . . One grain of water will require an electric current to be continued for three minutes and three quarters . . . to effect its decomposition, which current must be powerful enough to retain platina wire 1/104th of an inch in thickness, red hot.

Considering . . . that without decomposition transmission of electricity does not occur; and that for a given definite quantity of electricity passed an equally definite and constant quantity of water . . . is decomposed: considering also that the agent which is electricity, is simply employed in overcoming electrical powers in the body subjected to its action; it seems probable . . . that the quantity which passes is the *equivalent* of . . . that of the particles separated; i.e. that if the electrical power which holds the elements of a grain of water in combination, . . . could be thrown into the condition of a *current*, it would exactly equal the current required for the separation of that grain of water into its elements.

This view of the subject gives an almost overwhelming idea of the extraordinary quantity or degree of electric power which naturally belongs to the particles of matter. . . . I showed that two wires, one of platina and one of zinc . . . immersed in acid . . . yield as much electricity in three seconds . . . as a Leyden jar charged by thirty turns of a . . . powerful electric machine. This quantity, though sufficient . . . to have killed a rat or cat, as by a flash of lightning, was evolved by . . . so small a portion of zinc and water as to be inappreciable to our most delicate instruments. What an enormous quantity of electricity, therefore, is required for decomposing a single grain of water! The proportion is so high that I am almost afraid to mention it; 800,000 such charges of the Leyden battery as I have referred to . . . would be necessary . . . to decompose a single grain of water [65 mg.].

To liberate 1 gm. hydrogen by the electrolysis of 9 gm. water or to liberate one equivalent of any element, requires 96,540 coulombs. This quantity of electricity is now called 1 *faraday*. To realize the significance of Faraday's statement, let us calculate the work or energy required to separate 1 gm. hydrogen. The work done (W) by a waterfall depends on two factors, the quantity (Q) of water falling and the drop or difference of level (V), and W is directly proportional to both Q and W. Hence $W = QV$, if proper units are used. Similarly, in the electrolytic separation of 1 gm. hydrogen, for instance in a battery of electromotive force 2 volts (more accurately, 1.9), the work done $W = QV$ where Q is in coulombs, V in volts and W in joules. Then $W = 96,540 \times 2$ joules $= 96,540 \times 2 \times 0.75$ (1 joule $= 0.75$ ft. pd.) $= 140,000$ ft. pd. (approx.). This energy would be enough to lift a ton of coal 70 ft. or twice the height of a house. Since 9 gm. water is about 1/3 oz. (1 oz. $= 28.4$ gm.), $\therefore$ the energy of this kind latent in 1 gallon of water (10 lb.)

Fig. 314. The Royal Institution, London, where Faraday lived, performed his experiments, and gave his lectures.

would lift a ton of coal (70 × 3 × 16 × 10) ÷ 5,280 mi. = 7 miles! Was it any wonder that Faraday used strong expressions in discussing this topic?

According to [this view], the equivalent weights of bodies [substances] are simply those quantities of them which contain equal quantities of electricity . . . : it being the ELECTRICITY which *determines* the equivalent number, *because* it determines the combining force. Or if we adopt the atomic theory or phraseology, then the atoms . . . which are equivalents to each other . . . have equal quantities of electricity naturally associated with them.

The chemical action of a grain of water upon 4 grains of zinc can evolve electricity equal in quantity to that of a powerful thunder-storm. Is there not, then, great reason to hope and believe that . . . by a closer *experimental* investigation of the principles which govern the development and action of this subtile agent, we shall be able to increase the power of our batteries, or invent new instruments which shall a thousandfold surpass in energy those which we at present possess?[3]

Here Faraday peered into the future and foresaw interatomic energy yielding new degrees of power a thousandfold greater than those of his day. It is hardly probable, however, that even this farseeing philosopher had glimpsed the titanic magnitude of nuclear energy. The energy

[3]Ibid., pp. 111-172.

released by the nuclear fission of 1 gm. plutonium is about 200,000 times that involved in the electrolytic liberation of 1 gm. hydrogen. 2×10^5 in these matters is a multiplier of such magnitude that it taxes the imagination to grasp its practical meaning.

From the application of Faraday's fundamental discovery of induced currents arose a numerous group of new instruments, the generator, the motor, the transformer; also the *induction coil*, which can receive a unidirectional current of relatively low potential difference and produce an oscillating current of higher electromotive force. From Faraday's studies in electrolysis came the electrolytic cell and voltameter, and improved processes in electro-plating and electrolytic treatment of ores; also some exceedingly important theoretical results. Faraday, moreover, did pioneer work in the study of the condenser. The unit of electrical capacitance is named in his honour the *farad*. This great investigator's clearness of vision and freedom from bias were inspiring exemplars to all scientists; his prophetic insight was often astounding, and his careful distinction between fact and hypothesis needs to be more closely emulated in modern science. It is an inspiration and a benediction to contemplate the work and life of this great scientist and noble citizen.

SILK-COVERED WIRE

What a luxury a supply of insulated wire would have been to Faraday in his winding of coils! When Joseph Henry (1797-1878), professor of mathematics at Albany Academy, N.Y., and later, Secretary of the Smithsonian Institution, Washington, encountered the same difficulty in his experiments on improving the electro-magnet, he met it by inventing silk-covered wire:

The next improvement was made by myself. After reading an account of the galvanometer of Schweigger, the idea occurred to me that a much nearer approximation to the requirements of the theory of Ampère could be attained by insulating the conducting wire itself instead of the rod to be magnetized, and by covering the whole surface of the iron with a series of coils in close contact. This was effected by insulating a long wire with silk thread, and winding this around the rod of iron in close coils from one end to the other. The power of the instrument, with the same amount of galvanic force, was by this arrangement, several times increased.[4]

It is said that Henry's finances were so low at the time of this invention that the silk he used was taken from his wife's wedding-dress, which she tearfully but loyally sacrificed on the altar of science and her husband's career.

[4] *Smithsonian Miscellaneous Collections*, xxx, ii, 435; *Smithsonian Annual Report* (1857), p. 99.

OSCILLATORY DISCHARGE

In 1842, Henry showed that the discharge of the Leyden jar is oscillatory, a fact which proved important, for example, in the early stages of the study of electric waves, a certain range of which we use in radio. In his report, he accorded priority in the discovery of this fact to another scientist, Felix Savary:

M. Savary ... announced in 1826 that when several needles are placed at different distances above a wire through which the discharge of a Leyden battery is passed, they are magnetized in different directions. . . . This anomaly . . . which, at first sight, appears at variance with all our theoretical ideas of the connection of electricity and magnetism [Ampère's swimmer rule] was ... referred by the author to an action of the discharge, . . . never before recognized. The phenomena require us to admit the existence of a principal discharge in one direction, and then several reflex actions backward and forward, each more feeble than the preceding, until ... equilibrium is obtained.[5]

Fig. 315. Joseph Henry's monument in Washington, D.C.

SELF-INDUCTION

Although Faraday had the priority of publication for the discovery of induced currents, yet he gladly shared the honour of that contribution with Henry, who made the same discovery independently at the same time. Faraday's outspoken admiration of Henry's work knew no bounds. In fact, these two men were remarkably kindred spirits. The *self-inductive* effect which the turns of a coil have upon each other when the current strength varies, and the consequent "extra-current" when the circuit is broken, were discovered by Henry in 1832. Let us read a few of his statements:

[5]*Proceedings of the American Philosophical Society* (1842), II, 193; quoted in Magie, *Source Book of Physics*, p. 517.

When a battery is moderately excited, . . . and its poles . . . connected by a wire . . . thirty or forty feet long, . . . when the connection is broken, . . . a vivid spark is produced. . . .[6]

The following facts in reference to the spark and shock . . . from a galvanic battery, . . . were communicated by Prof. Joseph Henry and those relating to the spark, . . . illustrated experimentally: . . . A wire, coiled into a helix, gives a more vivid spark than the same wire when uncoiled. The property of producing an intense spark is induced on a short wire, by introducing at any point of a compound galvanic current, a large flat spiral. . . . The effect produced by an electro-magnet, in giving the shock, is due principally to the coiling of the long wire which surrounds the soft iron.[7]

Henry's scope was unusually broad, and his industry and productiveness immense. He was unquestionably the foremost American scientist of his day. In grateful recognition of his multitudinous services to science and humanity and particularly because of his work on self-induction, the International Congress of Electricians at Chicago in 1893 named the standard unit of inductance or inductive resistance, the *henry*.

LENZ'S LAW

As Ampère supplemented Oersted's discussion by enunciating the swimmer rule (or right-hand rule), so Faraday's demonstration of induced currents was supplemented by Emil Lenz (1804-65), professor of physics at St. Petersburg, Russia, who first stated the relation known as Lenz's law. It gives a method of predicting the sense of an induced current from the conditions which produce it.

In his *Experimental Researches in Electricity* which contains the discovery of . . . electrodynamic Induction, Faraday determines the sense of . . . currents, produced by this method . . . ; but even Faraday, indeed, confesses the difficulty of expressing clearly the relation between the sense of these currents and the factors that determine it. . . . Immediately upon reading Faraday's memoir, it occurred to me that all investigations in electrodynamic induction must lead back in a very simple manner to the laws of electrodynamic motion. . . . Since this opinion of mine has been confirmed by numerous experiments, I shall analyse it in what follows and test it partly by known experiments and partly by experiments performed expressly for the purpose. . . . The law is . . . as follows: If a metallic conductor moves in the field of an electric current or of a magnet, there is induced in it an electric current which has such a sense as would produce in the wire if stationary, a motion exactly opposite in sense to that actually given to it, provided that the wire when at rest were movable in the sense of this motion and in the opposite sense [fig. 316].

Lenz was, of course, a thoroughly trained physicist and his words show that he recognized immediately an analogy between electrodynamic induction and other instances of induced forces such as fric-

---

[6]*American Journal of Science* (1832), XXII, 403; Magie, op. cit., p. 514.
[7]*Journal of the Franklin Institute* (March 1835), XV, 169.

tion. These all come under Newton's third law of motion: reaction is equal and opposite to action. Hence we obtain a simple wording for Lenz's law: *The sense of an induced current is always such as to oppose the motion or change by which it is produced.* The third law of motion is implied qualitatively in the first or inertia law; and Lenz's law is an electrical analogue of the first law of motion as if electricity had a property corresponding to inertia. After deducing this law theoretically and before publishing his conclusions, Lenz tested them experimentally.

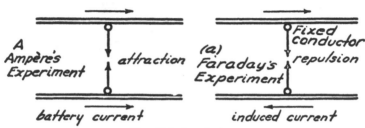

Fig. 316. Lenz's law.

Lenz quoted Ampère's law that parallel currents flowing in the same sense attract each other and in opposite senses, repel; then he proceeded with the corresponding case of induction:

If there are two straight conductors, parallel to each other, through one of which an electric current flows and the other conductor approaches it, remaining constantly parallel to it, then, during the motion, in the moving conductor, a current is induced, opposite in sense to that in the stationary conductor; if it is withdrawn, however, the induced current flows in the same sense as the inducing current. . . . By this analysis, I hope, the agreement between the experimental facts and all the implications of the law previously enunciated, has been sufficiently demonstrated.[8]

We can view this topic to advantage from the standpoint of energy. Let us distinguish the stationary wire of this paragraph and its current by the term *primary* (symbol, P), and the moving wire and its induced current, by the term *secondary* (S). If the sense of S were the same as that of P, then this induced current would induce current in P in the same sense as that of the primary current. Thus without S approaching P any further, the mutual actions of P and S would produce stronger and stronger currents in both wires. Then the system would yield more energy than we put into it. Such a machine is called a perpetual motion machine and, by the law of conservation of energy, is impossible—as Stevinus pointed out in 1585. By reductio ad absurdum, therefore, the induced current in S must be opposite in sense to the primary current.

[8]E. Lenz, "*Über die Bestimmung der Richtung der durch elektrodynamische Vertheilung erregten galvanischen Ströme*," *Poggendorff Annalen* (1834), xxxi, iii, 483.

THE WHEATSTONE BRIDGE

It is frequently necessary in electrical experiments to measure the resistance of a conductor. One of the best instruments for the purpose is the Wheatstone bridge, or a simple and convenient though less accurate modification of it, called the slide-bridge. Sir Charles Wheatstone described it in 1843, ascribing priority to Mr. S. H. Christie; but it has always been known by Wheatstone's name:

Fig. [317] represents a board on which are placed four copper wires, *Zb*, *Za*, *Ca*, *Cb*, the extremities of which are fixed to brass binding-screws. The binding-screws, *Z*, *C*, are for the purpose of receiving wires from the two poles of a "rheomotor" [source of current, e.g. battery *B*]; and those marked *a*, *b*, are for holding the ends of the wires of a galvanometer *G*. By this arrangement, a wire

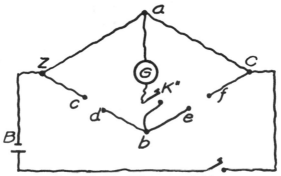

FIG. 317. The Wheatstone bridge.

from each pole of the rheomotor proceeds to each end of the galvanometer wire, and if the four wires be of equal length and thickness, and of the same material, perfect equilibrium is established. . . . The circuits *ZbaCZ* and *ZabCZ* are in this case, precisely equal, but as both currents tend to pass in opposite directions through the galvanometer, which is a common part of both circuits, no effect is produced on the needle. . . . But if a resistance be interposed in either of the four wires, the equilibrium of the galvanometer will be disturbed. . . . It may be restored by placing an equal resistance in either of the adjacent wires. For the purpose of interposing the measuring resistance and the resistance to be measured, the wires *Zb* and *Cb* are interrupted, and binding-screws *c*, *d*, and *e*, *f*, are fixed for the reception of the ends of the wires. The equilibrium, when once established, is not in any degree affected by fluctuations in the energy of the rheomotor. . . . Mr. S. H. Christie has described (*Phil. Trans.* 1833) a differential arrangement of which the principle is the same as that on which the instruments described in this section have been devised. To Mr. Christie, therefore, must be attributed the first idea of this useful and accurate method of measuring resistances.[9]

The Wheatstone bridge is principally an application of Ohm's law, and belongs to the fourth epoch in electricity.

[9] C. Wheatstone, *Collected Papers*, p. 127: *Phil. Trans. Roy. Soc. Lond.* (1843), cxxxiii, 303.

THE TRANSATLANTIC CABLE

For brilliant services rendered in several departments of science, and especially in connection with the installation of the first successful submarine cable and in discovering how to interpret its signals, Professor William Thomson of Glasgow (1824-1907), was knighted in 1866 and raised to the peerage as Baron Kelvin in 1892. Two scientists who had strong influence on Kelvin were Fourier, through his mathematical treatise on *Conduction of Heat*, and Regnault under whom Kelvin studied as a young graduate. An address which Kelvin delivered before the Royal Society in 1865 gave some description of the cable project:

FIG. 318. Lord Kelvin.

The forces concerned in the laying and lifting of deep submarine cables attracted much attention in the year 1857-8. An experimental trip to the Bay of Biscay in May 1858, proved the possibility not only of safely laying such a rope as the old Atlantic cable in very deep water, but of lifting it from the bottom without fracture. . . . The actual laying of the cable a few months later from mid-ocean to Valencia, Ireland, on the one side and . . . Newfoundland on the other, regarded merely as a mechanical achievement, took by surprise some of the most celebrated engineers of the day who had not concealed their opinion that the Atlantic Telegraph Company had undertaken an impossible task. . . . The electric failure, after several hundred messages . . . had been transmitted . . . was owing to electric faults existing in the cable before it went to sea. Under the improved electric testing, since brought into practice, such faults cannot escape detection . . . and the causes which led to the failure of the first Atlantic cable no longer exist. . . . But the possibility of damage being done to the insulation of the electric conductor before it leaves the ship . . . implies a danger which can be thoroughly guarded against only by being ready at any moment to back the ship and check the egress of the cable, and hold on for some time, or to haul back some length according to the results of electric testing. . . .

In the laying of the Atlantic cable, when the depth was two miles, the rate of the ship, six miles an hour, and the rate of paying out the cable, seven miles an hour, the resistance to the egress of the cable . . . measured by a dynamometer [spring balance], was only 14 cwt. But it must have been as much as 28 cwt. or the weight of two miles of the cable, hanging vertically down in water, were it not for the frictional resistance of the water. . . . In the event of a fault being indicated by the electric test, . . . the safe and proper course to be followed is, instantly . . . to stop and reverse the ship's engines and put on the greatest safe weight on the paying out brake. . . . The experience of the . . . expedition . . . not only verified the estimates of the scientific committee and of the contractors as to the strength of the cable, its weight in water . . . and its manageability, but it had proved that, in moderate weather, the *Great Eastern* could, by skilful seamanship, be kept in position and moved in the manner required. . . . The manageability of the *Great Eastern*, in skilful hands, has proved to be very much better than could have been expected. . . . She has both screw and paddle,—an advantage possessed by no other ship in existence. By driving the screw at full power ahead, and backing the paddles, to prevent the ship from moving ahead, . . . "steerage way" is created by the lash of the water from the screw against the rudder: and thus the *Great Eastern* may be effectually steered without going ahead. . . .[10]

One of Kelvin's footnotes gives some nautical help here for landlubbers: "A nautical mile, the length of a minute of latitude . . . is 6073 feet; for approximate statements . . . it may be taken as 6000 feet or 1000 fathoms." The *Great Eastern*, the ship that laid the cable, was the longest ever built up to that time; and she retained that preeminence for over half a century.

The instrument first used for receiving signals through a long submarine cable . . . was the mirror galvanometer. . . . The latest form of receiving instrument for long submarine cables, is that of the Siphon Recorder, for which the lecturer obtained his first patent in 1867.[11]

Some of Kelvin's most important work was done in collaboration with his friend Joule of Manchester. The labours of these two were fundamental in establishing the law of conservation of energy.

This is not our first gossip with Kelvin; but it requires much study to gain a true perspective of the magnitude and lustre of this man's work. When we say that few men are Kelvin's peers as to number and importance of researches, the story is only half told, for that leaves his indirect services unconsidered. Kelvin was a great teacher and an inspiring lecturer in spite of the fact that his native Scottish modesty amounted sometimes to diffidence. At times in his teaching, however, he would unfold enough to indulge in a quiet tease or even a flick of irony but always in the best spirit of jovial banter. He was always patient and encouraging to any student who was doing his best and Kelvin's patronage of young scientists was as fatherly as that of Franz Liszt toward aspiring musicians. According to the testimony of his old students and his protégés, progress in physics during the latter half of the nineteenth century was just as much owing to Kelvin's influence on others as to his direct contribution of discoveries and inventions. Higher praise a teacher could hardly receive.

[10]Kelvin, *Collected Papers*, ii, 153; *Proc. Roy. Soc. Edin.* (1865).
[11]Kelvin, *Collected Papers*, ii, 168; *Transactions of the Institute of Engineers and Shipbuilders* (Scotland, 1873), p. 168.

THE INDUCTION COIL

An important application of induced currents is the induction coil. It was a natural outcome of the work of Faraday and Henry. It was first given its modern form by C. G. Page of Salem, Massachusetts, in 1836. Its development, like that of the telephone, has consisted of a series of improvements introduced by a number of inventors. In structure it is essentially an electro-magnet surrounded by an insulated coil (secondary) which has more turns of wire than that of the electro-magnet (primary). The potential differences or electromotive forces in the two coils are in the same ratio as their numbers of turns of wire, or in algebra,

$$\frac{e''}{e'} = \frac{t''}{t'}.$$

Henry and Sturgeon improved the soft-iron core in 1835 by using a cylindrical bundle of straight wires insulated from each other by wax. Page and Foucault inserted in the primary circuit a circuit-breaker or interrupter but it had the disadvantage of a mercury-cup. It was in Germany that the circuit-breaker with platinum points like that of an electric bell or buzzer was first used by J. P. Wagner in 1839. In 1853, Fizeau placed a condenser in the base of the instrument connecting it in parallel with the primary spark-gap.

Between 1870 and 1900 makers vied with each other to see who could produce a coil that would yield the longest sparks. Notable among these technicians was Heinrich D. Ruhmkorff (1803-77), a Parisian instrument maker, whose coils became so famous that the induction coil came to be called the Ruhmkorff coil, whereas from priority it might fairly have been named after Page. A. Apps of London constructed a coil whose secondary contained 280 miles of fine insulated copper wire, with some 340,000 turns and a resistance of 100,000 ohms ($10^5\Omega$). With the current from 30 Grove cells in series in its primary, it gave a spark over a yard in length (42 inches or 105 cm.).

One of the many uses of this coil is to produce the spark in the spark-plug of a gasoline engine. Another is in operating X-ray machines; and we shall find it of crucial importance in the researches of Hertz, Lodge and other scientists.

CHAPTER TWENTY-THREE

Electro-Magnetic Waves
and Cathode Rays

THE most remarkable prophecy in the whole literature of science occurs in a treatise on electricity by James Clerk Maxwell. From calculations based on data obtained by himself and Faraday, Maxwell in 1865 inferred the existence of electric waves in air and in solar space. These waves may be taken as the sign manual of a sixth epoch in the story of electricity. In his calculations Maxwell drew important help from the work of some predecessors, notably Ampère, the latter's contemporary and fellow countryman Simeon D. Poisson of Paris, and the German physicist Karl F. Gauss of Göttingen. Prominent in his work also was Faraday's idea of lines of force or tensions in the aether.

Maxwell predicted that when these waves were discovered they would be found to have the following properties: they would travel through vacuum and transparent bodies with the speed of light and obey the same laws of reflection, refraction, and interference as light, differing from it, if at all, chiefly in wave-length and frequency. As we have seen, a hint of such waves was present in some of Galvani's frog experiments and it is also true that some experiments performed by Henry and Faraday fostered the same suggestion. It is proper that we should read these words of Maxwell with due respect:

> In several parts of this treatise, an attempt has been made to explain electro-magnetic phenomena by means of mechanical action transmitted from one body to another by means of a medium occupying the space between them. The undulatory theory of light also assumes the existence of a medium. We have now to shew that the properties of the electromagnetic medium are identical with those of the luminiferous medium. . . . We, therefore, obtain the numerical value of some property of the medium, such as the velocity with which a disturbance is propagated through it, which can be calculated from electromagnetic experiments and also observed directly in the case of light. If it should be found that the velocity of propagation of electromagnetic disturbances is the same as

the velocity of light, and this not only in air but also in other transparent media, we shall have strong reasons for believing that light is an electromagnetic phenomenon. . . .[1]

Fig. 319. James Clerk Maxwell.

Developing a set of equations that are known by his name, Maxwell showed the two velocities to be in the same order of magnitude and at least nearly equal:

In the meantime, our theory which asserts that these two quantities are equal and assigns a physical reason for this equality, is certainly not contradicted by the comparison of these results, such as they are.

[1]J. Clerk Maxwell, *Electricity and Magnetism*, ii, 383. Also, *Phil. Trans. Roy. Soc.* (1865).

It was one of Maxwell's fondest wishes that he might live to see his prophecy fulfilled, but his life was cut off untimely at the age of forty-eight, ten years before electro-magnetic waves were discovered by Hertz and Lodge. This was the culmination of two decades of intensive research on the part of these two investigators, and their striving was not without the features of a race. When Hertz succeeded in making the discovery, he also found that the Maxwellian waves agreed in close detail with Maxwell's predictions. So great can be the power of algebra in the hands of a master. Today these waves are in universal use. They are known as Hertzian waves and a certain range of them as radio waves.

Maxwell's major contribution to electricity was one of organization. His work brought together the various threads of the subject and wove them into one consistent pattern. Thus he rendered to electricity and magnetism the same service as Newton did to mechanics. Maxwell was the first of a trinity of great Cavendish professors at Cambridge, Maxwell, J. J. Thomson, and Rutherford, whose researches and leadership have been fundamental in the modern advancement in electricity.

When Maxwell realized the tragic fact that his life, which had been so full of signal service and which was still rich in promise, was to be ruthlessly cut down in mid-ascent, he accepted his fate with a fortitude and a quietness of spirit which betokened the great depth of his philosophy and his religious faith.

THE TELEPHONE

A brief account of the invention of the telephone in 1874 by Dr. Alexander Graham Bell is found in the records of a lawsuit, "To Annul the Bell Patents."

Salem, Mass., July 1, 1875.

Dear Mr. Hubbard,

The experiment to which I alluded when I saw you last, promises to be a grand success. . . . A membrane, M, is attached to a straining ring R. . . . The cross-bar B supports an electro-magnet HE. The leg C of the electro-magnet is covered with a coil H of insulated copper wire and to the uncovered leg E, . . . is pivoted one end of the steel armature A, . . . One of these instruments was placed in a room in the upper part of Mr. Williams' building . . . in Boston and the other in one of the work-rooms below. . . . While I was talking or singing into the instrument upstairs, I was interrupted by the sudden appearance of Mr. Watson who had rushed upstairs in great excitement, to tell me . . . he could hear my voice quite plainly and could almost make out what I said. . . . I recognize that the discovery of the magneto-electric current generated by the vibrations of the armature of an electro-magnet in front of one of the poles is the most important point yet reached. I believe that it is the key to still greater things. . . . I think . . . steps should be taken immediately towards obtaining a

Caveat or Patent. . . . When we can create a pulsatory action of the current which is the exact equivalent of the aerial impulses, we shall certainly obtain exactly similar results.

The telephone, of course, is fundamentally an application of induced currents and belongs to the fifth epoch. Bell was a specialist in pho-netics[2] but not specially trained in electricity. One astounding feature of the first telephone was its remarkable simplicity, especially in view of the fact that its inventor hoped that it would perform the infinitely complex feat of imitating the human voice. The phonograph had the same peculiarity. More than one expert electrician warned Bell that to expect a steel rod, a coil of wire, and the end of a fish-can to transmit the voice was madness. When the instrument proved a success, one great electrician said humourously of Bell that if he had known much electricity, he would not have invented the telephone. As an example of the irony of fate, one may consider the report that, in later years, Bell had the telephone removed from his house because he had found its persistent jingling an infernal nuisance.

FIG. 320. Alexander Graham Bell.

CONDUCTION BY AIR

The fact that the amber in Thales' experiment or any other charged body, no matter how carefully insulated, ultimately loses its charge, suggested to several scientists that air may be a conductor, though a relatively poor one. To get to the bottom of this matter would require an investigation of the conduction of electricity by air. Although the problem may seem rather simple, even insignificant, and certainly unspectacular, yet it is to be doubted whether in the whole history of science, a topic has ever been broached which led to a more wonderful series of beautiful and fascinating results. This cornucopia of novel and revolutionary advancements has brought a new epoch into the story of electricity and has even exerted a strong influence in other sciences, notably in the sister science of chemistry. In fact, important results

[2] D. C. Miller, *Anecdotal History of the Science of Sound* (Macmillan, 1935).

in this period follow each other in such profusion that the historian has difficulty in making a selection.

As early as 1710, Hauksbee was knocking at the door of the new subject by his study of electric discharge in vacuum. That door was flung wide open in 1869 by Heinrich Geissler, a master glass-blower and instrument-maker of Bonn, by Julius Plücker, professor of physics of the same city, and by Plücker's old student, J. W. Hittorf. To investigate the relation between the conductivity of air and its pressure, one obvious procedure was that of fig. 321, in which A and C are anode

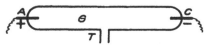

and cathode and G the gas whose pressure is altered by an air-pump attached to the tube T. This experiment was first performed in 1752 by William Watson of Lon-

FIG. 321. Aurora tube.

don, one of Stephen Gray's friends. The current used in Geissler's experiment was obtained either from an induction coil (because of its relatively high potential) or from an electric machine, e.g. a Wimshurst. This vacuum tube is called an *aurora tube* because the appearances seen in it remind one of the shimmering beauty of Northern Lights. Geissler devised many beautiful variations of this tube (fig. 322), which are named after him Geissler tubes.

Immediately a whole army of investigators were attracted into this intriguing field and an outburst of important results ensued such as the science of electricity had never before experienced. It seemed that

FIG. 322. Geissler tube.

every time anybody made any change in the experiments, he saw something new and beautiful, learned something interesting and useful, and was led into some new question requiring still other variations of the conditions. The aurora tube shows that reducing the pressure of air decreases the sparking-potential, i.e. the potential difference necessary to produce a spark between anode and cathode. Or we might say, increases the conductivity. Its colour displays suggest that Northern Lights are electrical discharges high in the atmosphere where the pressure is a small fraction of one atmosphere. By noting the appearance characteristic of each pressure, we obtain a new type of pressure gauge for gases at low pressure. When a Wimshurst was used, Geissler and others observed at the cathode a purple "cathode-glow." The minor glow at the other electrode, observed when an induction coil is used, is an ocular demonstration that the secondary current of an induction coil is oscillatory. Frequently the glass of a Geissler tube emits light and is said to *fluoresce*. Minerals placed in the tube also fluoresce, each with its own characteristic colour, and some continue to emit light after the current is shut off—a phenomenon called phosphorescence. This brief sketch may give some faint idea of the results but it is futile to try capturing in words the infinite beauty and variety of these phenomena which must be observed at first hand to be properly enjoyed.

The avenue which was thus opened up led directly to the electron theory, which is the earmark of the seventh epoch. This epoch followed the sixth so closely that they overlap and develop together with much mutual support. This explains why these two epochs are discussed in the same chapter.

Johann W. Hittorf (1824-1914), of Bonn and Münster, discovered in 1869 that if the anode is let in at the side of a vacuum tube, the glass at the end facing the cathode fluoresces, and that a piece of lead in the tube casts a shadow on that end of the tube. Therefore, there must be rays emanating from the cathode (fig. 323). Seven years later, Eugen Goldstein of Berlin very naturally named these "*cathode rays*."

FIG. 323. Cathode rays.

Any solid or fluid, whether insulator or conductor, which is placed in front of the cathode, cuts off the glow. . . . No bendings out of the straight line occur. . . . We see most clearly the rectilinear transmission of the glow if it goes out from a point cathode, and through a great length of the tube, brings the surface of the glass to fluorescence. . . . If in such conditions, any object is set in the space filled with the glow, it casts a sharp shadow on the fluorescing wall by cutting off from it the cone of rays, diverging from the cathode as origin.[3]

Hittorf and Goldstein observed that a cathode ray is emitted in general at right angles to the surface which produces it. One special form of vacuum tube is called a Hittorf tube.

[3] *Annalen der Physik und Chemie* (1869), CXXVI, 1.

DEFLECTION BY MAGNET

In much of his study of cathode rays Hittorf collaborated with his teacher, Julius Plücker (1801-68), also of Bonn. Plücker showed that *cathode rays are deflected by a magnetic field*.[4] The credit for this important discovery is shared by Hittorf and Sir William Crookes, who investigated the phenomenon in greater detail. At this time advances were taking place so rapidly that some priorities are disputed or shared by way of compromise. The sense of deflection of the cathode rays, Crookes found to be in accordance with Ampère's swimmer rule, so that a cathode ray behaves as though it were a "wireless current."

Plücker also invented a special form of vacuum tube called a Plücker tube. The central part of the tube is one or two millimetres in diameter. He used such tubes in making spectroscopic examination of the light produced with different gases in the tube. From these researches come the modern neon tube and its family of relatives.

Sir William Crookes made many valuable contributions to the science of electrical discharge through gases at low pressure. Faraday had observed a dark space that characteristically forms and grows near the cathode as the pressure is reduced. It is called the Faraday dark space. As the experiment proceeds another dark space forms near the cathode; of this Crookes made a detailed study and it is named the Crookes Dark Space. Several forms of vacuum bulbs that he invented and used in his researches are also given his name. A number of these were demonstrated in his Bakerian Lecture on this topic, delivered before the Royal Society of London in 1879:

> When the spark of a good induction coil traverses a glass tube containing a rarefied gas, certain phenomena are observed which vary greatly with the kind of gas and the degree of exhaustion. There is an appearance, however, which is constant in all gases which I have examined, and within very wide limits of pressure, viz., the well-known dark space round the negative pole [cathode].[5]

In the same lecture it was demonstrated that a magnetic field changes cathode rays from straight lines to curves (fig. 323):

> An electro-magnet is placed beneath the bulb. . . . The induction current being turned on, the shadow of the cross is projected along the tube. . . . The electro-magnet is now excited by one cell. The shadow is deflected sideways. The edges . . . when under no magnetic influence . . . are straight lines, but when deflected by the magnet, they are curved. . . . On reversing the battery current, passing round the magnet, the above-named deflections are obtained in the opposite direction.

Like most experiments in electricity, this one involves knowledge

[4]*Poggendorff Annalen*, CVII, 77; CXVI, 45.
[5]*Phil. Trans. Roy. Soc. Lond.* (1879), CLXX, i, 135.

and ideas from more than one epoch—conduction, battery current, induced current, and conductivity of gases at low pressure.

The biography of Crookes makes very inspiring reading. An excerpt from his lecture on "Radiant Matter" given at the Royal Institution in 1879 may help to convey some idea of his brilliance and power:

I have taken extraordinary pains to remove as much Matter as possible from these bulbs and have succeeded so far as to leave only about one-millionth of an atmosphere in them. It would seem that when divided by a million, so little matter will necessarily be left that we may justifiably neglect the trifling residue and apply the term vacuum. . . . To do so, however, would be a great error. . . . According to the best authorities, a bulb the size of the one before you (13.5 centimetres in diameter) contains a quadrillion molecules, (10^{24}). Now when it is exhausted to a millionth of an atmosphere, we still have a trillion molecules left in the bulb (10^{18}). . . . I perforate it by a spark from the induction coil. The spark produces a hole of microscopical fineness. . . . The inrush of air impinges against the vanes, and sets them rotating after the manner of a windmill. Let us suppose the

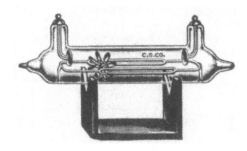

FIG. 324. Crookes tube with rotating wheel.

molecules to be of such a size that at every second, . . . a hundred millions could enter; how long, think you, would it take for this small vessel to get full of air? An hour? A day? A century? Nay almost an eternity! A time so enormous that imagination itself cannot grasp the reality.

Suppose that this exhausted glass bulb, endued with indestructibility, had been pierced at the birth of the solar system. Suppose it to have been present when the earth was "without form and void"; suppose it to have borne witness to all the stupendous changes evolved during the full cycle of geologic time, to have seen the first living creature appear and the last man disappear; suppose it to survive until the fulfilment of the mathematician's prediction that the sun, the source of energy, four million centuries from its formation, will ultimately become a burnt-out cinder, . . . at 100 million molecules a second, . . . this little bulb, . . . would scarcely have admitted its full quadrillion of molecules! . . .

In studying this Fourth State of Matter, we seem at length to have within our grasp and obedient to our control, the little indivisible particles which, with good warrant, are supposed to constitute the physical basis of the universe. We have seen that in some of its properties, Radiant Matter is as material as this table, whilst in other properties, it almost assumes the character of Radiant Energy. We have actually touched the border land where Matter and Force seem to merge into one another, the shadowy Realm between the Known and the Unknown, which for me has always had peculiar temptations. I venture to think that the greatest scientific problems of the future will find their solution in this Borderland and even beyond; here, it seems to me, lie ultimate realities, subtle, far-reaching, wonderful.

There is no doubt that the last paragraph was reaching out toward the electron theory. Other scientists of the same period, however, preferred to consider these phenomena as undular.

THE EDISON LAMP

Of all the analogues of von Guericke's classic bell-jar experiment, the commonest in everyday affairs is the ordinary electric incandescent lamp or bulb (fig. 325). It was invented in 1880 by Thomas A. Edison (1847-1931) of Orange, N.J., whose patent papers describe the instrument:

To all whom it may concern: Be it known that I, Thomas Alva Edison, of Menlo Park, in the State of New Jersey, United States of America, have invented an Improvement in Electric Lamps, and in the method of manufacturing the same . . . of which the following is a specification. . . . The invention consists in a light-giving body of carbon wire. . . . The invention further consists in placing such burner of great resistance in a nearly perfect vacuum to prevent oxidation and injury to the conductor by the atmosphere. The current is conducted into the vacuum bulb through platina wires sealed into the glass. . . . Only Platina can be used as its expansion is nearly the same as that of glass. . . . I have discovered that even a cotton thread, properly carbonized in a sealed glass bulb exhausted to one-millionth of an atmosphere, offers from one hundred to five hundred ohms resistance to the passage of the current. . . .

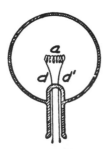

FIG. 325.
Edison lamp.

In fig. [325] *a* is the carbon spiral or thread, and *d*, *d'* are the platina wires. The spiral . . . is placed on the glass holder and a glass bulb is blown over the whole, with a leading tube for exhaustion by a mercury pump. This tube, when a high vacuum has been reached, is hermetically sealed. . . . I claim as my invention: 1. An electric lamp for giving light by incandescence, consisting of a filament of carbon of high resistance, . . . made as described. . . . 2. The combination of carbon filaments with a receiver, made entirely of glass and conductors, passing through the glass, and from which the air is exhausted for the purposes set forth.[6]

Edison's original lamp has undergone many modifications and improvements; but the spiral lamp-socket, which was also one of his many inventions, is not only still in use but is still standard and the best general-purpose socket. It combines extreme simplicity, durability, and efficiency. An invention which successfully withstands six decades of such testing declares the genius of its inventor.

[6]U.S. Patent Record, No. 223, 898; 1880. Quoted in *Science News Letter*, "Classics of Invention," August 10, 1929.

HERTZIAN WAVES

One of Helmholtz's greatest contributions to science was his training of Heinrich Hertz (1857-94), who became professor of physics at Bonn. Hertz's most famous achievement in electricity was the confirmation of Maxwell's electro-magnetic theory of light by a complete demonstration of electric waves and the properties which Maxwell had predicted and which we have taken as marking a sixth epoch in the story of electricity. In 1889, two decades after Maxwell's predictions were made and ten years after Maxwell's death, Hertz delivered, during the sixty-second meeting of the German Association for Advancement of Science at Heidelberg, a lecture which dealt with this subject:

I am here to support the assertion that light of every kind is itself an electrical phenomenon. . . . What then is light? Since the time of Young and Fresnel, we know that it is a wave-motion. We know the velocity of the waves and their length, and we know that they are transverse waves. . . . But transverse waves in fluids are unknown. They are not even possible. . . . Hence men are forced to assert that the aether . . . behaves like a solid body. But when they considered the unhindered course of the stars in the heavens, they found themselves forced to admit that the aether behaves like a perfect fluid. These two statements together land us in a painful and unintelligible contradiction which disfigures the otherwise beautiful development of Optics. Instead of trying to conceal this defect, let us turn to electricity, . . . we may perhaps make some progress towards removing the difficulty. . . .

What then is electricity? . . . Is there such a thing? The traditional conceptions of electricities which attract and repel each other and which are endowed with actions-at-a-distance as with spiritual properties—we are all familiar with these, and, in a way, fond of them. . . . The period at which those conceptions were formed was the period in which Newton's law of gravitation won its most glorious successes and in which the idea of direct action-at-a-distance was familiar. . . . Electric and magnetic attractions followed the same law. . . .

If the path indicated was a false one, warning could come only from an intellect of greatest freshness . . . from a man who looked at phenomena with an open mind and without preconceived opinions, who commenced from what he saw, not from what he had heard, learned or read. Such a man was Faraday. Among the questions that he raised, there was one that continually presented itself to him: Do electric and magnetic forces require time for their propagation? When we electrify and discharge a body in rapid succession, does the force vary at all distances simultaneously? Or do the oscillations arrive later the further we go from the body? In the latter case, the oscillation would propagate itself as a wave. . . . Are there such waves? . . . .

All the more must we admire the happy genius of the man who could connect together these apparently remote conjectures in such a way that they mutually supported each other. . . . This was a Scotsman, Maxwell. You know the paper that he published in 1865. . . . It is impossible to study his wonderful theory without feeling as if . . . the equations have an independent life and an intelligence of their own, as if they were wiser than ourselves, indeed, even wiser than their own discoverer, as if they give forth more than he put in them. . . . At this stage, I was so fortunate as to be able to take part in the work. To this I owe the honour of speaking to you to-day. . . .

If you give the physicist a number of tuning-forks and resonators, and ask him to demonstrate to you the propagation in time of sound waves, he will find no difficulty in doing so. . . . In exactly the same way, we proceed with our electric waves. In place of the tuning-fork, we use an oscillating conductor O, [fig. 326], and in place of the resonator . . . our interrupted wire R, which may also be called an electric resonator. We observe that in certain places, there are sparks at the gap G, but at others, none; we see that the dead points follow each other periodically in ordered succession [nodes and loops]. Thus the propagation in time is proved and the wave-length can be measured. . . . We multiply the wave-length by the calculated frequency of oscillations and find a velocity which is about that of light [*Eq.* 2, p. 189].

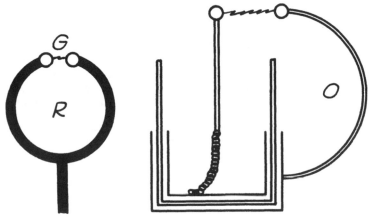

FIG. 326. Hertz's indicator.

After giving a complete and detailed demonstration of electrical shadows, rectilinear beams, reflection, refraction, and interference, Hertz concluded with these words:

The connection between light and electricity, . . . of which there were hints and suspicions and even predictions in the theory, is now established. . . . Optics is no longer restricted to minute aether waves, a small fraction of a milli-metre in length; its dominion is extended to waves that are measured in deci-metres, metres, and kilometres. And in spite of this extension, it appears merely . . . as a small appendage of the great domain of electricity. We see that this latter has become a mighty kingdom. . . . In every flame, in every luminous particle, we see an electrical process. Even if a body is not luminous, provided it radiates heat, it is a centre of electrical disturbances. Thus the domain of electricity extends over the whole of nature. It even affects ourselves closely: we perceive that we actually possess an electrical organ—the eye. . . . Just at present, physics is . . . inclined to ask whether all things have not been fashioned out of the aether.

These are the ultimate problems of physical science, the icy summits of its loftiest range. Shall we ever be permitted to set foot upon one of those summits? . . . We know not: but we have found a starting-point for further attempts which is a stage higher than any used before. Here the path does not end abruptly in a rocky wall; the first steps that we can see form a gentle ascent, and amongst the rocks, there are tracks leading upwards. There is no lack of eager and prac-tised explorers: how can we feel otherwise than hopeful of the sucess of future attempts?[7]

From this wonderful address, it is not difficult to see that the un-timely death of Hertz at the age of thirty-seven was a tragic blow to science, as the death of Mendelssohn was to music and that of Abel to mathematics.

LODGE'S EXPERIMENT

The search for electric waves became a race in which Hertz reached the goal first and Sir Oliver Lodge of Manchester (fig. 328) was a close second. As Hertz said, "If I had not anticipated him, he would have

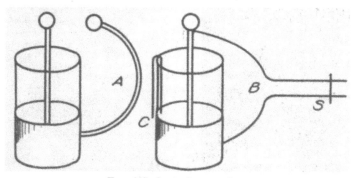

FIG. 327. Lodge's experiment.

succeeded in observing [electric] waves." One of Lodge's experiments is known by his name and is regularly used to show that dialing a receiving set is essentially a matter of adjusting the capacitance of the circuit to tune it into electrical resonance with the desired broad-casting station. An article that Lodge published in *Nature* in 1890 describes the experiment (fig. 327):

Two similar Leyden jars are joined up to similar fairly large loops of wire, one of the circuits having a spark-gap with knobs included (*A*), the other being completely metallic and of adjustable length *B*. The jar of this latter circuit has also a strip of tinfoil pasted over its lip so as to provide an overflow path complete with the exception of an air-chink *c*. . . . Then if the two circuits face each other at a reasonable distance, and if the slider *S* is properly adjusted, every discharge of *A* causes *B* to overflow. A slight shift of the slider puts them out of tune. . . .

It is needless to point out that the two jars constitute respectively a Hertz oscillator and receiver. . . .

[9]H. Hertz, *Miscellaneous Papers*, I, 313.

ROENTGEN RAYS

One morning in December, 1895, news flashed round the world that Dr. William Roentgen (1845-1923), professor of physics at Würzburg, had discovered a new kind of ray that could penetrate wood and other opaque or translucent substances. Obviously, for surgical reasons alone, such a discovery was front-page news. Roentgen published some of his results immediately:

When we send the discharge of a large induction coil through a . . . sufficiently evacuated . . . Crookes tube or similar apparatus, surrounded with a fairly close-fitting envelope of thin black cardboard, in a . . . darkened room,

FIG. 328. Sir Oliver Lodge.   FIG. 329. William Roentgen.

and bring near the apparatus a paper sheet, painted with barium platinocyanide, at each discharge, we see the sheet brightly luminous, fluorescing equally well whether the painted side or the reverse side is turned toward the discharge tube. . . . The first fact to strike us is that through the black cardboard which is opaque to visible and ultraviolet rays of the sun or the electric arc, an agent is transmitted which can readily cause fluorescence and one would naturally proceed . . . to investigate whether other bodies . . . have this property. We soon found that all substances are transparent to this agent but in very different degrees. Behind a book . . . of about one thousand pages, I saw . . . the screen light up brightly. . . . In the same way, the fluorescence appeared behind a double pack of cards. . . . Thick blocks of wood are also transparent. . . . Vulcanite, several centimetres thick, is transparent to these rays; for brevity's sake, I shall use the term "rays"; and to distinguish them from others, . . . I shall call them X-rays. Holding the hand between the discharge apparatus and the sheet, we see the darker shadow of the hand-bones within the shadow of the hand which is only slightly less dark. Lead 1.5 mm. thick is practically opaque and is much used because of this property.

After I had found the transparency of different substances, at various thicknesses, I forged ahead to discover how these X-rays behave on traversing a prism, whether they are refracted or not. Experiments with water and with carbon disulphide, enclosed in mica prisms . . . showed no deviation. . . .

Of special significance . . . is the fact that photographic dry plates have proved to be sensitive to X-rays. The photographic impressions can be obtained in a non-darkened room with the plates . . . either in the holders or wrapped in paper. On the other hand, from this property, it results as a consequence that undeveloped plates cannot be left for a long time in the neighbourhood of the X-ray bulb if they are protected by merely the ordinary covering.

Fig. 330
Roentgen-ray tube.

Thereby hangs a tale. It is said that Roentgen once happened to use a key to mark his place in a book. Later in the day, having loaded his plate-holders, he went out and took some landscapes. The first plate developed a good picture except that athwart the sky was a gigantic key. Roentgen saw that this key was an exact copy of his own in shape and size—but that key had been in his pocket when he was out snap-shotting. Thinking back, he remembered that the box of plates had been under the book when he was experimenting with the new cathode-tube. As suspicion always falls on a stranger, the discharge tube was suspect. The obvious procedure was to repeat the circumstances using a new box of plates. The key was placed in the book and this on a box containing a plate; then the vacuum bulb was brought near. When the plate was developed, it showed the key. Thus the X-rays were discovered and their ability to penetrate paper and cardboard and to affect a photographic plate.

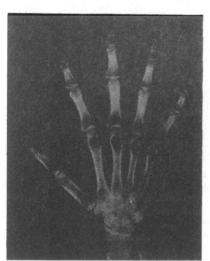

Fig. 331. X-ray photograph.

One of Roentgen's experiments was dangerous although he could hardly have been aware of the danger since he wrote: "The retina of the eye is not sensitive to these rays: the eye brought close to the discharge-tube [!] sees nothing." Roentgen's paper next opened the door of a new department of science, the X-ray examination of crystals; but he merely remarked on its possibilities, then closed the door:

It might be possible accordingly that the arrangement of the particles of a body would exert an influence on its transparency, e.g. a piece of calcspar of equal length and thickness might be unequally transparent depending on whether the rays were sent through in the direction of the axis or at right angles to it.

Roentgen also had some negative results to record:

"For interference phenomena of X-rays, I have diligently sought but unfortunately . . . without success."

These effects have since been found.

In distinguishing between cathode rays and X-rays, Roentgen naturally tried the effect of a magnetic field:

FIG. 332.
Guglielmo Marconi and his wife.

Another . . . noteworthy difference in the behaviour of cathode rays and X-rays, lies in the fact that, in spite of many endeavours, I have not succeeded in producing, even in very strong magnetic fields, a deflection of the X-rays by means of a magnet.[8]

The same is true of light rays; and in fact, X-rays are looked upon as being closely akin to light rays but with wave-lengths about one-thousandth that of violet light.

[8]W. Roentgen, *Sitzungsberichte der Physikalisch-medikalischen Gesellschaft zu Würzburg* (1895), XIX, 132. (From a copy in the library of the University of Göttingen, through the kindness of Dr. Leon Leppard.)

WIRELESS COMMUNICATION

The Nobel Laureate in Physics for 1909 was Guglielmo Marconi of Bologna (1874-1937).   As a boy his hobby was the study of electricity and his bent was toward applied physics. From a knowledge of the work of Maxwell, Hertz, and Lodge, gained with the help of his professor, Augusto Righi, this youth became intrigued with the idea of applying electromagnetic waves to achieve communication. In some of his experiments performed in 1895 at his father's country house, he effected wireless telegraphy over a distance of a mile. Next year, when he was only 22, he went to England and there continued his experiments and gave demonstrations involving even greater distances. That same year he took out the first patent in wireless telegraphy. In 1900 the Marconi Wireless Telegraph Co. was formed, and a year later Marconi succeeded in sending a wireless message across the Atlantic. He was given the title of Marchese in 1929 and made a senator in grateful recognition of the service he had rendered to the world and the honour he had brought to his native land.

CHAPTER TWENTY-FOUR

The Electron Theory

THE seventh epoch in the story of electricity is characterized by the electron hypothesis and its applications. The advent of this epoch was brought about by the work of so many able investigators that it is not easy to single out one man and accord him the laurels. If there is one name, however, that stands above all others, it is probably that of the gifted physicist and mathematician, Sir J. J. Thomson, who became Maxwell's successor as Director of the Cavendish Laboratory at Cambridge. A paper of his, published in 1896, proposed the assumption that atoms are composed of smaller particles—now called electrons, protons, neutrons, etc. Only a man with courage, much prestige, and irrefutable experimental data could have dared to stand up and tell the scientists of that day that what they called "atoms" (literally, "not divisible") should really be called "toms." Many of those scientists were such fanatical subscribers to the atomic theory that they resented an attack upon it as violently as they would have an attempt to undermine their religious faith—just like the peripatetics in Galileo's day. Science will have made an important advance when the fanatical support of any hypothesis is frowned out of court.

Professor Jean Perrin of Paris, a friend of Pierre and Marie Curie, had shown, by using a pierced anode, that cathode rays carry negative charges. By modifying Perrin's apparatus Thomson succeeded in making an ingenious series of measurements and calculations, and his conclusions from these brought the electron theory into the world— and a new and marvellous epoch into not only electricity but also some other sciences:

The apparatus used is represented in fig. [333]. The rays from the cathode C pass through a slit in the anode A, which is a metal plug . . . connected with the earth; after passing through a second slit B, . . . they travel between two parallel aluminium plates D, G; they then fall on the end of the tube E, and produce a narrow well-defined phosphorescent patch. . . . At high exhaustions, the rays were deflected when the two plates . . . were connected with the terminals of a battery of small storage cells . . . depressed when the upper plate was connected with the negative pole of the battery . . . and raised when it was . . . positive.

Let *m* be the mass of each of the particles, *e*, the charge carried by it. Let *N* be the number of particles passing across any section of the beam in a given

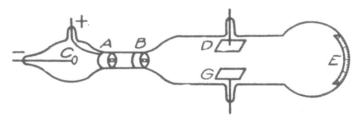

FIG. 333. Thomson vacuum tube.

time: then *Q*, the quantity of electricity carried by these particles, is given by the equation,

$$Ne = Q.$$

We can measure *Q* . . . with an electrometer.[1]

Thomson's paper continues:

From these determinations we see that the value of m/e is independent of the nature of the gas and that its value 10^{-7} [*e* in electromagnetic units] is very small as compared with the value 10^{-4}, which is the smallest value of this quantity previously known, and which is the value for the hydrogen ion in electrolysis.

The smallness of m/e may be due to the smallness of *m* or the largeness of *e* or to a combination of both of these two. . . . The explanation which seems to me to account, in the most simple and straightforward manner, for the facts is founded on a view . . . that the atoms of the different chemical elements are different aggregations of "atoms" of the same kind. . . . Thus, on this view, we have in cathode rays, matter in a new state, . . . a state in which all matter is of one and the same kind; this matter being the substance from which all the chemical elements are built up.[2]

The Thomson vacuum tube is the ancestor of the modern oscillograph and a group of related tubes.

FIG. 334. Sir J. J. Thomson.

Dr. G. Johnstone Stoney (1826-1911), professor of physics at the

[1]The calculations are given in S. G. Starling's excellent text, *Electricity and Magnetism* (Longmans, 1934), p. 326 (Intermediate ed.), p. 472 (Advanced ed.).
[2]J. J. Thomson, "Cathode Rays," in *Philosophical Magazine* (1897), S.5, XLIV, 293.

University of Dublin, drew attention in 1874 to what we found in Faraday's discussion of electrolysis, "the atoms which are equivalents to each other . . . have equal quantities of electricity naturally associated with them" (p. 465). In that sentence Faraday was standing on the threshold of the electron theory but instead of entering, he turned aside (as did Maxwell also) to consider stresses and strains in electromagnetic fields. This led to the discovery of electro-magnetic waves. Had Faraday and Maxwell chosen otherwise the sixth and seventh epochs might have been interchanged.

In spite of Stoney's clear reiteration his words received little attention—possibly because of Maxwell's preoccupation with other topics.

Attention must be given to Faraday's Law of Electrolysis which is equivalent to the statement that in electrolysis a definite quantity of electricity, the same in all cases, passes for each bond that is ruptured.[3]

His argument was that the relation between the mass of an atom (m) and its charge (e) must be the same as that between the mass of an ion liberated (M) in any experiment, and the quantity of electricity (Q) necessary to liberate it; or that their reciprocals are equal thus

$$\frac{e}{m} = \frac{Q}{M}.$$

Q and M were known; Faraday found that it requires 96,500 coulombs of electricity to liberate 1 gm. hydrogen ("or thereabouts").

$$\therefore \frac{e}{m} = \frac{96500}{1} \text{ (cl.)} \quad \therefore e = 96500\, m.$$

Hence if the mass of a hydrogen atom were known, e could be calculated. Kelvin and others had attempted to find m or, what amounts to the same thing, the number of molecules or atoms in a given volume of hydrogen. Their results varied over a wide range; but Stoney used an average of their values,

$$m = 1.04 \times 10^{-25} \text{gm.} \quad \therefore e = 96500 \times 1.04 \times 10^{-25} \times 3 \times 10^9$$
electrostatic units
$$= 3 \times 10^{-11} \text{ e.s.u.}$$

He stated the result thus:

The amount of this very remarkable quantity of electricity is about . . . 3 eleventhets [3×10^{-11}] of the electrostatic unit [of charge]. A charge of this amount is associated in the atom . . . with each bond. There may accordingly be several such charges in one atom . . . and there appear to be at least two in each atom [neutralization of opposite equal charges]. These charges, which it will be convenient to call "electrons" cannot be removed from the atom [?] but they become disguised when atoms unite chemically.

[3]*Scientific Transactions of the Royal Society of Dublin* (1891), S.11, IV, 563.

Thus in its original meaning the term electron stood for a certain charge of electricity either positive or negative and not for a particle.

In the work of Millikan we shall see e measured accurately and from it the value of m calculated. Hence N, the number of molecules in a molecular weight of any substance, is obtained. This number N is called Avogadro's (or Loschmidt's) constant and is of fundamental importance.

Stoney argued that since we assume matter to be corpuscular and since each atom has a definite constant charge, would it not be logical to assume that electricity also comes in parcels? If matter is granular, why not electricity? The atom of electricity he called an electron and proposed that it be adopted as a standard unit. In fact he urged that instead of the arbitrary units, centimetre, gram and second, we could to advantage base our system on three natural units, the velocity of light, the gravitational constant, and the electron.

With the sharpened backward vision of the historian, one can discern intimations of the seventh epoch in earlier records. The electron theory is, of course, a corpuscular hypothesis and so were all electrical hypotheses from Cabaeus (1620) to Hauksbee (1710). "All Bodies are electrics" said Dufay (1733) and "'tis probable that this Truth will lead us to the further Discovery of many other things." Franklin's prophetic insight was astonishing when he surmised that neutral bodies have a normal amount of "electric fire," some with a surplus (positive) and others a deficiency (negative). From influence experiments he inferred (1756) that "the electric matter consists of particles extremely subtle since it can permeate common matter . . . with such ease . . . as not to receive any appreciable resistance." Do you recall what Sir William Crookes said in 1879 about cathode rays? "We seem . . . to have within our grasp . . . the little indivisible particles which . . . are supposed to constitute the basis of the universe. . . . I venture to think that the greatest scientific problems of the future will find their solutions in this Borderland and even beyond" (p. 481).

THE FLEMING VALVE

The original of the modern radio bulb was invented in 1904 by Sir John A. Fleming (1849-1946), from whose entertaining and instructive little book, *Waves and Ripples in Water, Air and Aether*,[4] the following passages are quoted:

The signals are made in wireless telegraphy by creating long or short groups of these aetherial waves [Hertzian]. . . . If then, at a distant place, we erect another vertical wire, called the receiving aerial A, it will be struck by the

Sheldon Press, 1902.

advancing groups of waves, sent out by the transmitter. . . . The result of this movement of the electric ripple band across the aerial receiving wire is to create in that wire, . . . a feeble oscillatory current of electricity. . . .

In order to receive these signals, we have to contrive some method of detecting the presence of the corresponding feeble oscillations in the receiving aerial. . . . With this object in view, the receiving aerial A [fig. 335] has associated with it another circuit, called the condenser circuit, which comprises . . . a condenser C, and a coil of wire S, in series with it, which coil . . . is placed close to another coil R, inserted in the circuit of the receiving aerial. These two circuits are electrically tuned to each other so that electric vibrations in the receiving aerial . . . create other sympathetic and similar vibrations in the associated condenser circuit [Lodge's experiment]. . . . The most usual method of detecting these oscillations is by the use of an ordinary telephone receiver T, associated with some device called a rectifier or valve V.

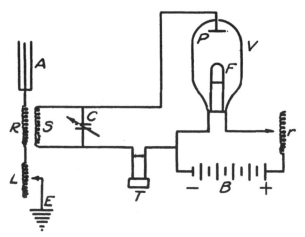

FIG. 335. The Fleming valve.

One such form of valve was devised by the author and is made as follows:—An ordinary small incandescent lamp . . . has a metal plate P included in the exhausted bulb, this plate being carried on a platinum wire, sealed through the glass. This arrangement is called a Fleming Valve or thermionic valve. When the filament F is made incandescent by a small battery, it gives off electrons . . . i.e. atoms of negative electricity, and the empty space between the hot filament and the cold metal plate can convey negative electricity from the filament to the plate but not in the opposite direction.[5] It therefore acts like a valve for electricity.

This valve is then joined to an ordinary telephone receiver T, one terminal of the telephone being connected to one terminal of the condenser C, the other end of the telephone to the filament in the bulb V, and the plate in the valve to the second terminal of the condenser. If, then, oscillations occur in the condenser, the valve permits movements of electricity through it in only one direction, and the telephone coil is traversed at each group of oscillations by a flow of elec-

[5]First observed by Edison in 1883; now called the Edison effect.

tricity . . . in one direction. Each gush of electricity through it corresponds to a single spark discharge at the transmitter. Hence, an observer listening to this telephone receiver, hears a sound which corresponds in pitch to the frequency of the spark discharge. If this spark sound is cut up by the sending key into short or long periods corresponding to the Morse Code, the observer hears these sounds as audible Morse Code signals.

The Fleming valve was invented about 1904 and belongs partly to the sixth epoch, for it has to do with the application of electro-magnetic waves in wireless telegraphy (and telephony). But the explanation of the valve's action which we have just read was published in 1912, twenty-nine years after the discovery of the Edison effect. Much had happened in the meantime. Advancements had crowded closely upon each others' heels and a new epoch had opened, with its electron theory. Thenceforth development in electric waves and in electron theory proceeded simultaneously, each in fact helping the other, as typified by Fleming's electron explanation of the action of a Fleming valve.

THE DE FOREST GRID TUBE

In 1908, the American inventor, Dr. Lee de Forest of Chicago, introduced an improvement in the Fleming valve. He placed a *grid* element between the filament and the plate. This led to the audion tube and to modern radio. De Forest's patent papers contain some description of these advances:

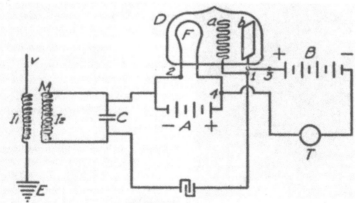

Fig. 336. De Forest grid tube.

D in fig. [336] represents an evacuated vessel, preferably of glass, having sealed therein three conducting members, *F*, *a* and *b* in the figure. The conducting member or electrode *F* is shown as consisting of a filament, preferably of metal, which is connected in series with the battery *A* or other source of electric current of sufficient strength to heat said filament, preferably to incandescence. The conducting member *b*, which may be platinum, has one end brought out to the terminal 3. Interposed between the members F and b, is a grid-shaped member a, which may be formed of platinum wire and which has one end brought out to the terminal 1. The local receiving circuit which includes the battery B, or other suitable source of electromotive force, and the signal indicating device T, which may be a telephone receiver, has its terminals connected to the plate b and the filament F at the points 3 and 4 respectively. The means for carrying the oscillations, to be detected, to the oscillation-detector, are the conductors which connect the filament F and grid a, to the tuned receiving circuit and, as shown, said conductors pass from the terminals, 2 and 1, to the armatures of the condenser C.

I have determined experimentally that the presence of the conducting member a, which as before stated may be grid-shaped, increases the sensitiveness of the oscillation-detector, and inasmuch as the explanation of this phenomenon is exceedingly complex and at best would be merely tentative, I do not deem it necessary herein to enter into the detailed statement of what I believe to be the probable explanation.[6]

CHARGE OF THE ELECTRON

About 1897, Dr. C. T. R. Wilson of Cambridge, England, made a determination of the value of e, the charge of an electron, by means of

FIG. 337. R. A. Millikan.

his now famous waterdrop experiment. A decade later Professor R. A. Millikan of Chicago, with the co-operation of some postgraduate students, H. Begeman, H. Fletcher and L. Gilchrist, modified the technique and determined e with greater accuracy, using suspended oil-drops. An account of this fundamental research is given in Millikan's book, *The Electron*, which should be read by every student of electronics.

While working with these "balanced drops," I noticed on several occasions on which I had failed to screen off the rays from the radium, that now and then one of the drops would suddenly change its charge and begin to move up or down in the field, evidently because it had captured, in one case, a positive and in the other, a negative ion.

It was only necessary to get a charged droplet entirely free from evaporation into the space between the plates of a horizontal air-condenser and then by alternately throwing on and off an electric field to keep this droplet pacing

6Lee de Forest, U.S. Patent, No. 879,532 (February, 1908).

its beat up and down between the plates until it could catch an atmospheric ion just the way I had already seen the water droplets do. The change in speed in the field would then be exactly proportional to the charge on the ion captured.

The final mean value of e obtained from the first 23 drops is 61.085×10^{-8}. This corresponds to

$$e = 4.774 \times 10^{-10} \text{ electrostatic units}$$
$$\text{and} \quad N = (6.062 \pm 0.006) \times 10^{23} \text{ [Avogadro's number]}.$$

It is only necessary to float such a body in the air, render it visible by reflected light in an ultra-microscopic arrangement of the sort we were using, charge it electrically by the capture of ions, count the number of electrons in its charge by the method described and then vary the potential applied to the plates or the charge on the body until its weight is just balanced by the upward pull of the field. This device is simply an electrical balance and it will weigh accurately and easily to one-ten-billionth of a milligram.[7]

For his fundamental contributions in this field, Dr. Millikan was awarded in 1923 the Nobel Prize in physics. On more than one occasion he has attributed a large share of his success to his association with A. A. Michelson, to whom his book is dedicated.

STRUCTURE OF ATOMS

For the radium which Millikan used in his oil-drop experiments, he was indebted to the researches of Pierre Curie of Paris and his wife Marie Curie. They discovered a new element in 1898 and named it radium to suggest that it emits rays. Such rays can affect electrometers and photographic plates and can make minerals fluoresce.[8] Further study brought out the fact that radium emits three kinds of rays, known as alpha, beta, and gamma rays.

In what follows, it will be necessary to select only a few names from the many who have contributed to the development of *electronics* and merely mark out a few mile-posts along the road. The name of Lord

Fig. 338. Lord Rutherford.

Rutherford, Sir J. J. Thomson's successor as Cavendish Professor, is outstanding. He proposed that the atom be considered similar to a

[7]R. A. Millikan, *The Electron* (Univ. of Chicago Press, 1917), p. 63.
[8]The wonderful but tragic story of the Curies is beautifully told in *Madame Curie*, written by their daughter Eve Curie (Doubleday, 1937).

small solar system with a central *nucleus* corresponding to the sun, and a cluster of electrons revolving about the nucleus somewhat like planets. These are sometimes called planetary electrons. Such an explanation is evidently an instance of using a mechanical picture or model. Rutherford published a paper on this subject in 1913 when he was professor of physics at Manchester University and a colleague of Sir Oliver Lodge. He studied the passage of α- and β-rays through gases. He refers to them as streams of α- and β-particles.

In comparing the theory outlined in this paper with the experimental results, it has been supposed that the atom consists of a central charge supposed concentrated at a point and that the large single deflexions of the α and β particles are mainly due to their passage through the strong central field. For concreteness, consider the passage of a high speed α particle through an atom having a positive central charge Ne, and surrounded by a compensating charge of N electrons [e is the positive or negative charge, equivalent to that of an electron and N is the number of such positive charges on the nucleus]. Remembering that the mass, momentum and kinetic energy of the α-particle are very large compared with the corresponding values for an electron in rapid motion, it does not seem possible from dynamic considerations that an α-particle can be deflected through a large angle by a close approach to an electron. It seems reasonable to suppose that the chance of single deflexions through a large angle due to this cause, if not zero, must be exceedingly small compared with that due to the central charge.

The general data available indicate that the value of this central charge for different atoms (Ne) is approximately proportional to their atomic weights.[9]

ATOMIC NUMBERS

It was one of the tragedies of science that young Harry G. J. Moseley[10] of Manchester was sent to the trenches at Gallipoli during the first World War and was there killed. Even at the early age of 27, he had already achieved one of the greatest advances in electronics that have ever been made. It was surely a grave error to place such rare human material in such jeopardy, although, on the other hand, young Moseley himself would have objected strenuously to any shielding from the dangers of serving in the army of his own country. Extremely complex and difficult is the problem of organizing society so that each may serve at the post which is best for him and for all concerned.

In his experiments, Moseley obtained X-rays from targets of different elements and examined the rays after reflection from crystals. In this work he was helped by the previous work of Professor Max Laue of Berlin and Sir William Bragg. Moseley's 1913 paper gives some of his results and inferences:

[9]*Philosophical Magazine* (May, 1913), p. 669.
[10]*Philosophical Magazine* (December, 1913), p. 1024.

We have here a proof that there is in the atom a fundamental quantity which increases by regular steps as we pass from one element to the next [in the periodic table]. This quantity can only be the charge on the central positive nucleus. . . . Now atomic weights increase on the average by about 2 units at a time, and this strongly suggests the view that the quantity N increases from atom to atom always by a single electronic unit.

We are therefore led by experiment to the view that N is the same as the number of the place occupied by the element in the periodic system. This atomic number is then for H, 1, for He, 2, for Li, 3, for Ca, 20, for Zn, 30 and so forth.[11]

THE RUTHERFORD-BOHR ATOM MODEL

Professor Niels Bohr of Copenhagen, Denmark, who had studied under Rutherford, showed that by making some assumptions about electrons, one could use Rutherford's atom-model to account for many facts then known about the spectra of elements. His 1913 paper, "On the Constitution of Atoms and Molecules," contains the following sentences:

Following the theory of Rutherford, we shall assume that the atoms of the elements consist of a positively charged nucleus surrounded by a cluster of electrons. The nucleus is the seat of the essential part of the mass of the atom, and has linear dimensions exceedingly small compared with the distances apart of the electrons in the surrounding cluster.

The total experimental evidence supports the hypothesis that the actual number of electrons in a neutral atom with a few exceptions is equal to the number which indicates the position of the corresponding element in the series of elements arranged in order of increasing atomic weight. For example, on this view, the atom of oxygen which is the eighth of the series, has eight [planetary] electrons and a nucleus carrying eight unit charges.

This paper is an attempt to show that the application of the above ideas to Rutherford's atom-model affords a basis for a theory of the constitution of atoms. It will further be shown that from this theory, we are led to a theory of the constitution of molecules.[12]

ISOTOPES

In 1886, E. Goldstein of Berlin, by drilling holes through a cathode plate, discovered rays emanating from the plate in opposite sense to the cathode rays which had been named by him ten years earlier. The new rays he called canal rays (*Kanalstrahlen*) because of the method of obtaining them. W. Wien of Berlin found that canal rays carry a positive charge. He found also that they are deflected by a magnet, in accordance with Ampère's right-hand rule, though to a less degree than cathode rays (by an equivalent field) but to the same degree as positive atoms of the gas in the tube. Sir J. J. Thomson called these

[11]An interesting account of Moseley is given in B. Jaffe, *Crucibles* (Tudor, 1934), p. 289.

[12]*Philosophical Magazine* (1913), p. 476.

rays "Positive Rays" and studied the masses of their constituent ions. With neon in the tube he found a surprising result—two deflections corresponding to two masses which differed by about 10 per cent. Presently two radioactive products were discovered which had different atomic masses but identical chemical properties; then several cases of this phenomenon were found. For any such pair or group, Dr. E. Soddy of Glasgow and Oxford proposed in 1913 the name "isotopes" which means "having the same position in the periodic table." In a paper in *Nature* (1917), he said, "When among the light elements we come across a clear case of large departure from integral value, such as magnesium 24.32 . . . , we may reasonably suspect the element to be a mixture of isotopes."

To decide this question, experimental methods with greater power of separating different masses were necessary. The first adequate method was developed in 1918 by Dr. Arthur J. Dempster of Chicago. Here is a snippet from his paper:

The charged particles . . . fall through a definite potential difference. A narrow bundle is separated out by a slit and is bent into a semicircle by a strong magnetic field; the rays then pass through a second slit and fall on a plate connected to an electrometer.[13]

FIG. 339. A. J. Dempster.

After an interruption owing to World War I, this research was resumed and in 1920 the examination of magnesium gave results which supported Soddy's statement:

After heating the magnesium slightly and pumping till a MacLeod gauge gave no pressure indication, the nitrogen molecule was the only particle present. The heating current was then increased by steps to vaporize the magnesium. . . . It was found that three strong new lines had appeared. The new lines which are undoubtedly due to magnesium were compared with the nitrogen rays which were still faintly present and found to have atomic weights 24, 25 and 26. . . . We conclude that magnesium consists of three isotopes of atomic weights 24, 25 and 26.[14]

[13]A. J. Dempster, "A New Method of Positive Ray Analysis," *The Physical Review* (1918), S.2, II, 316.
[14]A. J. Dempster, "Positive Ray Analysis of Magnesium," *Proceedings of the National Academy of Science* (February, 1921), VII, No. 2, p. 45.

DISCOVERY OF URANIUM-235

In 1920, Dr. F. W. Aston of Cambridge, by means of his mass spectrograph, demonstrated the existence of isotopes in ordinary elements. The masses of Thomson's two neon isotopes he found to be 20 and 22. During about two decades, Aston, Dempster, and others identified some 280 isotopes.

Special methods were devised to effect the analysis of some more difficult elements. One of these was developed by Dempster in 1935 and led to the discovery of over 30 isotopes. It is of exceptional interest because among these was the famous faint isotope of uranium of mass 235, which later was found to be fissioned by neutrons and was one of the two substances from which atomic bombs were made.

The new methods described in this paper are first, a new source of ions, and secondly, a new apparatus for use with these sources with the possibility of a greatly increased accuracy in the comparison of atomic masses. . . .

An ion beam always diverges slightly and contains ions of slightly different energies. . . . The arrangement [fig. 340] illustrates how an electric field may be combined with a magnetic field to produce a perfect focus. The ions are accelerated between S_1 and S_2 and a narrow ribbon of rays is separated by the slit S_3. The cylindrical condenser deflects them through a right angle and brings the velocity, v_1, to a focus at S_4. A bundle of slightly greater velocity, v_2, is reunited alongside those of velocity v_1. By making the radius of curvature in the electric field the proper ratio to the radius in the magnetic field, the bundles of varying energy will all reunite at one place on the photographic plate.

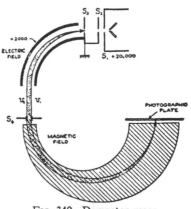

FIG. 340. Dempster mass spectrograph.

As uranium is of great importance for the subject of radioactivity, its analysis was tried with the new spectrograph. The spark was run with gold and uranium metal as electrodes, and also with an electrode made by packing a nickel tube with pitch-blende. . . .

It was found that a few seconds' exposure was sufficient for the main component at 238 reported by Dr. Aston, but on long exposures a faint companion of mass number 235 was also present. With two different uranium electrodes, the faint companion was observed on eight photographs and also on two photographs with the pitch-blende electrode. Pl. A, VI shows the two photographs with pitch-blende. . . So that I think we may safely consider the mass at 235 as an isotope of uranium.[15]

At that time, of course, Dempster could not know what an important isotope he had discovered.

[15]A. J. Dempster, "New Methods in Mass Spectroscopy," *Proceedings of the American Philosophical Society* (1935), LXXV, No. 8, p. 755.

THE ELECTRON MICROSCOPE

From a host of subsequent electronic developments only one will be discussed here—the electron microscope.[16] It belongs to the vacuum-tube family and opens a whole new world to investigation as did the (light) microscope three centuries earlier.

We have seen a corpuscular theory of cathode rays gaining preference in England. On the Continent, however, Roentgen and many others were inclined to favour a wave-theory and there the electron microscope was born. It is possible to assign a wave-length to cathode-rays, namely, about 1/100,000 that of light, i.e. about 1/100,000 of 6000 Å. or 0.06 angstroms. Louis de Broglie of Paris obtained the value 0.04 Å. at 85,000 volts. If two points are nearer to each other than about one-third (or even one-half) the wave-length of the light used in a microscope, they cannot be distinguished: the waves are too coarse to admit such fine distinction. In other words, 2000 Å. (or 3,000 Å.) is about the limit of the *resolving power* of the microscope. But since cathode rays have shorter wave-lengths they might enable us to see smaller particles and to obtain photographs' with greater detail. In fact the electron microscope under favourable conditions has given clear photographs of particles even less than 50 Å. in diameter.

As the action of a microscope depends on the focussing of light rays by lenses, so also that of the electron microscope. Cathode rays, of course, cannot traverse glass but they can be deflected by a magnet (Hittorf, Crookes). The focussing of cathode rays was effected in 1926 by H. Busch of Jena. His report says:

> The principle has been demonstrated [in this paper] that in a symmetrical magnetic field a sufficiently narrow beam of cathode rays diverging symmetrically from a point on the axis of symmetry, is always brought again to a focus.[17]

Accordingly the "lenses" of an electron microscope are electro-magnets.

Fig. 341 shows an electron microscope and three pioneers, apparently not working very hard at the moment—Professor E. F. Burton, of the University of Toronto, Dr. Albert Prebus of the Bell Telephone Laboratories, and Mr. William A. Ladd, M.A., electron microscopist of the Columbian Carbon Co., New York. One primary difficulty was to make an instrument sufficiently controllable to yield reproducible results. Burton and his assistants, J. Hillier, A. Prebus, W. Ladd, and others were first to produce an electron micro-

---

[16]Popular treatment by G. G. Hawley in *Seeing the Invisible* (Knopf, 1944).
[17]H. Busch, "*Berechnung der Bahn von Kathodenstrahlen im axialsymmetrischen elektromagnetischen Felde*," *Annalen der Physik* (1926), S. 993.

Fig. 341. The electron microscope. On the left, Professor Burton; in the centre, Dr. Prebus; on the right, Mr. Ladd.

scope which as one of them has put it, "would behave on Wednesday the same as on Tuesday."

A glowing cathode filament at the upper end of the tube emits cathode rays (thermionic emission) under a tension of some 70,000 volts. One of the most difficult problems is the control of this high potential. The beam by passing through a magnetic field is focussed on the specimen to be examined; thus the latter is "illuminated" by cathode rays instead of light. The rays are again brought to a focus farther down the tube on a fluorescent screen for ocular examination or on a photographic plate for permanent record.

A looser way of comparing the two microscopes is to give 3500 diameters as the limit of useful magnifications obtainable with the light instrument and from 40,000 to 200,000 diameters (final print) for the electron instrument up to the present. Electron pictures taken at 5,000 to 15,000 diameters can be enlarged optically 10 or 15 diameters and still give sharp detail. So the comparison really comes back to resolving power.

Dr. Vladimir K. Zworykin of Radio Corporation of America, New York, and his assistants, J. Hillier and others, have done much to put the electron microscope on a commercial basis. Already the instrument has given excellent service in the examination of pigment powders, abrasives, textile fibres, greases, carbon and rubber particles, and bacteria.[18] Toward the latter the electron microscope is a veritable Medusa, for it takes one glance at even the toughest bacterium and presto that "bug" is no longer alive. More power to the electron microscope.

SURVEY

And so we come to journey's end, although a journey of this kind could never really have an ending. We have made a rather hasty conducted tour through a broad and noble terrain. Many items of prime interest have been unavoidably overlooked. Important authors have been omitted from our visiting list and significant passages from the works of those interviewed. The reader could go on an unconducted tour of the same region and find endless new material equally interesting and profitable. But his general impression of the domain of science would probably not be altered essentially. It must be remembered that our attention has been restricted in general to such features as might appeal to beginners in physics. The fields which have thus been skirted are enormous in comparison with those which we have explored and examined.

If one thinks back to Thales' humble little experiment with amber, it can be seen that it contained in embryo the whole story of electricity.

> Flower in the crannied wall,
>   I pluck you out of the crannies;
> I hold you here, root and all, in my hand,
> Little flower—but *if* I could understand
> What you are, root and all and all in all,
> I should know what God and man is.

[18]G. G. Hawley, *Small Wonder, The Story of Colloids*. (Knopf, 1947), p.64.

The first epoch in the history of electricity extended from 600 B.C. to about A.D. 1729. During that period electricians such as Gilbert and Dufay were discovering and studying static charges, and their chief instruments were indicators called electroscopes. In every experiment from the beginning to the end of the story, an indicator of some kind has been necessarily present. An important general trend has been the invention of new indicators and the improvement of old ones, especially in sensitivity. Stephen Gray's discovery of conduction inaugurated the second epoch. The fact that the amber in Thales' experiment gradually loses charge is an indication of slow conduction. Thus the second epoch was potentially present in that protean experiment. The rediscovery of influence charges by Canton about 1754 opened the third epoch. It may be symbolized by the Leyden jar or the condenser. In stage II of von Guericke's pith-ball experiment, induction precedes attraction: but that experiment was really only a special form of Thales' amber "dust-test." Hence the third epoch was presaged by the ancient amber experiment. This is easier for the historian to realize than it was for the researchers of those times, for as an old Dutch philosopher said, "We advance forwards but we understand backwards."

Volta's invention of the voltaic cell in 1800 opened the fourth epoch, with its study and application of electric currents. All Volta's work grew directly from that of previous epochs, chiefly by the use of his condensing electroscope; hence this epoch was also adumbrated by the pristine amber experiment though it would have needed almost superhuman insight to foresee this development. That Volta glimpsed this advancement, though "as in a glass darkly," his writings prove beyond a doubt. An experiment of pre-eminent importance in this epoch was that of Oersted (1819), for it welded the sciences of electricity and magnetism.

The fifth epoch was ushered into the history of electricity by Faraday's discovery of induced currents (1831). This achievement consisted principally in reversing Oersted's experiment and was therefore implied by the amber experiment. It led to the invention of the generator and transformer and to the advent of the Electrical Age, in which people enjoy countless electrical conveniences and comforts not available to their ancestors—and encounter the concomitant dangers of luxury and decadence.

The sixth epoch was prophesied by Maxwell and inaugurated by Hertz. By the labours of a whole army of researchers, it has brought us rapid world-wide communication and symphony concerts in our homes (sometimes!). Such powers bring nearer to our reach the Parliament of Man, the universal brotherhood of mankind, and Isaiah's vision of a world without war.

The electron theory characterizes the seventh epoch; it was reached by the labours of Sir J. J. Thomson of Cambridge, Dr. J. Perrin of Paris, Sir William Crookes of London, Lord Rutherford of Nelson, New Zealand, Professor R. A. Millikan of Chicago, Professor Niels Bohr of Copenhagen, and a host of others. Its entry followed closely on the heels of its predecessor and the two have since developed together, rendering each other much mutual assistance. As we have seen, both these epochs originated in ideas reached in previous periods and were, therefore, foreshadowed in Thales' amber experiment.

At first the advancement was relatively slow: now it is so swift that it would be extremely difficult if not impossible for any person to keep abreast of all the detailed advances that are being made day by day in the various branches of the study and application of electricity. The more we learn about this mysterious and powerful agent, the more there is to learn; for in this science, as well as in others, increase of knowledge never compasses complete knowledge, but merely broadens one's horizon.

Most of the scientists whom we have met on our itinerary have proved to be worthy citizens. Priestley explained to us why this should be true. A large majority of them were great teachers and leaders, and throughout the story there has been a kind of apostolic succession, possibly not without the laying on of hands occasionally. Socrates was the teacher of Plato, Plato of Aristotle, Aristotle of Theophrastus. Dr. Joseph Black had famous students: Young, Watt, Leslie and Hope. Sir Humphry Davy taught Faraday from whom Maxwell gained inspiration. The mantle of greatness descended from Maxwell to Thomson and from him to Rutherford. We have seen many scientists helping each other, Dufay and Gray, Young and Fresnel, Kelvin and Joule.

One reason for the rapid advance of science has been the scientist's objective and impersonal or self-effacing attitude. Some scientists have lived a rather secluded life, retired in their laboratories, but applied science has remained socially vigorous; and now scientists are coming down from the ivory tower into the market-place to bring what help they can in aiding the community to solve its discouragingly difficult problems. We could mention Sir William Bragg, Sir Frederick Banting, and President James Conant of Harvard among others. If scientists can promote legislation which will prevent abuses of their inventions, e.g. the automobile, firearms, drugs, they will have rendered society a valuable service.

The goal of science is the comprehension and control of nature. Though much progress has been made, the goal is not yet reached. Successes have been won by means of a sword whose two edges are theory and practice.

In attempting to understand such a complicated entity as the physical universe, the only hope lies in organization. Hence the immediate purpose of physics is the organization of our experience of the physical world into a unified system. Organization effects economy of effort and time. One of the chief methods of organization is simplification as instanced by the principle of parsimony. Thus throughout the great epic portrayed in the story of physics one can see the advancement from the Many to the Few and if possible to the One.

In early physics most definitions and concepts were closely linked to the nature of man and may be described as anthropomorphic. To some degree this must always remain true: but there has been a marked tendency to make it less true. Although the physicist may sometimes think of force in subjective terms of muscular sensations, nevertheless, since Newton, his definition has been in objective terms of mass and acceleration measured in centimetres, grams and seconds. Hues were first defined in such physiological or psychological terms as red and blue; but now, in terms of wave-length. The replacement of fingers by thermometers was typical and significant. By the use of indicators, greater accuracy and precision have been secured and these, partly by elimination of personal peculiarities of observers—the personal factor.

It is the discarding of anthropomorphic features that has given modern physics an atmosphere which some laymen find alien and describe as inhuman. That the tendency has persisted indicates that it promotes progress. One aspect of such progress is seen in certain simplifications and unifications. Acoustics and thermics have become merged with mechanics, and optics and magnetism with electricity— an advancement from Several to Two. That the boundaries between the two fields are not everywhere sharply defined is suggestive. If and when they become sufficiently indistinct, that will signify an advancement to the One; and in this strain, one might be tempted to predict a probable future trend in physics.

It is hoped that these pages have helped the reader to gain some facts and principles of science which will bring him a better understanding of events that he observes, and will draw him into closer rapport with the scientific age in which he lives. More important perhaps than such facts and principles, to the layman, are the attitudes of scientists and their methods of attacking and solving problems. We have seen that although a scientist's attitude is largely impersonal and dispassionate, yet there is also a personal side to science; for every fact, every idea, every machine and device, every principle, is necessarily the discovery or invention and contribution of some person or group of persons.

Archimedes' Evaluation of π

THE diameter of the circle $D = 2R = 2.OB$ (fig. 342).

In the $\triangle OV_1B$, $V_1B : BO : V_1O = 1 : 2 : \sqrt{3}$ (Pythagorean theorem)

$\therefore V_1B = \frac{1}{2}R$ and $V_1O = \frac{1}{2}R(\sqrt{3}) = R\left(\dfrac{1.7320508}{2}\cdots\right)$

$$= R(0.866025\ldots).$$

The perimeter of the inner hexagon $P'_6 = 6.BE = 2.V_1B \times 6$

$$= 2R(\tfrac{1}{2})(6) = D(3.00000).$$

In the outer hexagon, $\because \triangle TGO \parallel\!\!\!\mid \triangle BV_1O$,

$\therefore TG/GO = V_1B/V_1O = 1/\sqrt{3}$

$\therefore TG = GO/\sqrt{3} = R/\sqrt{3} = R(0.577350\ldots)$

$\therefore$ perim. outer hex. $P''_6 = TT_1 \times 6 = 2.TG(6) = 2R(0.577350\ldots)(6)$

$$\therefore P''_6 = D(3.46410\ldots).$$

Intuitively it can be seen that $TG >$ arc $\overset{\frown}{BG} > V_1B$.

$\therefore$ the length of the circumference C is intermediate between P'_6 and P''_6 which differ by about 15%.

DODECAGONS

Archimedes next bisected $\angle V_1OB$ and $\angle GOT_1$ and obtained an inscribed and an escribed dodecagon as in the figure, with sides BG and KT_2 whose lengths must be found.

$\because \dfrac{A_1V_1}{A_1B} = \dfrac{V_1O}{OB}$ (Euclid, vi. 3) $\therefore \dfrac{A_1V_1}{A_1B + A_1V_1} = \dfrac{V_1O}{OB + V_1O}$

$\therefore A_1V_1 = \dfrac{(V_1B)\,R(0.866025\ldots)}{R(1 + 0.866025\ldots)} = R(0.232206\ldots).$

$A_1O = \sqrt{(A_1V_1)^2 + (V_1O)^2} = R\sqrt{(0.232206\ldots)^2 + (0.866025\ldots)^2}$

$$\therefore A_1O = R(0.89654\ldots).$$

By similar triangles, $V_2G/GO = A_1V_1/A_1O = \dfrac{0.232206\ldots}{0.89654\ldots} = 0.25882\ldots$

$$\therefore\quad V_2G = R(0.25882\ldots)$$

$\therefore$ perimeter of inner dodecagon P'_{12}

$$= 2.V_2G(12)$$
$$= 2R(0.25882\ldots)12 = D(3.10584\ldots).$$

By sim. $\triangle$'s, $KG/GO = A_1V_1/V_1O = \dfrac{R(0.232206\ldots)}{R(0.866025\ldots)}$

$$\therefore\quad KG = R(0.26795\ldots).$$

$\therefore$ perim. outer dodecagon $P''_{12} = 2.KG(12) = 2R(0.26795\ldots)(12)$
$$= D(3.21540\ldots)$$

As before, C is between P' and P'' and these differ now by about 4%.

The perimeter of the inner 12-gon is greater than that of the inner hexagon, for 6 corners have been added and $(BG + GE) > BE$ by Euclid, I.20. Also the perimeter of the outer 12-gon is less than

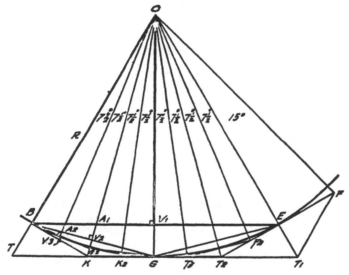

Fig. 342. Circumference of circle.

that of the outer hexagon, for 6 corners were cut off and $T_2F < (T_2T_1 + T_1F)$. Thus the inner perimeter has increased while the outer one has decreased. The two values have approached each other. Their difference has decreased from 15% to 4%. But ever the constant length of the circumference which we wish to measure is intermediate between P' and P'' no matter how closely they approach each other. In every case $P' < C < P''$ or $P'' > C > P'$. Not only are P' and P'' approaching each other but each is approaching the constant C.

A SQUEEZE-PLAY

Archimedes next bisected $\angle BOV_2$ and $\angle GOT_2$ and obtained an inner and an outer 24-gon and calculated their perimeters. Again the inner perimeter had increased and the outer one decreased. By this procedure, he had the circumference trapped between two calculable values which he could bring nearer and nearer to each other. He repeated this "squeeze-play" until he had 96-gons. Moreover, it is seen that he had contrived a repetitive plan of calculation by which he could obtain the data for each new polygon from those of the previous case. There is little doubt that the computations were laborious and tedious.

At this juncture we might well avail ourselves of the convenience of trigonometry with proper gratitude to Hipparchus and Ptolemy, the founders of trigonometry, and to their successors. In fig. 372, $BE = 2.V_1B = 2R. \sin 30°$; also $P'_6 = D(\sin 30)6$ and $P''_6 = D(\tan 30)6$. Similarly, $P'_{12} = D(\sin 15)12$ and $P''_{12} = D(\tan 15)12$. We need, therefore, a method for calculating $\sin A$ and $\tan A$ from $\sin 2A$.

$$\therefore \quad \cos 2A = \cos^2 A - \sin^2 A = 1 - 2.\sin^2 A = 2.\cos^2 A - 1,$$

$$\therefore \quad 2 \sin^2 A = 1 - \cos 2A \quad \therefore \quad \sin A = \sqrt{\frac{1 - \cos 2A}{2}}$$

$$\text{and } \cos A = \sqrt{\frac{1 + \cos 2A}{2}}$$

$$\therefore \quad \tan^2 A = \sin^2 A / \cos^2 A = \frac{1 - \cos 2A}{1 + \cos 2A} \quad \therefore \quad \tan A = \sqrt{\frac{1 - \cos 2A}{1 + \cos 2A}}$$

Thus from $\sin 15$ we can obtain $\sin 7\frac{1}{2}$ and $\tan 7\frac{1}{2}$ (24-gon).

$$\therefore \quad \sin 15 = 0.25882..., \quad \therefore \cos 15 = \sqrt{1 - (0.25882...)^2} = 0.96593...$$

$$\therefore \quad \sin 7\frac{1}{2} = \sqrt{\frac{1 - \cos 15}{2}} = \sqrt{\frac{1 - 0.96593}{2}} ... = 0.13053...$$

$$\text{and } \tan 7\frac{1}{2} = \sqrt{\frac{1 - \cos 15}{1 + \cos 15}} = \sqrt{\frac{1 - 0.96593}{1.96593}} = 0.13165$$

$\therefore$ perimeter inner 24-gon $P'_{24} = D(0.13053...)24 = D(3.13272...)$
and   "   outer 24-gon $P''_{24} = D(0.13165...)24 = D(3.15960...)$

This traps C between an upper and a lower boundary which differ from each other by about 1%.

96-GONS

Carrying on with dogged perseverance and patient attention to the accuracy of every detail, one reaches finally the result:

$$P'_{96} = D(\sin \tfrac{15}{8})96 = D(0.032715\ldots)96 = D(3.14064\ldots) \text{ and}$$
$$P''_{96} = D(\tan \tfrac{15}{8})96 = D(0.032735\ldots)96 = D(3.14256\ldots).$$

These differ by only 2 in 3,000 i.e. $\tfrac{1}{15}\%$. Their average is $D(3.1416\ldots)$ Now you can see why Archimedes reported his value for π as being between $3\tfrac{10}{70}$ and $3\tfrac{10}{71}$, i.e., in decimals, between 3.1428571... and 3.1408450... When comparing two fractions, the Greeks often made the numerators equal whereas our custom is to make the denominators the same. This seems to us another example of progress from the cumbersome and complex to the simpler and easier. The average of the two numbers is 3.1418..., which differs from the received value, 3.14159..., by only 2 in 30,000.

Two remarkable features of this astonishing achievement of Archimedes were, first, his ability to perform such calculations in Greek numerals and second, his use of the idea of a *variable approaching a limit*. This concept is fundamental in the infinitesimal calculus, for the invention of which nineteen centuries later Newton and Leibnitz became forever famous.

Index

513